Lecture Notes in Electrical Engineering

Volume 786

The book series *Lecture Notes in Electrical Engineering* (LNEE) publishes the latest developments in Electrical Engineering - quickly, informally and in high quality. While original research reported in proceedings and monographs has traditionally formed the core of LNEE, we also encourage authors to submit books devoted to supporting student education and professional training in the various fields and applications areas of electrical engineering. The series cover classical and emerging topics concerning:

- Communication Engineering, Information Theory and Networks
- Electronics Engineering and Microelectronics
- Signal, Image and Speech Processing
- Wireless and Mobile Communication
- Circuits and Systems
- Energy Systems, Power Electronics and Electrical Machines
- Electro-optical Engineering
- Instrumentation Engineering
- Avionics Engineering
- Control Systems
- Internet-of-Things and Cybersecurity
- Biomedical Devices, MEMS and NEMS

For general information about this book series, comments or suggestions, please contact leontina.dicecco@springer.com.

To submit a proposal or request further information, please contact the Publishing Editor in your country:

China

Jasmine Dou, Editor (jasmine.dou@springer.com)

India, Japan, Rest of Asia

Swati Meherishi, Editorial Director (Swati.Meherishi@springer.com)

Southeast Asia, Australia, New Zealand

Ramesh Nath Premnath, Editor (ramesh.premnath@springernature.com)

USA, Canada:

Michael Luby, Senior Editor (michael.luby@springer.com)

All other Countries:

Leontina Di Cecco, Senior Editor (leontina.dicecco@springer.com)

**** This series is indexed by EI Compendex and Scopus databases. ****

More information about this series at http://www.springer.com/series/7818

M. Mitra · Mita Nasipuri · Maitreyi Ray Kanjilal
Editors

Computational Advancement in Communication, Circuits and Systems

Proceedings of 3rd ICCACCS 2020

Springer

Editors
M. Mitra
Department of Electronics
and Telecommunication
IIEST Shibpur
Howrah, India

Mita Nasipuri
Department of Computer Science
and Engineering
Jadavpur University
Kolkata, India

Maitreyi Ray Kanjilal
Narula Institute of Technology
Kolkata, India

ISSN 1876-1100 ISSN 1876-1119 (electronic)
Lecture Notes in Electrical Engineering
ISBN 978-981-16-4034-6 ISBN 978-981-16-4035-3 (eBook)
https://doi.org/10.1007/978-981-16-4035-3

This Springer imprint is published by the registered company Springer Nature Singapore Pte Ltd.
The registered company address is: 152 Beach Road, #21-01/04 Gateway East, Singapore 189721, Singapore

Preface

This book gathers the proceedings of the International Conference on Computational Advancement in Communication Circuits and Systems (ICCACCS 2020), which was organized by Narula Institute of Technology under the patronage of the JIS group, affiliated to Maulana Abul Kalam Azad University of Technology. The book presents peer-reviewed papers that highlight new theoretical and experimental findings in the fields of electronics and communication engineering, advanced computing, pattern recognition and analysis, signal and image processing, etc. The respective papers cover a broad range of principles, techniques and applications in computing and communication, environment-friendly computing, reconfigurable computing, low-power nano technology and VLSI design. The proceedings reflect the conference's strong emphasis on methodological approaches and focus on applications within the domain of computational advancement in communication circuits and systems. It also addresses the emerging technologies in electronics and communication together with the latest practices, issues and trends.

Howrah, India — M. Mitra
Kolkata, India — Mita Nasipuri
Kolkata, India — Maitreyi Ray Kanjilal

Contents

About the Editors

Dr. M. Mitra did his B.Tech., M.Tech. & Ph.D. from the Department of Radiophysics and Electronics under Calcutta University in the year 1982, 1985–1986, and 1995, respectively. He joined as Lecturer in the Department of Electronics and Telecommunication Engineering of IIEST Shibpur (formerly B.E. College) in the year 1995 and became Professor in the same department in the year 2011. He served the department as Head from 2012–2014 and once again from 2016–2018. He had also served the institute at different capacity such as Head of the Department of Purabi Das School of Information Technology (PDSIT) of IIEST, Shibpur. His main area of research is on semiconductor microwave devices, especially IMPATT. He had fabricated and characterized IMPATT diode at Xband for the first time in India. He had published four books such as *Satellite Communication* (2nd Ed.), *Microwave Engineering* (3rd Ed.), *Microwave Semiconductor Devices*, and *Electronic Circuits* from the house of Prentice Hall. So far he had published about 80 research papers in different SCI indexed journals. Till now 15 students have got their Ph.D. degree under his supervision and 5 are continuing. He is Member of different prestigious societies like IEEE, Life Member of IETE, etc.

Prof. Mita Nasipuri received her B.E. and M.E. degrees in Electronics and Telecommunication Engineering from Jadavpur University, Kolkata, India, in 1979 and 1981, respectively, and did her Ph.D. degree in Engineering from Jadavpur University, in 1990. She is currently Professor in the Department of Computer Science and Engineering, Jadavpur University. From July 2004 to July 2006, she was Head of the Computer Science and Engineering Department, Jadavpur University. She has also served as Member of various UG/PG syllabus committees and the Doctoral committee of Engineering faculty of Jadavpur University. Since 2008, she is serving as Coordinator of Centre for Microprocessor Applications for Training Education and Research of Jadavpur University. Her research interest includes biomedical signal processing, image processing, pattern recognition, bioinformatics, medical image analysis, and multi-media system. She has authored or co-authored more than 500 research articles including several book chapters, out of which, more than 100 papers have been published in science citation indexed (SCI/SCIE) technical journals. Two

US patents have been granted on her work. She has supervised more than 22 Ph.D. students. She has been granted sponsored projects by the Government of India with a total amount of around INR 1.5 Crore. She is Fellow of Institution of Engineers (India) and West Bengal Academy of Science and Technology, India. She is also Senior Member of the IEEE, USA, since 1992.

Prof. Maitreyi Ray Kanjilal is currently holding the post of Principal of Narula Institute of Technology and also Professor of Department of Electronics and Communication Engineering. She did her UG and PG from Calcutta University. She has been awarded Ph.D. from Calcutta University in 2000. She has more than 22 years of teaching experience. She has more than 100 published papers in national and international journals and conferences. She has 6 Ph.D. scholars and has published two patents and one is filed. She is the author of two books on Basic Electronics and one book on Analog Electronics. Her subject interest is on high-power and high-frequency (MM-wave—THz region) operation of wide-bandgap semiconductor devices, Group IV-IV, III-V material-based devices, superlattice-based devices, applications of these devices in the medical arena, nanoscale devices: design and develop, heterostructure and heterojunction semiconductor devices, microelectronics fabrication, low power devices, VLSI circuits, nanodevices, spintronics, and quantum computing.

Proper Choice of a Machine Learning Algorithm for Breast Cancer Prediction

Arijit Das, Tanisha Khan, Subhram Das, and D. K. Bhattacharya

Abstract Breast cancer is the most common form of invasive cancer and after lung cancer, it is the second leading cause of cancer death in women. Many statistical models have been used to predict the malignancy of the tumor. Therefore due to the violation of the proportional hazard assumption, a statistical model may fail to predict breast cancer accurately. In the current epoch, machine learning algorithms play a decisive role in predicting the malignancy of a tumor with high accuracy. The primary purpose of this paper is to compare the performance of eleven different machine learning classification techniques for breast cancer prediction. Wisconsin Diagnostic Breast Cancer dataset is utilized to compare these established algorithms based on the k-fold Cross-Validation accuracy score. Additionally, three different feature selection methods have been incorporated to reduce the number of features on the dataset. After the reduction of the features, the same methods are applied again to compare performance based on their accuracy score. It is found that all the algorithms perform very well with more than 93% accuracy score; among these Logistic Regression, Support Vector Classification and Multilayer Perceptron get an accuracy score of over 98%. It is also observed that even after a drastic reduction in the number of features, the result remains satisfactory, and the accuracy score is more than 90% for all the applied algorithms.

Keywords Breast cancer · Machine learning · Feature selection · k-fold Cross-Validation

1 Introduction

Amidst developed nations, breast cancer happens to be the second most common cancer after lung cancer, from which most of the women across the world suffer

A. Das · T. Khan · S. Das (✉)
Computer Science and Engineering, Narula Institute of Technology, Kolkata, India

D. K. Bhattacharya
Pure Mathematics, University of Calcutta, Kolkata, India

M. Mitra et al. (eds.), *Computational Advancement in Communication, Circuits and Systems*, Lecture Notes in Electrical Engineering 786,
https://doi.org/10.1007/978-981-16-4035-3_1

most [1–3]. The percentage of occurrence and transience rate of breast cancer is very high [4–6]. This creates one of the significant and most pressing health problems in today's society.

With the help of mammography tools, breast cancer can be detected in the early stage even before any physical symptoms were developed. However, like any other screening tool mammography has got its loopholes, as it can both underdiagnose and overdiagnoses, subsequently leading to multiple diagnostic fallacies. To resolve this flaw, machine learning techniques and data mining are being used for the last few years to predict cancer diagnosis. Pathological and clinical data [7] are to be used for the computational purpose to assist conventional tools to predict this breast cancer. The more is the accurate data produced, the more is the accurate level of prediction under proper authenticated techniques. The use of computers has significantly improved technology along with the simultaneous advancement of medical database management systems leading to the creation of a huge volume of heterogeneous data and medical databases. Therefore, the intelligent healthcare system combined with various Machine Learning (ML) techniques is a crucial domain for cancer prediction. These current techniques also help doctors to diagnose cancer patients with better perfection [8]. Previously, basic software programs such as Microsoft Excel, SPSS, and STATA [9–11] were used by clinicians to analyze the factors influencing breast cancer survival rate. To demonstrate the advantages and potential of ML, a lot of work is done where different types of conventional statistical methods are compared with traditional ML classification methods. These traditional and statistical methods had various drawbacks, for which they subsequently failed to generate rational and creative visualizations. ML algorithms can be broadly classified into two categories—supervised learning and unsupervised learning. Whether to choose the first category or the second one depends on the types of data and their structures. A current review [12] shows that most of the ML algorithms, which are used in breast cancer diagnosis and prognosis, are of a supervised type. To establish that ML methods could be a reliable classification method for prediction of breast cancer, the classification methods DT and ANN are compared in 2004 with the statistical linear regression method based on a large dataset comprising of more than 200,000 cases. It is found that DT has an accuracy rate of 93.6%, ANN has an accuracy rate of 91.2% and linear aggression has an accuracy rate of 89.2% [13]. Thus both DT and ANN are better than linear regression. Moreover, DT is the best predictor and ANN is the next best. K-fold cross-validation is applied to the model in [14] based on Square Support Vector Machines (LS-SVM) concerning accuracy, sensitivity, and specify matrix. To solve classification problems under WBCD of [15], a new genetically optimized ANN (GOANN) algorithm is proposed in 2015 [16], which uses genetic programming (GP) based on ANN. In 2014, a classification accuracy of 94.56% is obtained on classifying WBCD [17]. This is achieved by applying the J48 decision tree method, which is originally developed by the WEKA project team [18]. Feature selection is also an important element in the field of machine learning as it reduces the number of features in the data, consequently decreasing the training time, noise, and complexity of the model giving us an upper hand of the set of all features. Thus the real challenge is to apply the feature selection method to find the minimum number of features, which can predict breast cancer as efficiently as possible on the WDBC data set using the K-fold method of accuracy. This is the motivation of the present paper. We validate our method on the Wisconsin Diagnostic Breast Cancer dataset. It is found that a combination of the K-fold cross-validation accuracy method and feature selection methods, breast cancer is predicted more efficiently by different ML algorithms. We also compared these different ML algorithms based on their accuracy score and choose the right alternatives. It is also proved that after a drastic reduction of the features results remains satisfactory.

2 Material and Methods

2.1 Data Sets

Wisconsin Diagnostic Breast Cancer (WDBC) data set [19] has been used in our current work. This data set contains information on 569 breast tumors, out of these 357 are benign classes and 212 are malignant classes. This data set consists of 10 different features (Radius, Texture, Perimeter, Area, Smoothness, Compactness, Concavity, Concave points, Symmetry, Fractal Dimension). All the features are computed from a digitized image of an acceptable needle aspirate (FNA) of a breast mass. All these 10 features are present with their 3 different measurements (mean, standard error, and worst measurement). Therefore, the data set comprised of 30 features in total.

2.2 Experiments

2.2.1 ML Algorithms

After pre-processing the dataset, the following eleven different machine learning algorithms have been applied: (a) Logistic Regression, (b) Support Vector Machine (Linear and Radial basis function kernel), (c) Multilayer Perceptron, (d) Adaboost, (e) Gradient Boosting, (f) K-Nearest Neighbors, (g) XG Boost, (h) Random Forest, (i) Linear Discriminant Analysis, (j) Decision Tree, (k) Naïve Bayes.

All the above-mentioned algorithms have been applied to the data set to calculate accuracy score, recall score, precision score, and f1 score. As of now, ample of articles on breast cancer prediction [20] has used test accuracy score for comparison of different machine learning algorithms whereas in this paper, instead of using test accuracy score for comparison, 10-fold cross-validation accuracy score has been used. In 10-fold cross-validation, data is divided into 10 number of folds of the same size. Now 9 folds of the data are used for training while the remaining onefold is used for testing and this process is then repeated 10 times. Thus, here the data is being used much more efficiently as every observation has been used for both training

Table 1 Measurement of performance using the different ML classification algorithms

	Accuracy	Recall	Precision	f1_score
Logistic Regression	0.982485	0.96688	0.986107	0.976053
SVC (RBF kernel)	0.982455	0.95735	0.995455	0.975339
MLP	0.982424	0.96212	0.990238	0.975717
Linear SVC	0.978975	0.95757	0.985627	0.970937
Adaboost	0.977189	0.96233	0.976605	0.969076
Gradient boosting	0.968509	0.94350	0.971605	0.956739
KNN	0.966694	0.91969	0.990097	0.952358
XG boost	0.963183	0.92943	0.971623	0.949542
Random forest	0.959734	0.93896	0.953338	0.945226
LDA	0.956227	0.89177	0.990238	0.93681
Decision tree	0.93169	0.91103	0.905976	0.906979
Naive bayes	0.931568	0.90064	0.919914	0.907504

and testing. After evaluating the accuracy score, we have manually selected all the hyperparameters for the algorithms. Eventually, all the algorithms have been sorted according to the accuracy score as shown in Table 1.

3 Feature Selection Methods

Feature Selection is the process to select the features which contribute most to predict the output variable. Having irrelevant features in the data set can negatively impact the performance of an ML algorithm. Thus, feature selection methods are used to remove unnecessary irrelevant features from the data set. Some recent studies [21] also show that a poor machine learning model with key important features works much efficiently than a low-error ML model that lack such important features. The following feature selection methods are incorporated into our study:

- Correlation-based feature selection

Correlation is a well-known method to measure the similarity between two features. For two linearly dependent features, their correlation is ± 1. When two features are uncorrelated, the correlated coefficient is 0. In this method, a correlation matrix is formed with the help of the Pearson correlation by which the features are filtered. Given a pair of features x, y, we can define Pearson correlation as:

$$\rho_{x,y} = \frac{\text{cov}(X,Y)}{\rho_X \rho_Y} \quad (1)$$

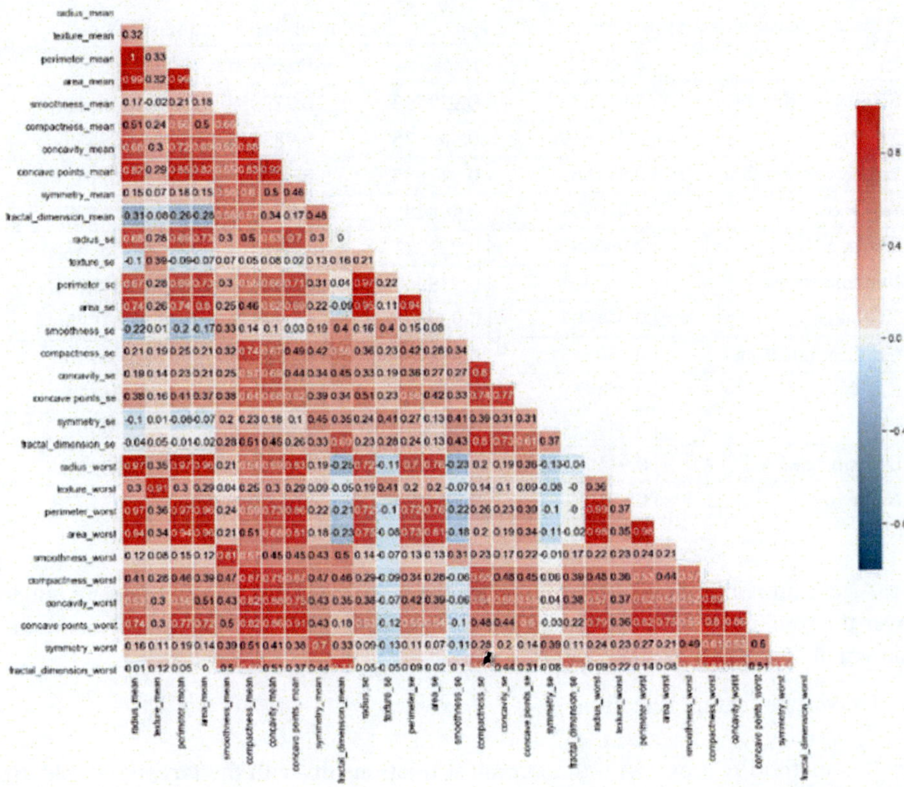

Fig. 1 Pearson correlation heatmap

where cov represents the covariance, σX represents the standard deviation of X, and σY represents the standard deviation of Y.

From this Pearson correlation heatmap in Fig. 1, it is observed that the following features are highly correlated:

(1). Radius_mean, perimeter_mean and area_mean, (2). compactness_mean, concavity_mean and concave points_mean, (3). radius_se, perimeter_se and area_se, (4). radius_worst, perimeter_worst and area_worst, (5). compactness_worst, concavity_worst and concave points_worst, (6). compactness_se, concavity_se, and concave points_se, (7). texture_mean and texture_worst, 8. area_worst and area_mean.

One feature is selected from a set of correlated features and remaining redundant features are removed from the data set. Therefore, in our experiment the following features are removed from the dataset: perimeter_mean, radius_mean, compactness_mean, concave points_mean, radius_se, perimeter_se, radius_worst, perimeter_worst, compactness_worst, concave points_worst, compactness_se, concave points_se, texture_worst, area_worst.

Table 2 Measurement of performance after applying Correlation-based feature selection

	Accuracy	Recall	Precision	f1_score
SVC(RBF Kernel)	0.975374	0.952597	0.982016	0.966019
MLP	0.973619	0.947835	0.982016	0.963458
Logistic regression	0.971865	0.947835	0.977016	0.96114
Adaboost	0.970201	0.948485	0.971536	0.958609
Linear SVC	0.966662	0.929221	0.981818	0.953567
Random forest	0.964907	0.938961	0.967452	0.95187
XG boost	0.964907	0.938961	0.967148	0.951273
Gradient boosting	0.959643	0.938961	0.954573	0.944716
KNN	0.954411	0.900649	0.976517	0.935511
LDA	0.947455	0.86342	0.995238	0.922951
Decision tree	0.941972	0.938745	0.913302	0.923911
Naive bayes	0.928091	0.886797	0.919804	0.901187

After removing those 13 features, machine learning algorithms have been applied over the remaining 17 features and the accuracy, recall, precision, and f1 score are shown in Table 2.

- Univariate feature selection

In this method features with the strongest relationship with the target-variable are selected. In our experiment chi squared (chi^2) statistical test has been used to select best n number of features, where n is a positive integer and then machine learning algorithms are applied on those selected feature sets. At first, we have chosen $n = 5$ to get the 5 most important features. The selected features are area worst, area mean, area se, perimeter worst, perimeter mean. Using these 5 features, again all the scores have been calculated and shown in Table 3.

Next, we have chosen $n = 20$, and selected features are area worst, area mean, area se, perimeter worst, perimeter mean, radius worst, radius mean, perimeter se, texture worst, texture mean, concavity worst, radius se, concavity mean, compactness worst, concave points worst, concave points mean, compactness mean, symmetry worst, concavity se, compactness se. Similarly calculated Scores using these 20 features have been shown in Table 4.

- Tree-based feature selection

This feature selection method can be applied to tree-based methods as in these methods computing importance of individual features is possible. The importance of a feature can be computed by calculating how much that feature is decreasing the impurity. Features that can decrease the impurity more are more important. For classification problems, the impurity can be either Gini impurity or the information entropy. The Gini impurity and the information entropy can be defined as:

Table 3 Measurement of performance with 5 best features obtained from univariate feature selection

	Accuracy	Recall	Precision	f1_score
XG Boost	0.938679	0.901948	0.934503	0.915624
Random forest	0.937017	0.89697	0.932019	0.913451
Gradient boosting	0.933506	0.901948	0.920883	0.909626
Logistic regression	0.928181	0.869264	0.937902	0.898672
KNN	0.92815	0.85	0.954636	0.897001
Linear SVC	0.926551	0.836147	0.962058	0.892617
Adaboost	0.926427	0.873377	0.926985	0.897516
MLP	0.926426	0.878571	0.923856	0.897936
SVC (RBF kernel)	0.924674	0.854762	0.941671	0.892958
LDA	0.921041	0.830952	0.953482	0.88608
Naive bayes	0.912421	0.822078	0.939639	0.8723
Decision tree	0.910387	0.882468	0.881863	0.88035

Table 4 Measurement of performance with 20 best features obtained from univariate feature selection

	Accuracy	Recall	Precision	f1_score
Adaboost	0.971986	0.957792	0.967619	0.962203
SVC (RBF kernel)	0.971958	0.952814	0.971755	0.961515
Logistic regression	0.971927	0.957576	0.96721	0.961628
Linear SVC	0.971895	0.948268	0.976339	0.96118
MLP	0.970202	0.95303	0.966972	0.959081
XG boost	0.968601	0.948268	0.967708	0.957166
Random FOREST	0.966784	0.938961	0.971345	0.954209
Gradient boosting	0.963276	0.943506	0.958423	0.950198
KNN	0.963184	0.929437	0.971006	0.948468
LDA	0.958014	0.891558	0.995455	0.938591
Decision tree	0.938741	0.934416	0.903869	0.918418
Naive bayes	0.935078	0.895887	0.929886	0.910582

$$\text{Gini} = 1 - \sum_{i=1}^{n} p^2(c_i) \tag{2}$$

$$\text{Entropy} = \sum_{i=1}^{n} -p(c_i)\log_2(p(c_i)) \tag{3}$$

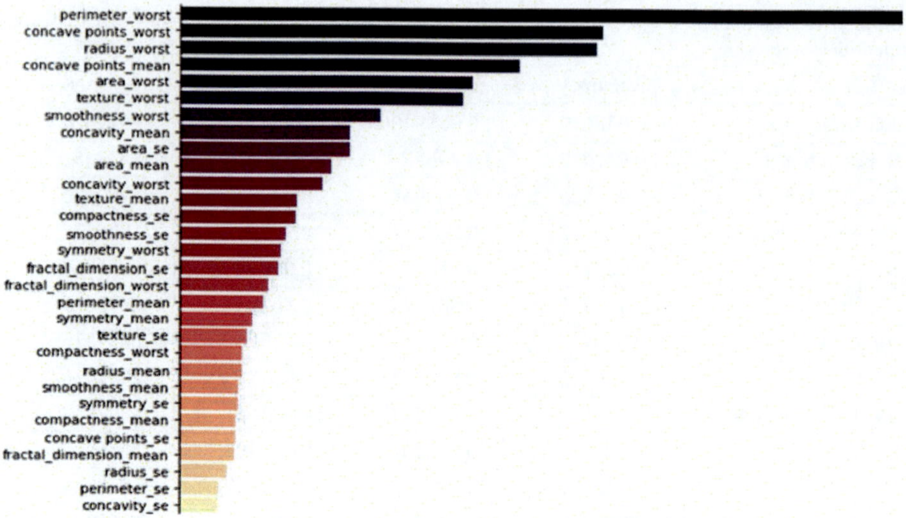

Fig. 2 List of important features based on the tree-based feature selection method

where $p(ci)$ is the probability of choosing a data point with class i. After calculating the impurity decrease from each feature, the average impurity decrease for the whole tree is calculated and the final importance of the feature is calculated. In this way, the importance of each feature has been calculated for Decision Tree, Random Forest, Gradient Boosting, AdaBoost, XGB algorithm. For the comparison purpose, all these important values have been scaled in the range of 0 and 1 using the following formula:

$$z = \frac{x - \min(x)}{\max(x) - \min(x)} \tag{4}$$

Finally, the sum of the value of features importance has been calculated for each model and has been plotted in Fig. 2.

Now, the top 5 features from Fig. 2 has been selected, and after applying all the algorithms on those features, calculated scores are shown in Table 5.

Similarly, the top 20 features from Fig. 2 have been selected and the calculated scores are shown in Table 6.

4 Results and Discussion

From Table 1 it is observed that when all the 30 features are taken, Logistic Regression, Support Vector Classification and Multilayer Perceptron got an accuracy score over 98% and all the other algorithms have performed well with more than 93% accuracy score. After applying correlation-based feature selection by considering 17 features, the accuracy score of XG boost, Random Forest, and Decision tree algorithm

Table 5 Measurement of performance with top 5 features obtained from tree-based feature selection

	Accuracy	Recall	Precision	f1_score
SVC (RBF kernel)	0.949175	0.905628	0.96087	0.930479
Gradient boosting	0.947545	0.92013	0.940549	0.929096
Logistic regression	0.947452	0.924675	0.938608	0.929537
Linear SVC	0.94742	0.900866	0.96087	0.927645
Naive bayes	0.943973	0.92013	0.93276	0.924973
XG boost	0.943942	0.915368	0.935812	0.92434
MLP	0.943912	0.91039	0.94337	0.923914
LDA	0.942219	0.872944	0.971292	0.917983
KNN	0.940464	0.89632	0.943743	0.918403
Decision tree	0.936954	0.911039	0.921575	0.914616
Random forest	0.936894	0.905628	0.927818	0.914293
Adaboost	0.931567	0.89632	0.921741	0.906558

Table 6 Measurement of performance with top 20 features obtained from tree-based feature selection

	Accuracy	Recall	Precision	f1_score
Linear SVC	0.978945	0.957576	0.985213	0.970836
Adaboost	0.97725	0.962554	0.976583	0.968966
Logistic regression	0.977222	0.957576	0.980668	0.968511
SVC (RBF kernel)	0.977191	0.957576	0.9811	0.968383
MLP	0.977191	0.957576	0.980451	0.968513
Random forest	0.970232	0.95303	0.967276	0.959542
XG Boost	0.966693	0.948052	0.963342	0.95473
KNN	0.966601	0.919481	0.990652	0.952724
LDA	0.963276	0.906061	0.995238	0.947488
Gradient boosting	0.96146	0.938745	0.957514	0.947209
Naive bayes	0.943943	0.919697	0.932772	0.924323
Decision tree	0.940277	0.929437	0.914695	0.920899

increased a bit, whereas the accuracy of all the other algorithms decreased by around 1%. Then univariate feature selection has been used to select the top 5 and top 20 important features. Now with 20 features, the accuracy score for XG boost, Random Forest, Decision tree, LDA, and Naive Bayes has increased; and the accuracy of the remaining algorithms has decreased by around 1%. And with the top 5 features, the accuracy score of all the algorithms are in the range from 91 to 93%. Similarly, the top 20 and top 5 features have been selected with tree-based feature selection. However, after obtaining the performance with 20 features, the accuracy of Adaboost, Random Forest, XG Boost, LDA, Naive Bayes, and Decision Tree has subsequently increased whereas the accuracy score of Linear SVC and KNN is almost same, and the accuracy score of Logistic Regression, SVC (RBF kernel), MLP, Gradient Boosting has decreased. With the top 5 features, the accuracy score of all algorithm ranges between 93 and 94%. In Fig. 3, we have compared the results which were obtained without feature selection with those three methods obtained after applying feature selection. In Fig. 4, we have compared univariate and tree-based feature selection method with 5 features. So, by comparing univariate feature selection using tree-based feature selection using the same number of features, tree-based feature selection has given

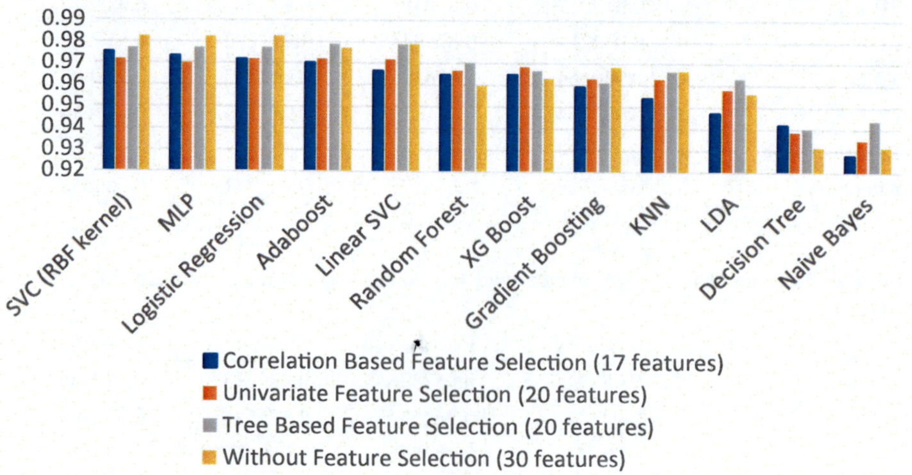

Fig. 3 Comparison of different feature selection methods

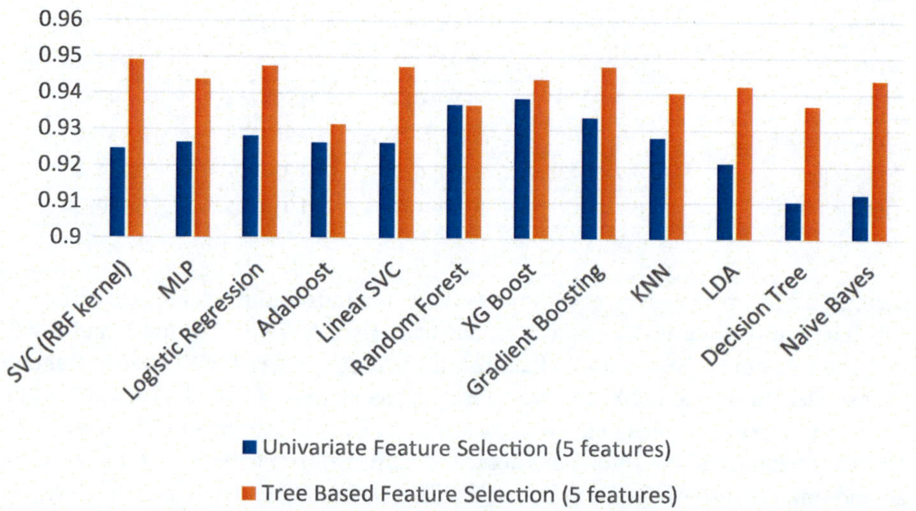

Fig. 4 Comparison between univariate and tree-based feature selection with 5 features

comparatively better results. Another observation is that the Tree-based feature selection has given better results than correlation-based feature selection though only 17 features are used in the case of correlation-based feature selection method. Finally, if we compare all the algorithms in terms of accuracy score, Logistic Regression, Support Vector Classification and Multilayer Perceptron are giving better results in most of the cases.

5 Conclusion

In the healthcare industry, early-stage breast cancer detection is an important issue. Detecting it at a preliminary stage is extremely crucial to prevent the huge death toll. In this paper, eleven machine learning algorithms and three feature selection methods are compared for breast cancer prediction to find the best algorithm and feature selection method in terms of accuracy. Here, we have used k-fold cross-validation for the comparisons which have facilitated efficient data utilization and guarantees a much accurate measurement of the model's performance than the traditional test accuracy. In most cases, it is observed that Logistic Regression, Support Vector Classification, and Multilayer Perceptron outweigh in terms of accuracy than other machine learning algorithms. It is also clearly visible that, tree-based feature selection is giving better results in terms of accuracy score when compared to other feature selection methods.

References

1. Ponnuraja CC, Lakshmanan B, Srinivasan V, Prasanth BK (2017) Decision tree classification and model evaluation for breast cancer survivability: a data mining approach. Biomed Pharmacol J 10:281–289
2. Malehi AS (2014) Diagnostic classification scheme in Iranian breast cancer patients using a decision tree. Asian Pac J Cancer Prev 15:5593–5596
3. Shrivastava SS, Sant A, Aharwal RP (2013) An overview on data mining approach on breast Cancer data. Int J Adv Comput Res. 3(4):256–262
4. Chen W, Zheng R, Baade PD, Zhang S, Zeng H, Bray F, Jemal A, Yu XQ, He J (2016) Cancer statistics in China, 2015. CA Cancer J Clin 66:115–132
5. Siegel RL, Miller KD, Jemal A (2016) Cancer statistics, 2016. CA: Cancer J Clin 66:7–30. https://doi.org/10.3322/caac.21332
6. Torre LA, Bray F, Siegel RL, Ferlay J, Lortet-Tieulent J, Jemal A (2015) Global cancer statistics, 2012. CA Cancer J Clin 65:87–108
7. Pavlopoulos SA, Delopoulos AN (1999) Designing and implementing the transition to a fully digital hospital. IEEE Trans Inf Technol Biomed 3:6–19
8. Yue W, Wang Z, Chen H, Payne A, Liu X (2018) Machine learning with applications in breast cancer diagnosis and prognosis. Designs 2(2):13
9. Bhoo-Pathy N, Verkooijen HM, Tan E-Y, Miao H, Taib NAM, Brand JS et al (2015) Trends in presentation, management and survival of patients with de novo metastatic breast cancer in a southeast Asian setting. Sci Rep 5:16252

10. Yip CH, BhooPathy N, Uiterwaal CS, Taib NA, Tan GH, Mun KS et al (2011) Factors affecting estrogen receptor status in a multiracial Asian country: an analysis of 3557 cases. Breast 20:S60–S64
11. Ng CH, Pathy NB, Taib NA, Ho GF, Mun KS, Rhodes A et al (2014) Do clinical features and survival of single hormone receptor positive breast cancers differ from double hormone receptor positive breast cancers? Asian Pac J Cancer Prev 15:7959–7964
12. Cruz JA, Wishart DS (2006) Applications of machine learning in cancer prediction and prognosis. Cancer Inform. 2:59–77
13. Delen D, Walker G, Kadam A (2005) Predicting breast cancer survivability: a comparison of three data mining methods. Artif Intell Med 34:113–127
14. Mangasarian OL, Setiono R, Wolberg WH (1990) Pattern recognition via linear programming: theory and application to medical diagnosis. In: Coleman TF, Li Y (eds) Large-scale numerical optimization, SIAM Publications, Philadelphia, pp 22–30
15. Bhardwaj A, Tiwari A (2015) Breast cancer diagnosis using genetically optimized neural network model. Expert Syst Appl 42:4611–4620
16. Koza JR, Rice JP (1991) Genetic generation of both the weights and architecture for a neural network. In: IJCNN-91-Seattle international joint conference on neural networks vol 2, pp 397–404. https://doi.org/10.1109/IJCNN.1991.155366
17. Witten IH, Frank E, Trigg L, Hall M, Holmes G. Cunningham SJ (1999) Weka: practical machine learning tools and techniques with Java implementations. (Working paper 99/11). Hamilton, New Zealand: University of Waikato, Department of Computer Science
18. Sumbaly R, Vishnusri N, Jeyalatha S (2014) Diagnosis of breast cancer using decision tree data mining technique. Int J Comput Appl 98:16–24
19. Frank A, Asuncion A (2010) UCI machine learning repository [http://archive.ics.uci.edu/ml]. Irvine, CA: University of California, School of Information and Computer Science
20. Agarap AFM (2018) On breast cancer detection: an application of machine learning algorithms on the wisconsin diagnostic dataset. In: Proceedings of the 2nd international conference on machine learning and soft computing (ICMLSC '18). Association for computing machinery, New York, NY, USA, pp 5–9. https://doi.org/10.1145/3184066.3184080
21. Balachandran PV, Xue D, Theiler J, Hogden J, Gubernatis JE, Lookman T (2018) Importance of feature selection in machine learning and adaptive design for materials. In: Lookman T, Eidenbenz S, Alexander F, Barnes C (eds) Materials discovery and design. Springer Series in Materials Science, vol 280. Springer, Cham. https://doi.org/10.1007/978-3-319-99465-9_3

Word Boundary Detection Using Convolutional Neural Network (CNN) and Decision Tree Method

Kaushik Sarkar, Arnab Sadhukhan, Atreyee Mukherjee, Shramana Guchait, and Sudipta Banerjee

Abstract Speech recognition is a vast area for research. In speech recognition word boundary detection is an important part. Understanding and fixing the problems of efficiently detecting where the words are present in a signal is still challenging. In this paper, we have discussed our work on word boundary detection using two different approaches: (1) Convolutional Neural Network and (2) Decision Tree Method. CNN is efficiently used in areas like face recognition, object detection, image classification, etc. whereas a decision tree is broadly used in decision analysis. This study can further help with speech recognition.

Keywords Signal processing · Word boundary detection · Convolution neural network · Decision tree

1 Introduction

Speech signal consists of the words spoken along with some silent parts comprising random noises. During continuous speaking, the consecutive words have very small pauses between them. Thus, it becomes difficult to separate them within a signal. In this paper, the first approach to detect the word boundary is Convolutional Neural Network (CNN).

CNN model is a type of neural network which is used for visual imagery analysis. It recognizes the pattern and based on them it classifies the images. It finds the relation between two nearest pixels and tries to recognize any pattern or higher dimensional features which is present in an image.

In CNN approach, first, we have created word profiles of the audio signals. Then we have done some preprocessing to reduce the input and output shape. Finally, Mel spectrogram transformation is done to create the training dataset. The next

A. Sadhukhan (✉) · A. Mukherjee · S. Guchait · S. Banerjee
Narula Institute of Technology, Kolkata, India

K. Sarkar
Springer Heidelberg, Tiergartenstr. 17, 69121 Heidelberg, Germany

M. Mitra et al. (eds.), *Computational Advancement in Communication, Circuits and Systems*, Lecture Notes in Electrical Engineering 786,
https://doi.org/10.1007/978-981-16-4035-3_2

approach Decision tree method is a type of algorithm that uses a tree-like structure for classification purpose. It creates the branches from the input data and in every branch, it applies some operations based on a comparison among some quantities.

In this method, some transformations on the raw audio signal are done to get the features from the audio signals. Here we have used MFCC matrix as the input to the model. Word profiles are used as the labels for the model.

2 Methodology

For this total work, first, we have to trace that raw audio signal's portion where the words are present. We can get the word boundaries by following two techniques:

- Convolutional Neural Network (CNN)
- Decision Tree Method

We will further discuss these techniques. Before that, we will explain few signal processing techniques which we used in this project.

2.1 *Mel Spectrogram*

Spectrogram is a powerful visualization tool to represent the signal strength of a signal. It is basically a two-dimensional graph where horizontal axis is the time axis, vertical axis is frequency axis, and amplitude is represented by color intensity where the dark color implies lower amplitude and the bright color implies higher amplitude.

In case of a speech signal, we transform the frequency axis into log scale and color (amplitude) axis into Decibels.

MFCC [3] The human auditory system interpretation of pitch is not in a linear manner. To represent this in a linear scale, the Mel Scale was developed by Stevens, Volkmann, and Newman in 1937 by experimenting with human ears interpretations of pitch [2]. This scale is constructed by assigning a perceptual pitch of 1000 Mels to a 1000 Hz frequency, 40 dB above the listener's threshold and above 1000 Hz the scale shows a logarithmic nature. The formula for Mel scale is:

$$Mel(f) = 2595\log(1 + \frac{f}{700})$$

Mel scale spectrogram is a spectrogram where the y axis is the Mel scale.

2.2 Mel Frequency Cepstral Coefficients (MFCC)

MFCC [3] Speech is the sound that is generated by a human by changing the shape of the vocal tract which includes tongue, teeth, etc. The shape of the vocal tract is responsible for the variation of sound [1]. If we can determine the shape accurately, then we can predict which phoneme is being produced. The shape of the vocal tract manifests itself in the envelope of the short time power spectrum. Here MFCC comes to accurately get the envelope from the audio signal.

Mel Frequency Cepstral Coefficients (MFCCs) is a mostly used feature in automatic speech and speaker recognition.

3 Word Boundary Detection Using Convolutional Neural Network

We are supposed to create a model which can take an audio signal as input and can tell what are each word's starting and ending position in that signal.

Let us consider the following audio signal.

Just by looking at this raw audio signal, we can't get any clue about where the word breaks will be and another problem is different audio signals have different lengths. The required model is supposed to work for audio signals of any length. The following audio signal shown above has five words.

3.1 Create Word Profile

We have used 150 audio signals as our dataset. We have manually taken each word's start time and its duration and saved it as a csv file. Then we have created a profile of words from the csv files.

The csv file looks like this:

Name	Start	Duration
marker 01	0:00.202	0:00.491
marker 02	0:00.718	0:00.517
marker 03	0:01.537	0:00.532

```
Steps:
length = length (audio signal)
word profile = zeros (shape = (length))
for each marker in (no. of markers):
  end time = start time + duration
```

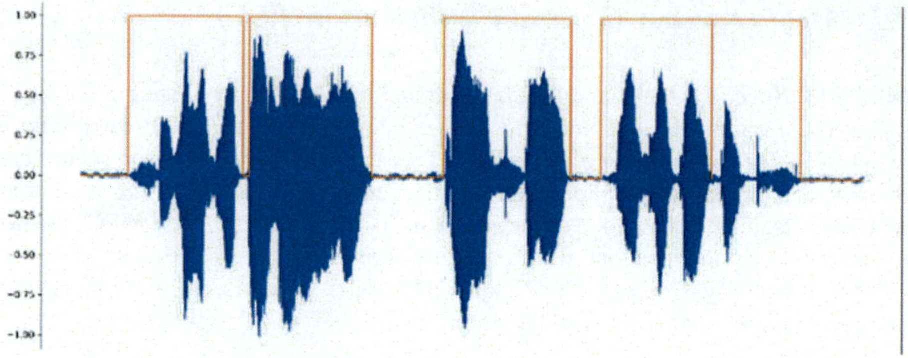

Fig. 1 Raw audio waveform with word boundary

```
    word profile [start time: end time] = 1
  end for
```

In this way, we have created a word profile for the signals which looks like the orange profile shown in Fig. 1.

Note: The word profile has the same length as the audio signal.

Now the word profile holds all the information:

- where the words are
- how many words are there
- every word's starting and ending time or position

Now the main idea is, we're giving audio signal to a model as input and it will give the signal's word profile as output.

But this is difficult because suppose a normal audio length is 3 s and if its sampling frequency is 22,050/s, then we have a length of audio signal of 3 × 22,050 = 66,150 samples and the length of word profile is also the same, i.e., 66,150 samples and feeding raw audio signal to train a model is not a good idea. So, we have done some preprocessing on the audio signals.

3.2 *Data Preprocessing*

We have divided the signal into 0.5 s segments. We have done the same thing to the word profile (Fig. 2).

Now we have 0.5 s input and 0.5 s output. If sampling frequency = 22,050/s then input shape = 22,050 * 0.5 = 11,025 samples and output (word profile) shape = 11,025 samples.

The red dots in Fig. 3 are the 5 ms interval points. The sequence is 500 ms, so we have taken one value for every 5 ms interval which is 0 or 1. These are enough

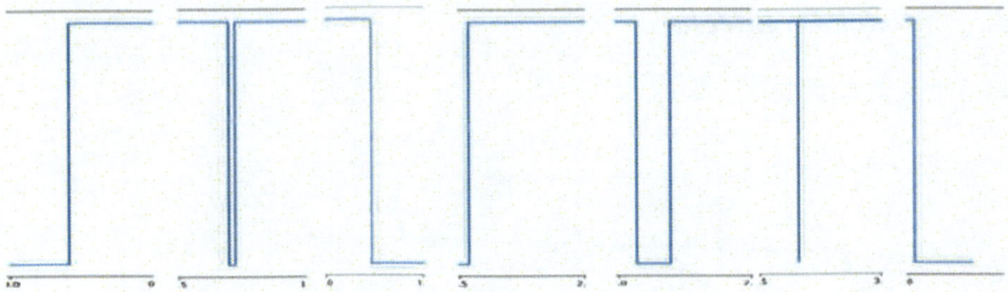

Fig. 2 0.5 s segment of a complete word profile

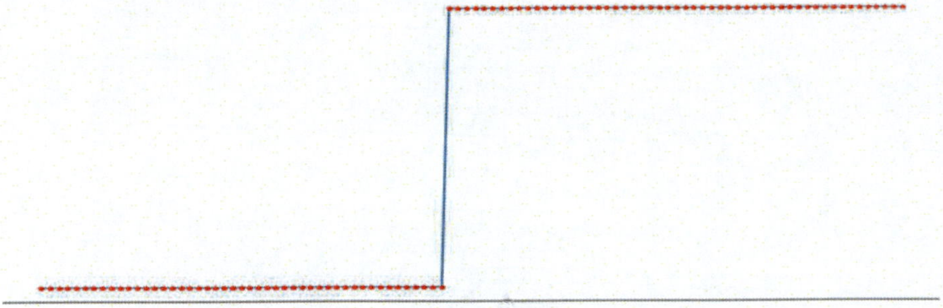

Fig. 3 Word profile with 5 ms interval points

values to represent the profile. So now, the number of points will be 500 ms /5 ms = 100 samples.

Finally, we have 11,025 numbers of points for the input signal for the model and 100 output points.

3.3 Model Architecture

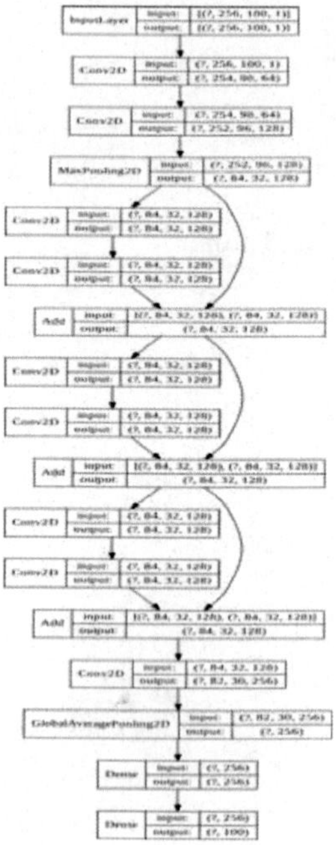

3.4 Create Training Dataset

Now the model can be trained but there is still one problem left. Training the model with raw audio signals is not a good idea. We have to transform the audio from a time domain to any other domain.

The transformation options are: (1) FFT, (2) STFT, (3) Spectrogram, (4) Mel Spectrogram, (5) MFCC.

Among these Mel spectrogram gives the best result, so we have performed this transformation.

To get the 0.5 s Mel spectrogram we need to set the parameters:

window = 0.005 s = 5 ms, fs = 22,100.

We've used python library librosa to get the mel spectrogram.

Librosa [2] librosa.feature.melspectrogram(y = audio_samples_array, sr = fs, n_mels = 256, fmax = 8000, hop_length = int(fs*window), n_fft = 1024).

This gives a 256 × 100 matrix for every 0.5 s. Now we have a [256 × 100] matrix. This matrix is the input to the model.

4 Word Boundary Detection Using Decision Tree

This is also the same approach as CNN model. We have performed some transformations on the raw audio signal to get the features from the signals and we have used the word profiles as the labels for our model. In CNN based model the disadvantages are, this model takes too much time to train and lots of computation is required for the prediction. So instead of using CNN based model, we can use a Decision tree. In this method, we use the MFCC transformation for the raw audio signals. We have created training data from this MFCC matrix and given this matrix as input to the model and used the word profile as the output.

4.1 Create MFCC Matrix from Audio Signals

To get the MFCCs from the audio signal, we've performed the following steps:

- First, we break the audio signal into small 5 ms frames. So, if we take the signal which is sampled at 22050 Hz, we get 22,050*0.005 = 110 samples, If the speech file does not divide into an even number of frames, pad it with zeros so that it does.
- Then we take each frame and apply complex DFT to it. We take the absolute value of the complex Fourier transform and square the result. We would generally perform a 1024-point FFT and keep only the first 512 coefficients (this gives only the positive frequencies). This is known as periodogram power spectral.
- Compute the Mel-spaced filter bank. This is a set of 20–40 (26 is standard) triangular filters that we apply to the periodogram power spectral estimate from step 2. Our filter bank comes in the form of 26 vectors of length 512. Each vector is mostly zeros but is non-zero for a certain section of the spectrum. To calculate filter bank energies, we multiply each filter bank with the power spectrum, then add up the coefficients. Once this is performed, we are left with 26 numbers that give us an indication of how much energy was in each filter bank.
- Then take the log of each of the 26 energies from step 3. This leaves us with 26 log filter bank energies.
- Then we take the Discrete Cosine Transform (DCT) of the 26 log filter bank energies to give 26 cepstral coefficients. In our case, we take only the lower 20 of the 26 coefficients.

4.2 Creating Training Data for Decision Tree

Now we have the MFCC matrix as input and for the output we have a word profile, which is in binary form (0 or 1).

Each of the frames is taken at a 5 ms interval. So, the first frame output should be first value of word profile, 2nd frame output should be 2nd value of word profile and so on but if we take each frame of feature matrix as input and each frame of word profile (which is 0,1), the model doesn‘t know which features are present in the previous slot and which will come in future, it just predicts on current frame features. As a result, the model always tells that every silence part is 0 and each power part is 1.

To remove this problem, we have given the model a few previous frames and a few future frames slot along with the present frame (we have taken the previous 5 frames and future 4 frames along with the current frame, each frame is 5 ms long, so the model gets a $10*5 = 50$ ms slot of the signal to predict the word profile).

Now the model gets the full information about previous and future frames. This significantly improves the model output.

To create training data:

- Pad the total MFCC matrix with 5 frames at the beginning and the ending.
- Take first 10 frames (frame 0 to 9) as input and first value of word profile as output.
- Then shift one frame right and take next 10 frames (frame 1 to 10) and take next value of word profile as output.
- Continue the process for all audio signals and finally train the model with this training data.

5 Result Analysis

5.1 Predict Word Profiles Using Convolutional Neural Network Approach

From the results, we can conclude that the first three figures in Fig. 4 the model is able to predict all the word profiles correctly but in the last figure first two words are predicted as one word but the rest of the words are predicted correctly. In the final figure, the model predicts some individual words together but there is no wrong prediction if we consider all the word boundaries.

5.2 Predict Word Profiles Using Decision Tree Approach

From the results, we can conclude that the first three figures in Fig. 5 the word profiles are predicted correctly but in the last figure, the model predicts some individual words together.

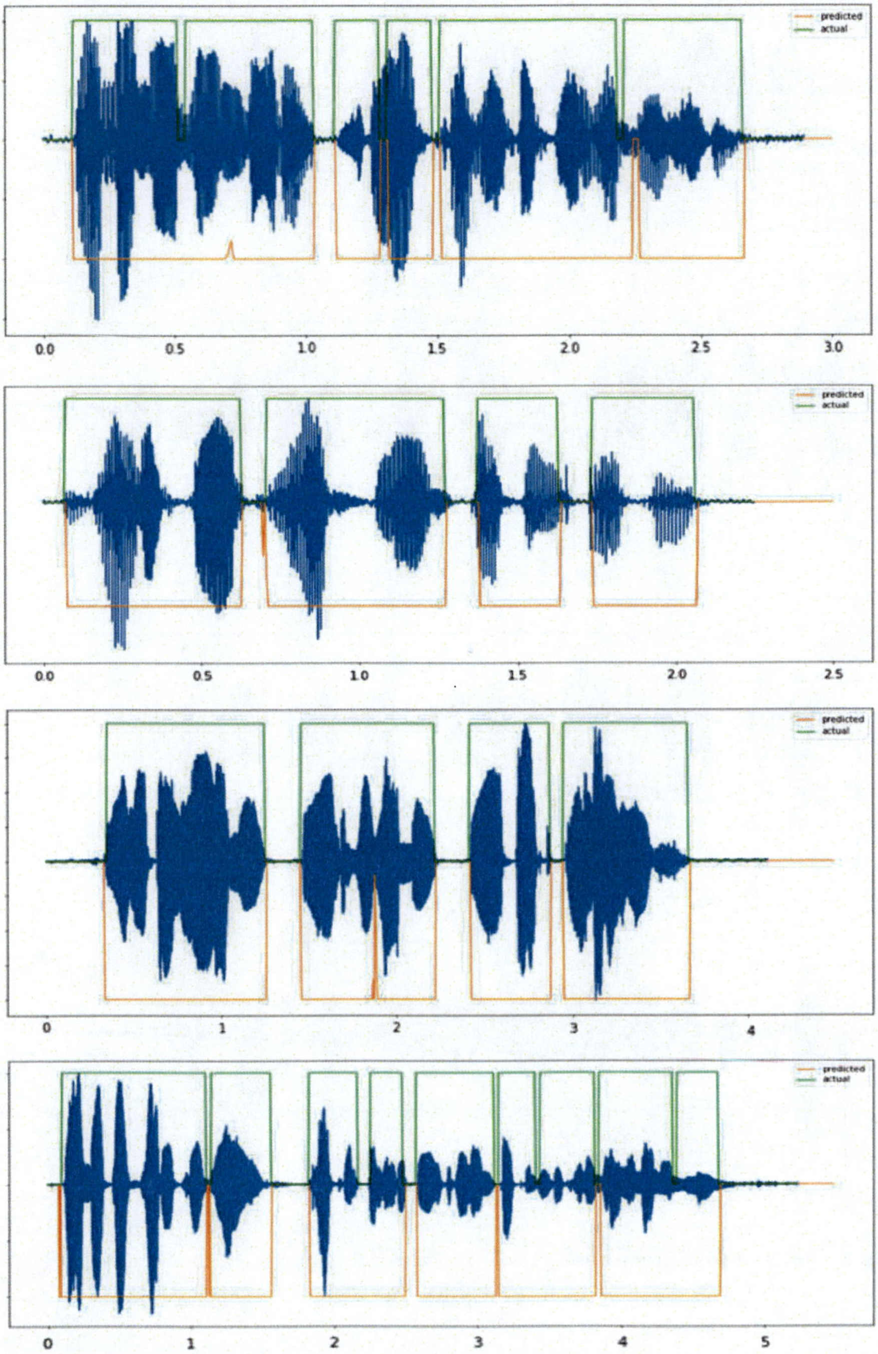

Fig. 4 Predicted word profiles using convolutional neural network model

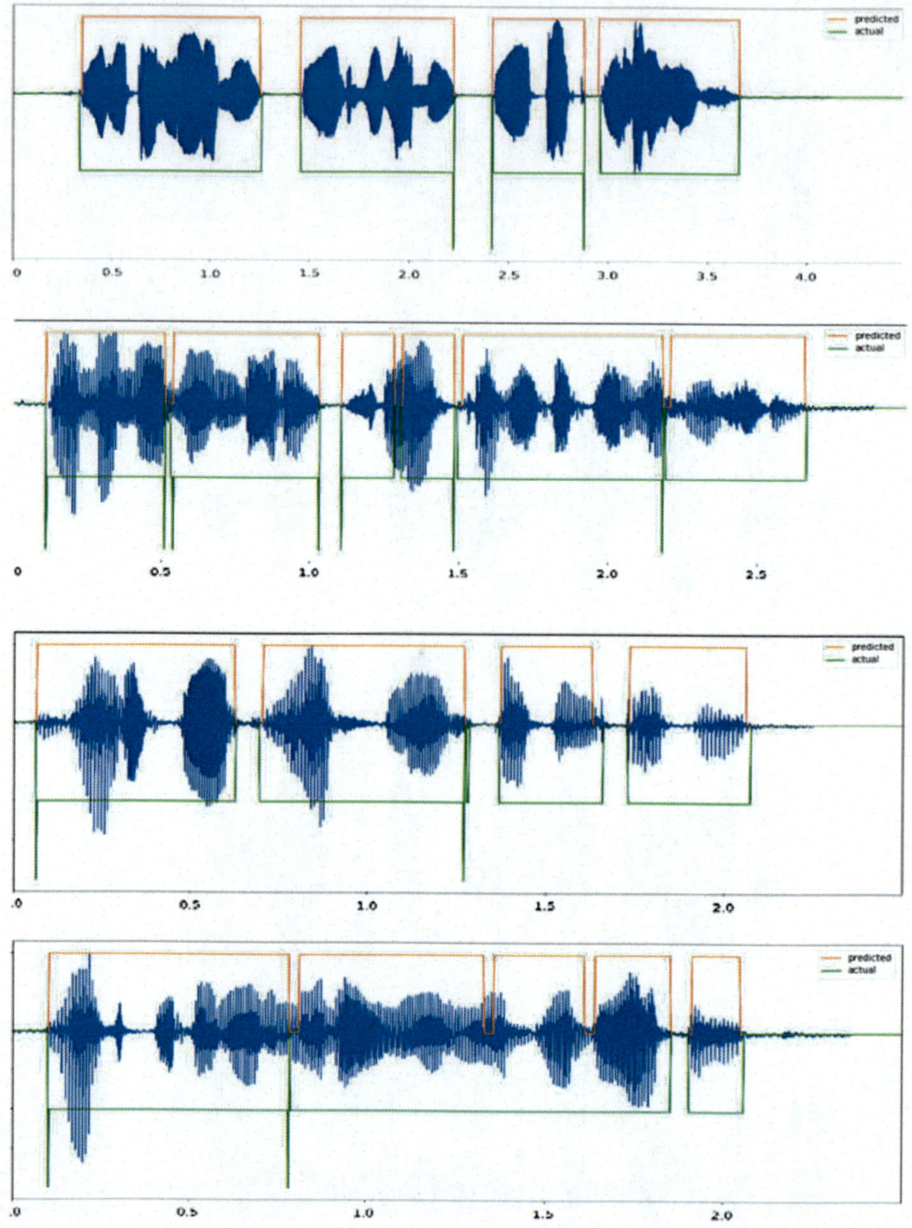

Fig. 5 Predicted Word Profiles using Decision tree model

Table 1 Accuracy table of Convolutional Neural Network (CNN) model with different threshold values

Threshold	Fully accurate	Mostly accurate	Overall accuracy
0.5	0.171	0.4	0.571
0.6	0.214	0.471	0.685
0.7	0.200	0.514	0.714
0.8	0.341	0.4	0.741
0.9	0.285	0.471	0.756

6 Evaluation

6.1 Accuracy of Convolutional Neural Network (CNN) Model

We have to binarize the model's predicted profiles with some threshold value. Different threshold value gives different accuracy. So we have calculated the accuracy with different threshold values and selected the threshold value which gives high accuracy.

The model's prediction is not correct always, sometimes it predicts some individual words together but if we consider only the word boundaries, the model's output is not wrong. So, to measure the accuracy based on the number of words predicted by the model and actual number of words present in that audio signal, if the model's predicted words and actual number of words are the same then we call it fully accurate and if the model predicts some two words together but predict the other words correctly then we call it as mostly correct. Finally, we have calculated the overall accuracy by adding the two accuracy values (Table 1).

From the table, we can see that the threshold value of 0.9 gives the highest accuracy so we take the threshold value as 0.9.

6.2 Accuracy of Decision Tree Based Model

We have used different Decision tree based model and this model's accuracy changes with respect to the depth or number of estimators. So we have built different models and chosen the model which gives highest accuracy.

We have calculated the accuracy of the decision tree based model the same as we calculated in case of CNN model. The overall accuracy is calculated by adding the fully accurate and the most accurate values (Table 2).

From the table, we can see that the random forest model with 10 estimators gives the highest accuracy so we can use that model for final predictions.

Table 2 Accuracy table of Decision Tree based models

Model type	Fully accurate	Mostly accurate	Overall accuracy
Decision tree depth = 5	0.185	0.442	0.627
Random forest estimators = 5	0.285	0.457	0.742
Random forest estimator = 10	0.242	0.514	0.756
Random forest estimator = 20	0.142	0.428	0.57

7 Conclusion

In this project, we have used two methods to get the word boundaries from an audio signal. The CNN method output profile is noisy but it produces good output for unseen data. If we apply some threshold to remove the noise, it will show a perfect output. Sometimes it failed to find some intermediate words, but in most cases it produces good output.

In case of Decision tree, its output profiles are nearly accurate but it doesn't give good results for all data. In case of unseen data, sometimes both the model failed to recognize the intermediate word or predicting two words together.

CNN model's performance is good but a lot of computation is required for a single audio file prediction, but Decision tree model doesn't require that much computation. its prediction is quite fast. This is an easier method to find the word boundary and it can be improved by using adaptive boosting algorithms like ADABOOST method.

References

1. Acoustics of Bangla Speech Sounds, Asoke Kumar Datta, Springer Publications
2. Librosa: a python audio library. https://medium.com/@patrickbfuller/librosa-a-python-audio-libary-60014eeaccfb
3. Mel frequency cepstral coefficient (MFCC) tutorial. http://practicalcryptography.com/miscellaneous/machine-learning/guide-mel-frequency-cepstral-coefficients-mfccs/

Brain Computer Interface: A Review

Debrupa Pal, Sujoy Palit, and Anilesh Dey

Abstract Brain-computer interface (BCI) enables their users to use brain signals instead of the brain's normal peripheral nerve and muscle output paths to communicate or control external devices. Several methods can be used to obtain data from the brain sensors that basically monitor physical processes Brain computer interface technology is an emerging area of research with several applications in medical fields. In this review, we discuss the current status and future prospects of BCI technology and its applications in several fields. We will define BCI, examine BCI-related signals from the human brain, and describe the functional components of BCI. We will also review the different applications of BCI technologies in the field of medicine, in entertainment and games, safety and security and in biomedical. Finally, we will discuss the current restrictions of BCI technology, obstacles to its widespread clinical application, and expectations for the future.

Keywords Brain computer interface · BCI · EEG

1 Introduction

For generations, the greatest desire of mankind is to conquer every nook and corner of this universe. There is no secret hidden in this universe. Earlier human brain was considered as complicated and humans were inquisitive to explore it for a long time [1]. With the explosive growth of technology, partition between humans and central machines have begun to shrink. In the year 1970, research on Brain Computer

D. Pal
Department of Computer Application, Narula Institute of Technology, Kolkata, India
e-mail: debrupa.pal@nit.ac.in

S. Palit
Ericsson India Global Services Private Limited, Kolkata, India

A. Dey (✉)
Department of Electronics and Communication Engineering, Narula Institute of Technology, Kolkata, India
e-mail: anileshdey@ieee.org; anilesh.dey@nit.ac.in

M. Mitra et al. (eds.), *Computational Advancement in Communication, Circuits and Systems*, Lecture Notes in Electrical Engineering 786,
https://doi.org/10.1007/978-981-16-4035-3_3

Interface was started at the University of California Los Angeles (UCLA) under a grant from National Science Foundation followed by a contract from DARPA. Brain-computer interface (BCI) is the direct communication path between brain and computer which can be considered as replacement of natural connection between nervous system (CNS) and musculoskeletal system [2]. The research community initially considered biomedical applications when developing BCI, they help to restore physical disabilities or lock the user's athletic ability, and make up for lost athletic functions [3].

In spite of that, the scope of research has been further extended to include non-medical applications. Recent research is aimed at normal people by exploring the use of BCI as a new input device and investigating the generation of hands-free applications [4, 5]. On the other hand, several advantages of BCI for healthy users have been discussed in [6].

The research paper will guide beginners to fully understand what is BCI, how and why. In addition, BCI applications, challenges, possible solutions and the future was also discussed.

1.1 Functions of BCI

Invasive BCIs are those that are implanted directly into the brain and have the highest quality signal. The function of the application program of the brain computer interface is based on observing the user status or allowing the user communicates his/her thoughts. BCI system records brain waves and send them to the computer system to complete the expected task. The transmitted waves are used to obtain an idea or control an object.

Some of the functions of BCI is described as.

1.1.1 Communication Bridge

The Brain Computer Interface (BCI) system establishes a communication bridge between the human brain and the outside world, thereby eliminating the need for typical information transmission methods. They manage the.

sending of information to the human brain and interpret their silent thoughts. Therefore, they can help people with disabilities to speak and write down their opinions and ideas through multiple methods, such as in silent speech communication [7], spelling application [8], categorization based on semantics [9].

BCI assistive robots can provide support for people with disabilities in their personal and professional life, strengthening their cooperation in building communities [10].

1.1.2 User Status Monitoring

Early BCI applications have been targeted at disabled users with limited mobility or speech problems. Their purpose is to provide these users with alternative communication channels. With large explosion in this technology BCIs may become useful for healthy users in specific situations. It is used as a physiological measurement tool to retrieve and use an individual's emotional, cognitive state. The goal of brain signal utilization has been extended to control certain objects or provide alternatives for certain objects Function, which is known as passive BCI [11].

The BCI user status monitoring function is considered to be a useful in the human–machine interface, and it is adjusted according to the estimated user emotion or cognitive status [6, 9]. It works in shared controls environment and determine the best possible control type used in some cases. The next section will focus on some applications that use the brain computer interface.

2 Applications of BCI

Various forms of brain-computer interfaces have proven to be widely used in almost all research areas. The technology was originally introduced to help people with physical disabilities is now used in medical, neuroergonomics and smart environment, neuromarketing and advertisement, educational and self-regulation, games and entertainment and security and authentication fields. The BCI area of applications can be briefly summarized in Fig. 1 [12].

2.1 Medical Application

Healthcare field has multiple applications that takes advantages of brain signals in all relevant stages including prevention, discovery, diagnosis, rehabilitation and

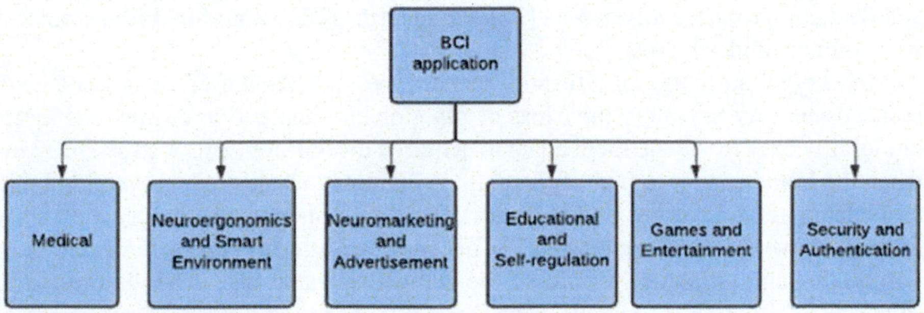

Fig. 1 BCI application fields

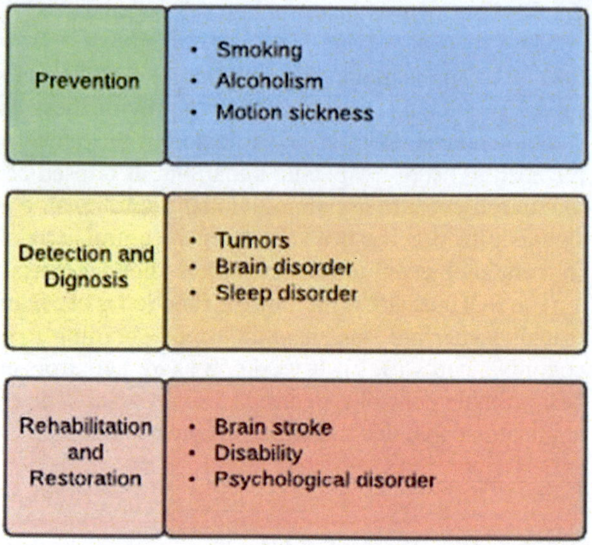

Fig. 2 BCI applications in medical fields

recovery as shown in Fig. 2 [13]. The importance of medical prevention lies in the possible loss of function and reduction in level of alertness caused by smoking and/or drinking. The effect of smoking and alcohol on attentiveness of brain waves have been enlightened in [14–18]. Traffic accidents are contemplated the main cause for death or some serious injuries as stated in [19, 20]. Analyzing the reason for prevention in the future has always been the focus of research in various fields. Thus, concentration level for those suffering from motion sickness, especially drivers, has been studied [21, 22]. The mental state monitoring function of the BCI system also helps to predict and detect health problems, such as Abnormal brain structures (such as brain tumours), seizures (such as epilepsy), sleep disorders (such as narcolepsy), and Swelling of the brain (such as encephalitis).

Using EEG as a cheap secondary alternative to MRI and CT-SCAN, it is possible to find tumours that are caused by uncontrolled cell division. Brain tumours detection system based on EEG have been the main theme of researches in [23, 24] while [25] has been focused on breast cancer identification using EEG Signals. Sharanreddy and Kulkarni proposed a system in [22] that Identifies EEG abnormalities associated with tumours and seizures.

Mobility rehabilitation is a form of physical rehabilitation used for patients with reduced mobility to restore their loss of function and restoration of previous levels of mobility or at least help them adapt to acquired disabilities [26]. A stroke happens when the blood supply to part of the brain is disrupted or reduced, preventing brain tissue from getting oxygen and nutrients. The patient may suddenly lose the ability to speak, have memory impairment, or become paralyzed on one side of the body. Disabilities and stroke have become the subject of many researches interested in solutions Involving brain signals. It has been established in [27] that brain structure related to stroke could be reorganised and the impaired motor function could be restored through neuroplasticity [28, 29]. BCI based prosthetic limbs, also called neuroprosthetic devices can be used to restore normal functionality for patients who cannot recover previous levels of mobility or communication as discussed in [30–33].

2.2 Neuroergonomics and Smart Environment

Smart environments, such as smart houses, workplaces or vehicles, could also uses the brain computer interface to provide further safety, luxury and physiological control for human daily life. It monitors the user's mental state and adapts to the surrounding environment accordingly. They are also expected to see collaboration between Internet of Things (IOT) and BCI technologies as stated in [34].

2.3 Neuromarketing and Advertisement

The field of marketing has also been an area of interest for BCI researches. The benefits of using EEG evaluation of TV advertisements related to both commercials and the political realm is discussed in [35]. The researchers had considered the effect of another cognitive function in the field of neuromarketing. They had been keen in estimating the memorization of TV commercials, which provide another way to evaluate advertising.

2.4 Educational and Self-Regulation

Neurofeedback is a promising way to strengthen the brain performance by adjusting to human brain activity. It invades the education system, which uses EEG signals to determine the clarity of the studied information. Individualised interaction to each learner is entrenched according to the resultant response experienced [36]. In [37] EEG based emotional intelligence has been applied in sports competitions to control the stress associated with it. BCI technology has been explained in self-regulation and skill learning via functional Magnetic Resonance Imaging (fMRI) neurofeedback in [28].

2.5 Games and Entertainment

Nonmedical brain computer interfaces are used extensively in entertainment and gaming applications. Player's physiological functions like brain's signals, heartbeat and facial expression are used in this kind of neuro gaming. In [38] several games are discussed where helicopters are made to fly to any point in either a 2D or 3D virtual world. In [6], Tan and Nijholt described a brain game designed to reduce stress levels of the players. The players can only move the ball by relaxing, therefore calm players are more likely to win, so they would learn to control their stress while being entertained.

2.6 *Security and Authentication*

Biometrics based, knowledge based and object-based authentications are used in security system. There are several applications in this field; detection of signal distortions through EEG and eye movement, detection of irregular behaviour and suspicious objects is discussed in [39]. In a scenario, several testers and viewers are observing the recording of a doubtful event, only EEG and precise eye movement can recognize the potential targets that cannot be determined by any other method [39]. Various researches have considered the authentication of EEG signals generated from driving behaviour as a part of smart navigating systems. To verify driver's identity on demand, a simplified driving simulator with mental tasked condition is used in [40, 41].

3 Components of BCI System

As shown in Fig. 3 [42], a BCI system consist of the following components: signal acquisition, preprocessing, feature extraction, classification and application interface. The signal acquisition component is responsible for recording the electrophysiological signals and transmitting it for signal enhancement. Two techniques of brain acquisition methods are invasive and non-invasive method as shown in Fig. 4 [13]. Preprocessing component is responsible for enhancing the signal-to-noise ratio. The goal of feature extraction is to find discriminative characteristics for the improved signal, reducing the size of the data applied to the classification component. Translation of produced feature into device commands is carried out by classifiers [3, 43].

4 Challenges in BCI System

Usage of brain signals in establishing the communication interface has faced several challenges. System obstacles specially those related to EEG features characteristics are technical challenges. Limitations affecting the level of human acceptance are described as usability challenges [44].

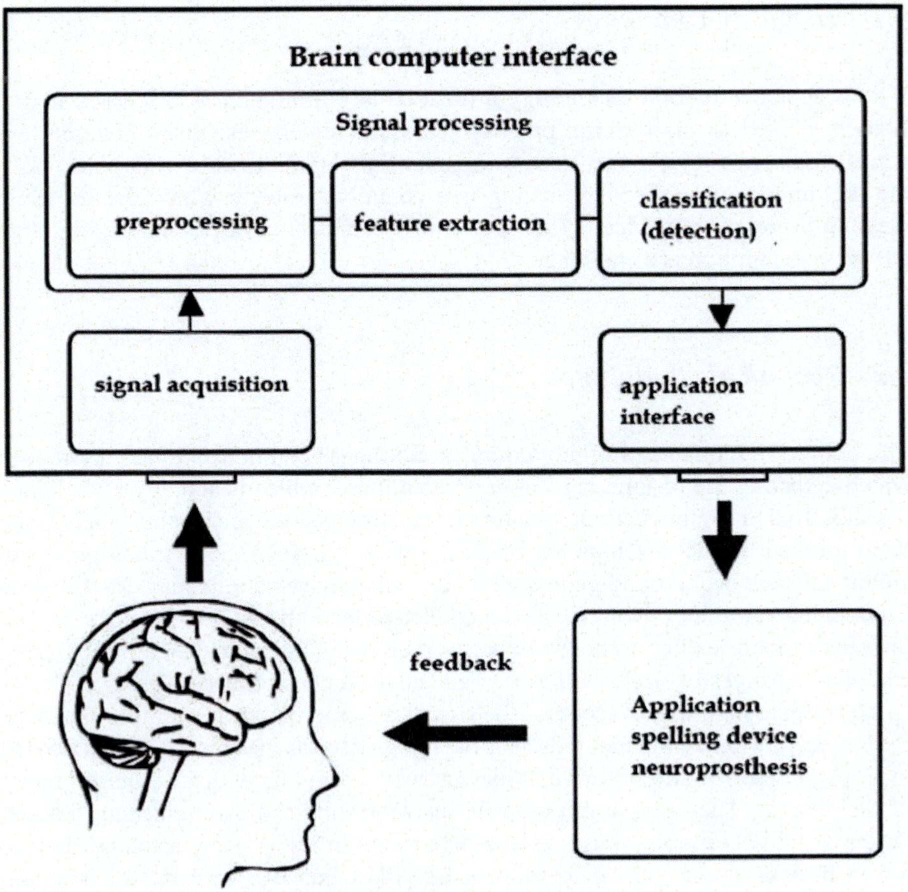

Fig. 3 Components of a BCI system

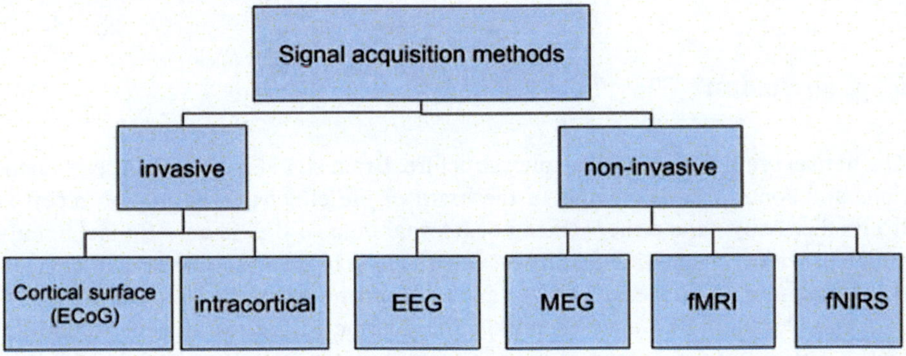

Fig. 4 Brain acquisition methods

4.1 Usability Challenges

User acceptance of BCI technology utilization is a limitation as discussed in [4]. Matters related to the training process necessary for classes' discrimination. are considered. Training the user occurs in either the classifier calibration phase or in the preliminary phase [45]. Sweating is a common problem which occurs while wearing a prosthetic device. When a person wears such a device managing large energy consumption is a challenge.

4.2 Technical Challenges

The brain of a human is a highly complex, nonlinear and nonstationary system in which detecting the chaotic behaviour of neural ensembles is actually a challenge faced by BCI. The non-stationary nature of electrophysiological brain signals is the main problem in the development of BCI systems [5, 46]. The psychological and emotional state background generated by different conversations may contribute to EEG signal variability. Noise is also a challenge faced by BCI technology and an important factor leading to nonstationarity problem. This includes harmful signals caused by changes in electrode position and environmental noise [47].

To preserve high spatial accuracy the signals are recorded from multiple channels. Several feature extraction methods have been suggested as the amount of data needed to properly describe different signals increases exponentially with the dimensionality of the vectors. They play a critical role in identifying the distinguishing feature. Preferably it is desirable to use, at least five to ten times as many training samples for each class as the number of dimensions [44]. For BCI system, this solution cannot be sustained in a highly dimensional environment causing the extension of the dimensionality curse [48].

5 Conclusion

The human brain is a highly complex structure. Brain signals reflect the user's intentions and controlling behaviour of the brain or the effect of information received from other body parts either sensing or internal organs. BCI is a useful technology which provide a channeling facility between human brain and computers or devices. Applications of BCI have encouraged and attracted the researchers around the world.

Applications of BCI have encouraged and attracted the researchers around the world. This paper presented a study of BCI with growing interest in several sectors such as medical, organizational, transportation, gaming and entertainment, and security and authentication fields. It also demonstrates five stages of BCI. These stages are signal acquisition, preprocessing and signal enhancement, feature extraction, feature classification and finally the application interface. It also illustrates several devices used for capturing brain signals. The study of BCI applications reflects that the user can perform any difficult work or any impossible work for a paralyzed person, with the help of only his/her thoughts and without involving any paralyzed organ. There are few challenges and issues posed as a result of utilizing brain signals for example understanding human brain activity, usability, capturing minute details, wearing problems and hardware problems.

According to the review presented in this paper, it can be acknowledged that once the challenges are resolved, we will be able to control the power of decision making and manipulation of human body to act and react to certain situations in an entirely different way. In the more distant future, developments in BCI will have a huge impact for the interaction between human beings. Concisely, we are advancing towards connecting people through neural signals.

References

1. Umair A, Ashfaq U, Khan MG, Recent trends, applications, and challenges of brain-computer interfacing (BCI). Department of electrical engineering, National University of Computer and Emerging Sciences, Chiniot-Faisalabad
2. Luz María A-V, Víctor Rodrigo M-G (2017) Enrichment of human-computer interaction in brain- computer interfaces via virtual environments. Comput Intell Neurosci 12. Article ID 6076913
3. Bi L, Fan X-A, Liu Y (2013) Eeg-based brain-controlled mobilerobots: a survey. Human-Machine Syst IEEE Trans 43(2):161–176
4. van Erp J, Lotte F, Tangermann M (2012) Brain-computer interfaces: beyond medical applications computer 45(4):26–34
5. Rao R, Scherer R (2010) Brain-computer interfacing [in the spotlight] Signal Process Mag IEEE 27(4):152–150. Google Scholar
6. Tan DS, Nijholt A (2010) Brain-computer interfaces: applying our minds to human-computer interaction. Springer
7. Brumberg JS, Nieto-Castanon A, Kennedy PR, Guenther FH (2010) Brain-computer interfaces for speech communication. Speech Commun 52(4):367–79
8. Lelievre Y, Washizawa Y, Rutkowski TM (2013) Single trial BCI classification accuracy improvement for the novel virtual sound movement-based spatial auditory paradigm. In: Signal and information processing association annual summit and conference (APSIPA), 2013 Asia-Pacific. IEEE, pp 1–6
9. Wang W, Degenhart AD, Sudre GP, Pomerleau DA, TylerKabara EC (2011) Decoding semantic information from human electrocorticographic (ecog) signals. In: 2011 Annual international conference of the IEEE, engineering in medicine and biology society, EMBC. IEEE, pp 6294–98
10. Prataksita N, Lin Y-T, Chou H-C, Kuo C-H (2014) Brain-robot control interface: development and application. In: 2014 IEEE international symposium on bioelectronics and bioinformatics (ISBB). IEEE, pp 1–4
11. Brouwer A-M, van Erp J, Heylen D, Jensen O, Poel M (2013) Effortless passive BCIs for healthy users in universal access in human-computer interaction design methods. Springer, Tools and Interaction Techniques for eInclusion
12. Hanan AR, Akkar, Faris Ali (2017) Brain computer interface technology and applications, a survey review. Int J Comput Appl 7(3):1797–2250

13. Sarah NA, Atia A, Mostafa MSM (2015) Brain computer interfacing: applications and challenges. Egypt Inf J 16(2):213–230
14. Hanafiah ZM, Taib MN, Hamid N (2010) Eeg pattern of smokers for theta, alpha and beta band frequencies. In: 2010 IEEE student conference on research and development (SCOReD). IEEE, pp 320–23
15. Di D, Zhihua C, Ruifang F, Guangyu L, Tian L (2010) Study on human brain after consuming alcohol based on eeg signal. In: 2010 3rd IEEE international conference on computer science and information technology (ICCSIT), vol 5. IEEE, pp 406–409
16. Eksi Z, Akgu¨l A, Recep Bozkurt M (2013) The classification of eeg signals recorded in drunk and non-drunk people. Int J Comput Appl 68
17. Shri P, Sriraam N (2012) Eeg based detection of alcoholics using spectral entropy with neural network classifiers. In: 2012 International conference on biomedical engineering (ICoBE). IEEE, pp 89–93
18. Malar E (2011) A novel approach for the detection of drunken driving using the power spectral density analysis of EEG. Int J Comput Appl 21
19. Fan X, Bi L, Wang Z (2012) Detecting emergency situations by monitoring drivers' states from eeg. In: 2012 ICME international conference on complex medical engineering (CME). IEEE, pp 245–248
20. http://www.eip.gov.eg/Documents/StudiesDetails.aspx?id=522. Accessed on 05 Oct 14
21. Fan X, Bi L, Wang Z (2012) Detecting emergency situations by monitoring drivers' statesfrom EEG. In: 2012 ICME international conference on complex medical engineering (CME). IEEE, pp 245–248
22. Sharanreddy M, Kulkarni P (2013) Automated EEG signal analysis for identification of epilepsy seizures and brain tumour. J Med Eng Technol 37(8):511–519
23. Selvam VS, Shenbagadevi S (2011) Brain tumor detection using scalp eeg with modified wavelet-ica and multi-layer feed forward neural network. In: 2011 Annual international conference of the IEEE engineering in medicine and biology society, EMBC. IEEE, pp 6104–6109
24. Sharanreddy M, Kulkarni P (2013) Detection of primary brain tumor present in eeg signal using wavelet transform and neural network. Int J Biol Med Res 4(1)
25. Poulos M, Felekis T, Evangelou A (2012) Is it possible to extract a fingerprint for early breast cancer via eeg analysis? Med Hypotheses 78(6):711–716
26. Ang CS, Sakel M, Pepper M, Phillips M (2011) Use of brain computer interfaces in neurological rehabilitation. Brit J Neurosci Nurs 7(3):523–528
27. Tan H, Kong K, Shee C, Wang C, Guan C, Ang W (2010) Post-acute stroke patients use brain-computer interface to activate electrical stimulation. In: 2010 Annual international conference of the IEEE engineering in medicine and biology society (EMBC), pp 4234–4237
28. Birbaumer N, Ruiz S, Sitaram R (2013) Learned regulation of brain metabolism. Trends Cognitive Sci 17(6):295–302
29. Ruiz S, Buyukturkoglu K, Rana M, Birbaumer N, Sitaram R (2014) Real-time fmri brain computer interfaces: self-regulation of single brain regions to networks. Biol Psychol 95:4–20
30. Ang KK, Guan C, Sui Geok Chua K, Ang BT, Kuah C, Wang C, Phua KS, Chin ZY, Zhang H (2010) Clinical study of neurorehabilitation in stroke using eeg-based motor imagery brain-computer interface with robotic feedback. In: 2010 Annual international conference of the IEEE engineering in medicine and biology society (EMBC). IEEE, pp 5549–5552
31. King CE, Wang PT, Mizuta M, Reinkensmeyer DJ, Do AH, Moromugi S, Nenadic Z (2011) Noninvasive brain-computer interface driven hand orthosis. In: 2011 Annual international conference of the IEEE engineering in medicine and biology society, EMBC. IEEE, pp 5786–5789
32. Meyer T, Peters J, Brtz D, Zander TO, Scholkopf B, Soekadar SR, Grosse-Wentrup M (2012) A brain–robot interface for studying motor learning after stroke. In: 2012 IEEE/RSJ international conference on intelligent robots and systems (IROS). IEEE, pp 4078–4083
33. Jones CL, Wang F, Morrison R, N.Sarkar N, Kamper DG (2014) Design and development of the cable actuated finger exoskeleton for hand rehabilitation following stroke. IEEE Syst J

34. Domingo MC (2012) An overview of the internet of things for people with disabilities. J Netw Comput Appl 35(2):584–596
35. Vecchiato G, Astolfi L, De Vico Fallani F, Salinari S, Cincotti F, Aloise F, Mattia D, Marciani MG, Bianchi L, Soranzo R et al (2009) The study of brain activity during the observation of commercial advertising by using high resolution EEG techniques. In: 2009 Annual international conference of the IEEE engineering in medicine and biology society EMBC. IEEE, pp 57–60
36. Sorudeykin KA (2010) An educative brain-computer interface. arXiv preprint arXiv:1003.2660
37. Marquez BY, Alanis A, Lopez MA, Magdaleno-Palencia JS (2012) Sport education-based technology: stress measurement in competence. In: 2012 International conference on e-learning and e-technologies in education (ICEEE). IEEE, pp 247–252
38. Royer AS, Doud AJ, Rose ML, He B (2010) Eeg control of a virtual helicopter in 3-dimensional space using intelligent control strategies. Neural Syst Rehabilitation Eng IEEE Trans 18(6):581–589
39. Mathan (2008) Feature-Image search at the speed of thought. Interactions 15(4):76–77
40. Nakanishi I, Baba S, Li S (2011) Evaluation of brain waves as biometrics for driver authentication using simplified driving simulator. In: 2011 International conference on biometrics and Kansei engineering (ICBAKE). IEEE, pp 71–76
41. Nakanishi I, Baba S, Ozaki K, Li S (2013) Using brain waves as transparent biometrics for on–demand driver authentication. Int J Biometrics 5(3):288–305
42. Pfurtscheller G et al (2005) Human brain-computer interface. In: Vaadia E, Riehle A (eds) Motor cortex in voluntary movements: a distributed system for distributed functions (frontiers in neuroscience). CRC Press, Boca Raton, FL, USA, pp 405–441
43. Thorpe J, Van Oorschot P, Somayaji A (2005) Pass-thoughts: authenticating with our minds. In: SN Abdulkader et al (eds) Proceedings of the 2005 workshop on new security paradigms, new security paradigms workshop: 228, vol 20, no 23, pp 45–56
44. Lotte F, Congedo M, Le´cuyer A, Lamarche F, Arnaldi B, et al. A review of classification algorithms for eeg-based braincomputer interfaces. J Neural Eng 2007;4.
45. Panoulas KJ, Hadjileontiadis LJ, Panas SM. Brain–computer interface (BCI): Types, processing perspectives and applications. In: Multimedia Services in Intelligent Environments. Springer; 2010. p. 299–321.
46. Samek W, Muller K-R, Kawanabe M, Vidaurre C (2012) Braincomputer interfacing in discriminative and stationary subspaces. In: Engineering in medicine and biology society (EMBC), 2012 annual international conference of the IEEE. IEEE, pp 2873–876
47. Sosa OP, Quijano Y, Doniz M, Chong-Quero J (2011) BCI: A historical analysis and technology comparison. In: Health Care exchanges (PAHCE), 2011 Pan American. IEEE, pp 205–09
48. Soria-Frisch A (2013) A critical review on the usage of ensembles for bci. Springer, Towards practical brain-computer interfaces

COVID-19 Economic Tracking and Assistance System (CETAS)

Tamajit Biswas, Pranab Hazra, Baishali Sarkar, Debdas Mondal, Deepali Kumari, and Niladri Mallik

Abstract Amidst the current pandemic situation which has been going on for a significant period of time now, although vast numbers of people are safe in their homes, this crisis has rendered millions of people unemployed, mainly the daily wage labourers, etc., who are now struggling with not only the gravity of the coronavirus but also with hunger and unemployment. We are witnessing a rich and poor divide that has crossed boundaries that were never explored before. We are all aware of the hardships of the people who are daily struggling just to manage a day's food for their family, and so as responsible engineers, we plan to develop an Android application which aims on directing financial aid to those who are actually needy by gathering funds from millions of donors across the nation who share the same ideology and wish to help these people out but are unable due to the pandemic related restrictions. People who are deserving of this financial aid and also the people who wish to donate will be able to register themselves on the application, post which verification will be conducted for both the parties after which a designated amount of money will transferred to the needful directly. All cyber security protocols are aimed to be implemented so as there is no presence of any anomaly.

Keywords Unemployment · COVID-19 · Pandemic · Economic · Android application · Assistance · Livelihood

1 Introduction

Approximately, 7 months ago in the month of April, the world as we knew it suddenly entered into a pandemic crisis because of a virus, named coronavirus or COVID-19 which originated in Wuhan, China, last year in the month of December. This has resulted in the death of millions of people around the world. Moreover, due to the

T. Biswas (✉) · P. Hazra · B. Sarkar · D. Mondal · D. Kumari · N. Mallik
Narula Institute of Technology, 81 Nilgunj Road, Agarpara, Kolkata, India

P. Hazra
e-mail: pranabhazra2017@nit.ac.in

M. Mitra et al. (eds.), *Computational Advancement in Communication, Circuits and Systems*, Lecture Notes in Electrical Engineering 786,
https://doi.org/10.1007/978-981-16-4035-3_4

lockdown restrictions imposed in various countries to contain the spread of the virus, several businesses have been forced to come to a standstill especially in India. People working in menial jobs such as daily wage workers, farmers, small business owners, etc. are the ones those have been the most affected in the wake of this pandemic. This has further widened the gap between the rich and the poor. The poor are struggling more than ever to make their daily ends meet and to feed their family. Due to import and export restrictions, there has been a decline in economics in multiple countries.

This has resulted in a situation that ultimately affects the poor consensus of mankind the most. So, as engineering students, we might not be capable of directly improving the economy of the country momentarily, but we can definitely contribute and help to reduce its effect on the common masses of people who belong to the BPL category. This can be done in the form of donations from like-minded people who wish to contribute towards this cause so that the ones who do not have a source of income can at least manage to keep their families fed and safe. We wish to develop an Android-based application that aims to become a bridge between these two sections of people. These donations would be collected from willing donors who would register themselves on this application, and the collected funds are transferred to the corresponding receivers who are registered on the same platform.

2 Genesis

The International Labour Organization [1] in its report describes the coronavirus pandemic as the worst global crisis. About **400 million people (76.2% of the total workforce)** functioning in the informal economy in India are under threat of being further pulled towards poverty because of catastrophic circumstances caused by the COVID-19 pandemic. Since most of the world has been subjected to extensive lockdowns, it has resulted in the loss of 195 million full-time jobs or 6.7 per cent of working hours globally which has surely increased the unemployment drastically. Several thousands of people work menial jobs where sudden loss of income is disastrous to livelihood **(International Labor Organization, 2020)** [1] (Fig. 1).

Similarly, to cope up with the COVID-19 pandemic, the Indian government had imposed lockdown on 25th March, 2020, in India. Consequentially, millions of people, especially migrant workers and daily wage labourers lost their jobs. Till April, already **4 million people have lost their jobs,** and the number is keeping on increasing. The most critical and immediate impact of COVID-19 pandemic is unemployment while lower to none economic growth and rise in financial gap in people would be the long-term effects, according to a survey by the **Indian society of Labour Economics (ISLE)** [3] (Fig. 2).

During lockdown, the exports and imports were entirely put a halt, which resulted in minimal cash inflow into the nation thus all the more reducing any and all scopes of employment. This not only affected the India's economy but also affected millions of people who struggled to manage their basic expenses for sustaining. Other sectors such as tourism, food and beverage industry were also completely shut down during

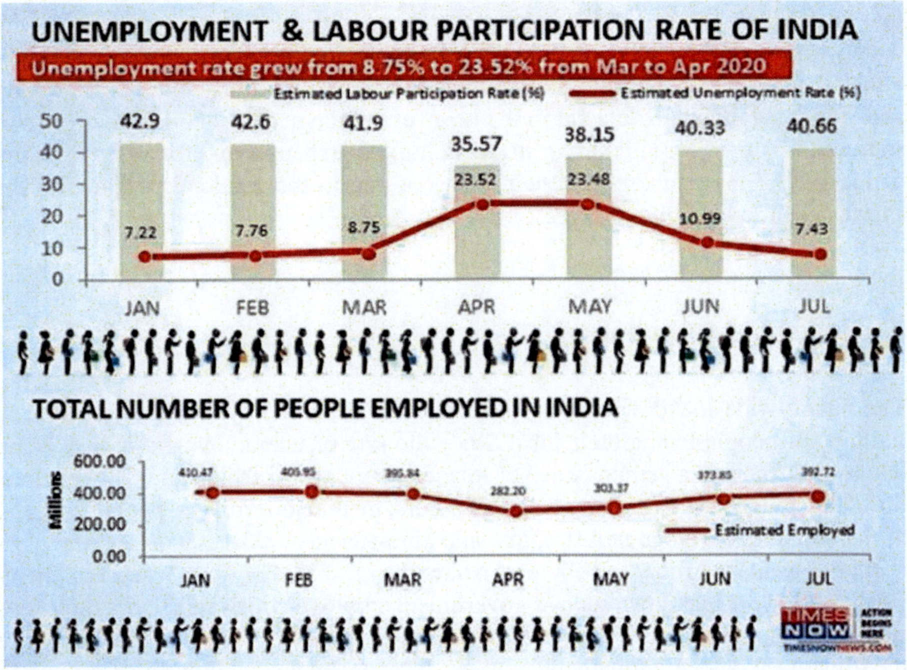

Fig. 1 Unemployment and Labour Participation Rate of India [2]

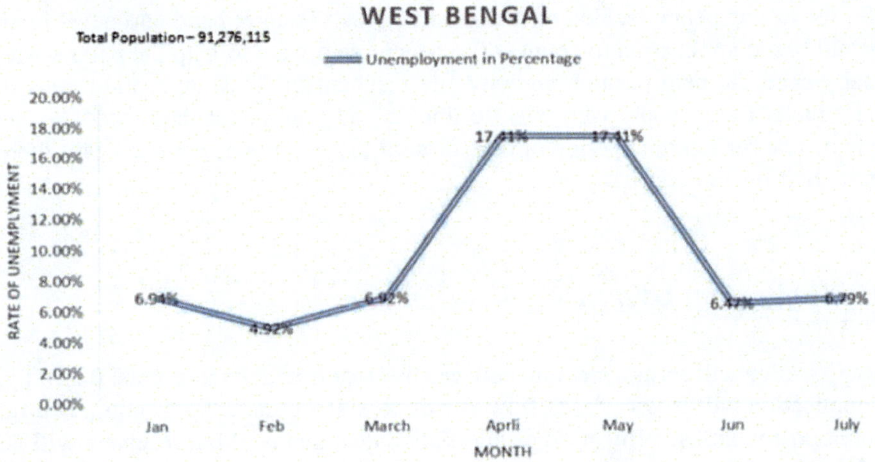

Fig. 2 Unemployment rates in the state of West Bengal in the last 6 months [2]

that time which resulted in the unemployment of several thousands of people who were working in this sector. Due to all these reasons, India and its people are badly affected economically. While the nation would deal with this situation differently and would surely come out of it in the near future, the people who lost their jobs or sources of income are still struggling with hunger, expenses and for an opportunity of income. Many of them lost their lives not due to the pandemic but due to the shortage of food and money.

3 Solution Methodology

The nation-wide lockdown greatly affected India's economy and thus resulted in millions of people losing their jobs. Since the rate of unemployment is increasing day by day, there is a genuine need of financial support for the people who were the most affected by it such as daily wage labourers, menial job workers, people working in the food and beverage industry, taxi and auto-rickshaw drivers, coolies, etc.

There are also millions of people who have the means to comfortably sustain themselves during the lockdown such as government employees, private sector employees such as engineers, doctors, chartered accountants, etc. These people are all aware of the situation and were sympathetic towards the people who are struggling for survival financially but were unable to reach out their helping hands due to the restrictions of lockdown.

So we as engineers decided to address this grave issue at hand and develop an Android-based application to connect the people who want to help the people who actually need the help through an erstwhile mechanism. This application aims to collect funds from people who wish to donate and then direct these funds to the needy people for a certain period of time or until the situation is conducive for them to earn their livelihoods again.

3.1 Application Overview

The application will act as a medium between the donors and the acceptors. Basically, the application will collect funds from the donors and transfer them into a neutral account post which a verification of all relative documents of the receivers will be conducted to ensure authenticity. After successful verification, a database of genuine receivers will be created while maintaining priority for each receiver based on number of heads in family, and a certain amount of money will be directed to the accounts of the receivers monthly for a certain period of time.

3.2 Features

- Basic Information:

After successful installation of the application, the user has to complete the basic registration for signing up in the application. Basic registration includes providing information of the user such as their email address, personal mobile phone number and then they have to create their own password. After successfully signing up in the application, they can utilize the application's services.

- Creating an Account of Donor/Acceptors:

After a user has successfully registered himself/ herself, there will be two sections to choose from: one for offering to donate funds as a donor and the other for receiving the financial aid as an acceptor. For either of the sections, the user has to provide their Aadhar Card number, upload their photograph, and also their PAN card number (optional). Only after these documents are uploaded or produced, they would be treated as a donor or acceptor account.

Donor: **For a donor account, firstly, they have to provide some basic information about them such as name, address, age, occupation, email address and phone number.**

After that, they will have to select their method of payment for contributing any amount that they desire to the fund. There will also be the provision of donating multiple times. After successful payment, they get an e-receipt as a proof of their successful donation.

Acceptor: For an acceptor account, the acceptor has to give their basic information much similar to the donor such as their name, address, age, email address (optional) and phone number. Apart from these, they would also have to fill a form wherein they have to select their profession before unemployment or their current employment, daily income, income certificate, number of members in family, mention family members' name and age. They would also have to select if the user and/ or the family members are ration card holders, if the children are registered in government schools, the bank account number and the bank name for receiving the payment.

3.3 Salient Aspects

- Holding Funds:

All the funds that would be received from the donors will be deposited in a bank account which would be registered under the name of the application and would be treated as an NGO body. The transaction details of this account will be available to all the users that are registered in the application as donors.

- Fund Management:

It is the most crucial part of the application which will decide the amount that is to be transferred and also would hold the database of acceptors who wish to receive the financial aid on the basis of priority which would be set as per their chosen parameters.

- Acknowledgements:

An e-acknowledgement receipt will be issued to the people who make donations to through the application complimented with a dedicated message for their service.

- Help Section:

In the application, there will be a dedicated help section where all the users would be able to learn about the aspects of the application and would also be made aware of using the application ethically. This section will also include tutorials for the application and the option to contact a customer support representative.

4 Conclusion

The COVID-19 Economic Tracking and Assistance System or the CETAS android application aims to become the bridge between the people who have more than their means and the people who are becoming and living below their means so that in the next phase of pandemic, the society does not have to deal with the additional impacts of hunger and poverty. This android application will help in stabilizing the condition of millions of people across the nation so that we can all fight together with this pandemic and come out strong and victorious.

References

1. https://www.ilo.org
2. https://www.timesnownews.com/business-economy/economy/article/how-the-covid-19-outbreak-has-affected-the-joblessness-rate-in-india-explained-in-4-charts/634284
3. http://www.isleijle.org/

Use of Convolutional Neural Network (CNN) to Detect Plant Disease

Navoneel Moitra, Akanksha Singh, and Subhram Das

Abstract The automatic and accurate detection of diseased leaves is a challenging job for researchers. It offers a promising step towards food security and agricultural growth. On contrary, the conventional manual interpretation is time-consuming and expensive. In this paper, it proposes a new approach to detect plant diseases using the deep learning Convolutional Neural Network. We have used 1900 images, taken from a public dataset to train our model. This deep learning model is designed to consist of 25-layer for plant disease classification. The trained model achieved 96.64% accuracy to detect the plant disease. The proposed deep learning convolution neural network model may have great potential in disease detection for current cultivation on large scale.

Keywords Plant disease · Deep learning · Convolutional neural networks

1 Introduction

The recent trends of Computer Science involve the usage of Artificial Intelligence up to a great extent which has significantly reduced the labor and time for various tasks which were once predominantly manual. Like other fields technology has found its way to agriculture as well. The recent trend of technological revolution that agriculture is going through is mainly automation. Institutions both profit and non-profit have come up with a digital trend in agriculture. It is a non-debatable point that agriculture with the help of automation will increase farmer yield to a great extent. Humans now meet the criteria for food production to cater to the demands of almost all the world's population. Even though food security is threatened by several factors

N. Moitra
DXC Technology, Bangalore, India

A. Singh
Wipro, Kolkata, India

S. Das (✉)
Computer Science and Engineering, Narula Institute of Technology, Kolkata, India

M. Mitra et al. (eds.), *Computational Advancement in Communication, Circuits and Systems*, Lecture Notes in Electrical Engineering 786,
https://doi.org/10.1007/978-981-16-4035-3_5

like climate change [1, 2] and different diseases [3]. Plant diseases are not only a threat to the availability of food security around the world, it can have devastating consequences for small scale farmers whose income depends on healthy crops. It has been found out that 80% of agricultural produce is generated by small scale farmers [4], this section of farmers has a loss percentage of around 50% due to pests and diseases [5]. Moreover, this is the largest section of people plagued by hunger in a farmer's household [6]. Efforts have been taken to minimalize the loss of crops due to diseases. Widespread application of pesticides has been used in the past 10 years through means of integrated pest management approaches [7]. Although this proves to be effective to a certain extent, the effectiveness can be increased if the diseases are identified early due to the efficient disease management steps which can be later taken. The present advent of smartphones offers an efficient way of detecting plant diseases, each of the devices has a significant processing power with high definition cameras coupled with a high-resolution display. This is further aided by the fact that mobile broadband access is accessible almost all around the world and can be used to capture data for processing [8]. The advancement of computer vision and object recognition has made tremendous progress in the last few years. A few of the notable works [10–12] are set as standards for image recognition particularly Artificial Intelligence enabled vision including object classification. In 2012, a deep convolutional neural network achieved the task of classifying images in 1000 categories having a top 5 error of 16.4% [13]. Further works in this technology have lowered the error rate to 3.57% [13–17]. Deep learning in the present is utilized in various diverse regions of technology. A proper rendition includes the image of a plant as input and hence gives a crop-disease pair as output [18]. Deep Learning neural networks involve mapping the input layer and output layers over a series of nodes. The implementation involves correctly mapping the input layers and output layers by a proper structure and correct edge weights. The entire process is challenging in computational terms, though in recent times has improved drastically through a breakthrough in technology [19, 20]. Progressive research [21, 22] shows that leaves photographed in natural environments do not give the same results as a controlled environment and hence the conventional dataset [9] is modified with images from the natural environment as well as controlled environments to include diversion and achieving the desired results. Bringing all into a common roof to identify the diseases of a variety of plants with high accuracy is a part of the primary reason for carrying out the work and hence implementing it from a broader perspective. Convolutional Neural Networks using Keras is a feature that can be used to greatly solve the mentioned problem domain as it yields high accuracy with the ability to combine a diverse variety of plant species, such that in the near future scope it can be implemented real-time to provide a sustainable and effective solution in the agricultural domain.

```
└── PlantVillage
    ├── Pepper__bell___Bacterial_spot
    ├── Pepper__bell___healthy
    ├── Potato___Early_blight
    ├── Potato___Late_blight
    ├── Potato___healthy
    ├── Tomato_Bacterial_spot
    ├── Tomato_Early_blight
    ├── Tomato_Late_blight
    ├── Tomato_Leaf_Mold
    ├── Tomato_Septoria_leaf_spot
    ├── Tomato_Spider_mites_Two_spotted_spider_mite
    ├── Tomato__Target_Spot
    ├── Tomato__Tomato_YellowLeaf__Curl_Virus
    ├── Tomato__Tomato_mosaic_virus
    └── Tomato_healthy
```

Fig. 1 File structure of data set used

2 Material and Methods

2.1 *Data Sets*

The major challenge that pertains to the detection of the plant leaf is the fact that classification results are poor when done in real-time environments as compared to the one in lab environments. One of the primary reasons is that the non-availability of datasets in correspondence to this work, most of these data sets are mainly derived from the plant village data set. Plant Village data sets (Fig. 1) consist of 54,323 images divided into 38 classes of various diseased and healthy pairs. There are 14 different crop species in total in this dataset, each crop has is grouped under a class with a crop-diseased pair. The images are resized into 256 × 256 pixels in the preliminary step as model optimization and predictions are performed on these images, which are downscaled. The source of this dataset is available in Kaggle (a crowd sourced platform by Google). All the images are taken on a solid color background which is labeled only by using a class name. Figure 2 to adapt the Plant Village for practical usability the plant disease dataset is introduced.

2.2 *Convolutional Neural Network*

2.2.1 Model Overview

Convolutional neural networks are used to train the image dataset by using the pre-processed images as it emulates human-retinal functionality by building an Image Classifier model. This process, involves learning which is complex by nature and classification process and hence uses a GPU, which is available via Kaggle Notebooks for faster execution. This like other neural networks also functions on neurons arraigned

Fig. 2 Images of leaves in the dataset

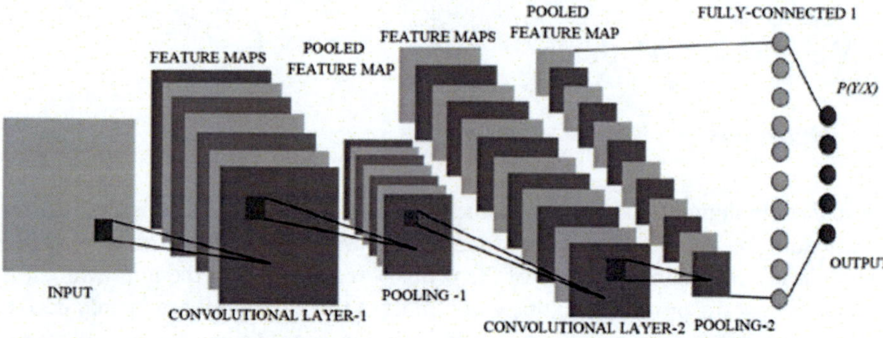

Fig. 3 Illustration of a CNN model

in layers, which contribute to efficiency. The efficiency can be also increased by the increase of the depth of layers and/or can be increased by altering the image parameters of the input image.

A convolution multiplies (Fig. 3) a matrix of pixels with a kernel (filter matrix) and adds up the multiplied values, this procedure is repeated into every pixel and is repeated until all the image pixels are hence covered. Visualization is shown below.

2.2.2 Preprocessing of Data

- *Resizing of the dataset images*

Preprocessing of the images is needed to achieve resizing the images into a fixed side of (256 × 256) this is done to decrease the processing load on the neural network. This process is achieved by using OpenCV (Open Computer Source Vision) which is a library of programming functions mainly used for image processing. This is mainly supported by multiple platforms and is supported by deep learning modules such as

TensorFlow, PyTorch, Torch, etc. The image and the directories in which the images are stored are kept in two distinct lists mainly by using tools from OpenCV.

- *Assigning Binary Values to images*

The leaves are needed to classified whether they are healthy or diseased, this is achieved by using LabelBinarizer, particular leaves are assigned binary values to assign its state, conventionally using the transform method, this process assigns one binary label to multiple classes (whether it belongs or doesn't belong to a particular class).

- *Conversion to a NumPy list*

The features extracted from the leaves are converted to NumPy arrays, with all the features extracted and hence stored into matrices. A NumPy array consists of a grid of values that is indexed by a non-negative tuple. The rank of the array represents the dimension of the array, the shape being the tuple of integers giving the size of the array along each dimension, the value of each pixel is stored in the NumPy array list.

- *Splitting up the dataset into Test and Train*

The dataset is then split up into Test and Train which has the same predictors or variables. The difference between the two is the observation and the specific values in the variables, by fitting the model on the training dataset errors are implicitly minimized and correct responses are found. The model which is fitted hence provides a good prediction of the training dataset. The test dataset is then used for testing the model and checks the validity and efficiency of the model. Test and train are randomly split into subsets from the original dataset.

2.3 Training the Model

- *Image Data Generation*

Image data generator classes are used to configure random transformation and normalization on the image dataset, this is mainly used by the Keras model which is used for the classification. The process is done on augmented image batches via the data-labels or directories. The model then accepts the generated data as inputs, which are used to fit the data generators, evaluate, and finally predict them.

- *Building the model*

The model is built using Sequential layers. As it has a plain stack of layers specifically one input and one output tensor. Each layer has a weight attached to it, which corresponds to the layer following it. Keras provides simplicity in doing this process. There are 25 layers and output layers in the model. ReLU (Rectified Linear Unit) is used as an activation function, which is defined mathematically as

$$y = \max(0, x). \tag{1}$$

The function is linear for values that are greater than zero, and hence has a lot of desirable characteristics which are achieved in linear activation function can be obtained, while backpropagating in training the model and yet is nonlinear as values lesser than zero is always considered as zero. This makes ReLU an automatic choice while choosing the activation function.

- *Accessing Model Training History*

  Training matrices for each Epoch can be accessed by using Keras' Model Training History, which includes the loss and accuracy for the classification problems as well as for the validation dataset. Either one is set then the history object is returned as the fit function is used to train the model. The loss is used to train the gradients of a neural network and which in turn is used to update the weight of the neural net.

- *Evaluation of Model*

  The accuracy score is the number of true positives and true negatives divided by the number of true positives, true negatives, false positives, and false negatives and since this is a classification-based model hence, Accuracy is found out for this particular model as well.

3 Results and Discussion

The results presented in this section are related to training with the whole database containing both original and augmented images of leaves that are both healthy and diseased. As it is known that convolutional neural networks (CNN) can learn features when trained on larger datasets (pre-processed training image) and tries to replicate the human-retinal functionality, and hence building an image classifier that consists of only original images is limited in scope. After fine-tuning the model to a certain extent an accuracy of 96.64% is achieved in the process. The whole process consists of about twenty-five epochs or hyperparameters or iterations consisting of a forward pass and a backward pass (Table 1).

The plot between the training accuracy and validation accuracy is also plotted, hence it is evident that with each epoch or iteration the accuracy level increases in the training accuracy (Fig. 4).

The plot between the training and validation loss is also plotted which shows that the loss level decreases with each epoch or iteration (Fig. 5).

Table 1 Accuracy and loss percentage of epochs

Loss	Accuracy	Validation loss	Validation accuracy
0.1276	0.9717	0.1693	0.9724
0.1084	0.9723	0.4877	0.9490
0.1117	0.9718	0.2886	0.9605
0.1049	0.9720	0.2649	0.9724
0.0966	0.9734	0.2252	0.9673
0.0943	0.9737	0.1999	0.9679
0.0869	0.9740	0.2717	0.9520
0.0805	0.9752	0.1514	0.9663
0.0786	0.9755	0.1767	0.9618
0.0767	0.9762	0.0940	0.9723
0.0703	0.9775	0.1532	0.9659
0.0678	0.9781	0.1157	0.9687

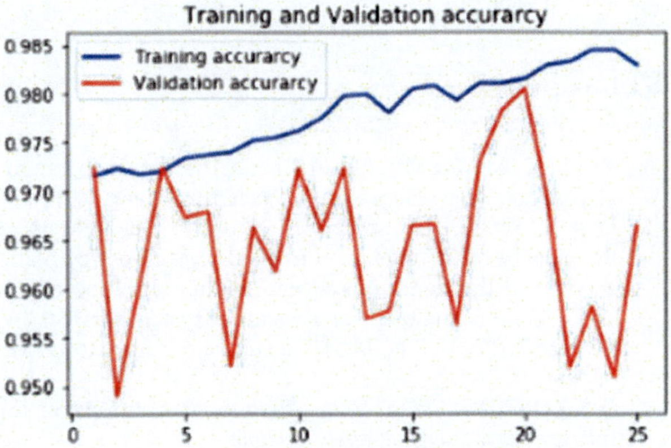

Fig. 4 A plot of the training and testing accuracy

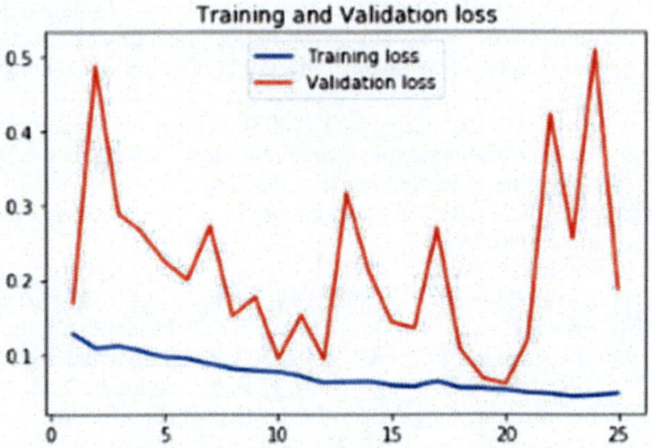

Fig. 5 A plot of the training and validation loss

4 Conclusion

In this paper, image classification is done using a deep learning convolution neural network to detect plant disease automatically. A distinctive and detailed study has been carried out by previous implementations that have been tested but have been insufficient in properly dealing with the exact problem domain and providing an accurate analysis. This paper explicitly discusses and tests Deep Learning technology implementation in plants. There has been a recent surge in the studies of Artificial Intelligence in the field of agriculture, nevertheless mostly all studies converge specifically to a species or two, and hence to provide results to a broader array of plants this study has been carried out. Our model provides an accuracy of 96.64% to detect plant disease which is also fair reasoning for its adoption. This deep learning technology is a new smart technology to detect plant disease without human interactions. Overall, the existing method works practically well with many different crop species to detect diseases and with more training data is likely to improve performance significantly.

References

1. Tai APK, Martin MV, Heald CL (2014) Threat to future global food security from climate change and ozone air pollution. Nat Clim Chang 4(9):817–821
2. Diaz S, Settele J, Brondízio E, Ngo HT, Guèze M, Agard J, Zayas CN (2019) Report of the plenary of the intergovernmental science-policy platform on biodiversity and ecosystem services on the work of its seventh session. In: Proceedings of the intergovernmental science-policy platform on biodiversity and ecosystem services (IPBES), Paris, France, 7
3. Strange RN, Scott PR (2005) Plant disease: a threat to global food security. Ann Rev Phytopathol 43:83–116
4. IFAD U (2013) Smallholders, food security and the environment, vol 29. International Fund for Agricultural Development, Rome
5. Harvey CA, Rakotobe ZL, Rao NS, Dave R, Razafimahatratra H, Rabarijohn RH, Rajaofara H, MacKinnon JL (2014) Extreme vulnerability of smallholder farmers to agricultural risks and climate change in Madagascar. Philos Trans Roy Soc B Biol Sci 369(1639):20130089
6. Sanchez PA, Swaminathan MS (2005) Cutting world hunger in half. Science 307(5708):357–359
7. Nesci AV, Etcheverry MG (2006) Control of Aspergillus growth and aflatoxin production using natural maize phytochemicals under different conditions of water activity. Pest Manage Sci Formerly Pestic Sci 62(8):775–784
8. Facts ICT (2015) Figures-the world in 2015. Geneva: the international telecommunication union (ITU)
9. Hughes D, Salathé M (2015) An open access repository of images on plant health to enable the development of mobile disease diagnostics. arXiv preprint arXiv:1511.08060
10. Everingham M, Van Gool L, Williams CKI et al (2010) The Pascal visual object classes (VOC) Challenge. Int J Comput Vis 88:303–338. https://doi.org/10.1007/s11263-009-0275-4
11. Russakovsky O, Deng J, Su H, Krause J, Satheesh S, Ma S, Huang Z, Karpathy A, Khosla A, Bernstein M, Berg AC Fei-Fei L (2015) Imagenet large scale visual recognition challenge. Int J Comput Vision 115(3):211–252
12. Deng J, Dong W, Socher R, Li LJ, Kai L, Li F-F (2009) ImageNet: a large-scale hierarchical image database. In: 2009 IEEE conference on computer vision and pattern recognition pp 248–255. https://doi.org/10.1109/CVPR.2009.5206848

13. Krizhevsky A, Sutskever I, Hinton GE (2012) Imagenet classification with deep convolutional neural networks. Adv Neural Inf Process Sys 25:1097–1105
14. Simonyan K, Zisserman A (2014) Very deep convolutional networks for large-scale image recognition. arXiv preprint arXiv:1409.1556
15. Zeiler MD, Fergus R (2014) Visualizing and understanding convolutional networks. In: European conference on computer vision, pp. 818–833. Springer, Cham
16. He K, Zhang X, Ren S, Sun J (2016) Deep residual learning for image recognition. In: Proceedings of the IEEE conference on computer vision and pattern recognition (CVPR), pp 770–778. https://doi.org/10.1109/CVPR.2016.90
17. Szegedy C, Liu W, Jia Y, Sermanet P, Reed S, Anguelov D, Erhan D, Vanhoucke V, Rabinovich A (2015) Going deeper with convolutions. In: 2015 IEEE conference on computer vision and pattern recognition (CVPR). pp 1–9. https://doi.org/10.1109/CVPR.2015.7298594
18. Mohanty SP, Hughes DP, Salathé M (2016) Using deep learning for image-based plant disease detection. Front Plant Sci 7:1419
19. LeCun Y, Bengio Y, Hinton G (2015) Deep learning. Nature 521(7553):436–444
20. Schmidhuber J (2015) Deep learning in neural networks: an overview. Neural Netw 61:85–117
21. Sun Y, Liu Y, Wang G, Zhang H (2017) Deep learning for plant identification in natural environment. Comput Intell Neurosci, vol. 2017, Article ID 7361042, 6p. https://doi.org/10.1155/2017/7361042
22. Es-Saady Y, Massi IE, Yassa ME, Mammass D, Benazoun A (2016) Automatic recognition of plant leaves diseases based on serial combination of two SVM classifiers. In: 2016 International Conference on Electrical and Information Technologies (ICEIT), pp 561–566, https://doi.org/10.1109/EITech.2016.7519661

Prediction of Blended Fuel Characteristics Through Regression Modelling

Sumit Nandi, Debopriya Dey, and Rupa Bhattacharyya

Abstract Blended fuel attracted considerable attention for the environmental sustainability, mitigation of scarcity of non-renewable fuels and enhancement of property modification for the last few decades. Jatropha Curcas oil (JCO), a non-edible vegetable oil, can be utilized for the preparation of non-conventional alternative energy sources like biodiesel which may be blended with diesel fuel for better environmental sustainability. Initially, biodiesel is prepared from JCO with methanol through transesterification reaction maintaining optimized reaction parameters in the presence of biocatalyst. After that mathematical relationships between fuel properties and blended fuel have been established through regression analysis method for the prediction of fuel properties like density, kinematic viscosity, cloud point and flash point. The blended samples are prepared ranging from 10 to 60% (B10 to B60) for biodiesel-diesel fuel. From the experimental results, graph of each fuel property has been plotted and mathematical equation of each fuel property for biodiesel-diesel blends are approximated with their respective coefficient of determination (R^2). The results of estimation show that blended fuel properties have linear relationships regarding density, kinematic viscosity, cloud point and flash point. The equations identified for the properties of blended fuels are prerequisites as input data research findings. From the estimation of mathematical regression equation based on experimental findings, prediction can be done for any fuel properties for any ratios of biodiesel-diesel blends. So mathematical understanding contributes a better pathway for finding out the properties of blended fuels which may help to reduce the scarcity of conventional fuels.

Keywords Biodiesel · Blended fuel · Regression modelling · Jatropha Curcas oil

S. Nandi (✉) · D. Dey · R. Bhattacharyya
Department of Basic Science and Humanities, Narula Institute of Technology, Kolkata 700109, India
e-mail: Head_bshu.nit@jisgroup.org

M. Mitra et al. (eds.), *Computational Advancement in Communication, Circuits and Systems*, Lecture Notes in Electrical Engineering 786,
https://doi.org/10.1007/978-981-16-4035-3_6

1 Introduction

Blended fuel or biodiesel-diesel blends may be the most suitable option in the future energy sector due to the depleting conditions of fossil fuels along with the degradation of environmental conditions. Mathematical understanding for the estimation of blended fuel properties is one of the most suitable options for the identification of the best process technology. Different raw materials are used for the preparation of alternative fuel but JCO is probably the most appropriate raw material in this respect due to its easy availability, non-edible nature and low cost [1–4]. Blended fuels are prepared with biodiesel prepared from different edible and non-edible sources by different researchers and their characteristics are also analyzed [5–8]. Chen et al. analyzed the fuel properties and combustion characteristics of blended fuel prepared from JCO [9]. Mofijur et al. also evaluated the blending performance of Jatropha curcas methyl ester along with engine performance and emission characteristics [10]. Present author also studied different aspects for the preparation of biodiesel from JCO, e.g. by varying alcohols [11], with recycling of enzymes [12] and through comparative studies between chemical and biocatalytic methods [13]. Roy et al. analyzed the mass transfer kinetics for biodiesel preparation from JCO through mathematical techniques [14]. For the prediction of blended fuel properties, different mathematical techniques and methodologies have been applied for the appropriate outcomes. Adnan Berber analyzed the fuel flow performance of diesel engines through mathematical techniques [15]. El Galad et al. made an economical study for blended fuel through empirical equations [16]. Acharya et al. studied the blending performance from mahua biodiesel through regression models and showed that 30% blending is the optimum blending ratio taking Cold Filter Plugging Point into account [17]. Gulum and Bilgin also measured and predicted the density and viscosity of different blended fuels by using the exponential model [18]. Different researchers applied different mathematical techniques for the prediction of properties of different blended fuels like biodiesel-diesel blends, vegetable oil-diesel blends, ethanol–diesel blends, etc. [19–23]. But few studies have been made for the analysis of fuel properties made of biodiesel-diesel blends prepared from JCO and methanol through transesterification reaction in the presence of enzyme as a catalyst. So, in the present research investigation, blended fuel properties like density, viscosity, cloud point and flash point have been analyzed through regression model.

2 Materials and Methods

The JCO used in this study was collected by M/s. Arora Oils Ltd., Burdwan, West Bengal, India. The enzyme used for the transesterification reaction was a kind gift of Novozyme South Asia Pvt. Ltd. Bangalore, India. The chemicals used in this work such as methanol (99.8% pure) and hexane were purchased from S.D. Fine

Chemicals (Mumbai, India). Except otherwise specified all other chemicals used were A.R. Grade.

Transesterification reaction was carried out by taking JCO and methanol in an appropriate proportion in an Erlenmeyer flask (250 ml) fitted with a water condenser and stirred by a magnetic stirrer at different temperatures for 8 h in the presence of immobilized enzyme. Samples were taken from the reaction mixture at specified time intervals for analysis. Biodiesel characterization was done according to the American Standard Testing Method. After biodiesel preparation, blending was done with diesel fuel in definite proportions (20–80%) for the estimation of viscosity, density, cloud point and flash point. For each property of blended fuel, a mathematical equation and regression coefficient (R^2) has been developed using regression analysis through mat lab software.

3 Results and Discussions

The different properties of JCO, JCO biodiesel and mineral diesel are presented and compared in Table 1. The JCO is found to have a higher density and higher kinematic viscosity than mineral diesel. The higher density is due to the higher molecular weight of triglyceride molecule. Its calorific value is also lower than mineral diesel which is attributed to its difference in the chemical composition with that of diesel. So direct use of JCO in engines is difficult for these issues. After transesterification of JCO to JCO biodiesel, the kinematic viscosity reduces to 1/5 of its original value and comes closer to that of mineral diesel. The properties like density, kinematic viscosity, flash point, refractive index, cloud point, calorific values and pour point of biodiesel almost satisfy according to the test methods. The JCO biodiesel has a

Table 1 Comparative properties

Properties	JCO	JCO biodiesel	Biodiesel standard	Diesel fuel	Test method
Kinematic Viscosity @ 40 °C (mm^2/s)	20.65	4.35	1.9–6.0	1.3–4.1	ASTMD-445
Flash point (°C)	213	180	>120	60–80	ASTMD-93
Pour point (°C)	7.4	3.8	−15 to 10	18	ASTMD-93
Density @ 15 °C kg/m^3	930	872	860–900	825	–
Refractive index at 40 °C	1.61	1.46	1.34	1.45	–
Calorific value (MJ/kg)	37.56	36.9	33–40	45	ASTM-6751
Cloud point (°C)	15	13	–	–	ASTMD-2500

higher flash point which ensures less chance of flammability hazard which is better for safety purposes.

Variation of density with blending ratio:

The density is an important fuel property which affects the engine performance due to changes in fuel injection capacity. As the fuel injection system of an engine measures the fuel by volume so any change in the density will directly influence the engine output. So for the relationship of variation, a mathematical approach has been applied and the least square regression line of density (*D*) versus biodiesel percentage (*X*) may be represented as.

$D = a_1X + a_0$ where a_0 and a_1 are obtained from the normal equations

$$\sum D = a_0 N + a_1 \sum X;$$
$$\sum XD = a_0 \sum X + a_1 \sum X^2 \text{ and}$$
$$a_0 = \frac{(\sum D)(\sum X^2) - (\sum X)(\sum XD)}{N \sum X^2 - (\sum X)^2} \quad \text{and}$$
$$a_1 = \frac{N \sum XD - (\sum X)(\sum D)}{N \sum X^2 - (\sum X)^2}$$

So, the linear relationship between density and blended fuel can be represented as

$$D = 0.3686\,X + 832.93 \tag{1}$$

From the above understanding, linear correlation coefficient (*r*) of regression line can be represented as

$$r = \frac{N \sum XD - (\sum X)(\sum D)}{\sqrt{\left[N \sum X^2 - (\sum X)^2\right]\left[N \sum D^2 - (\sum D)^2\right]}}$$

and the coefficient of determination (R^2) can be represented as

$$R^2 = \frac{\text{Explained variation}}{\text{Total variation}} = \frac{\sum (D_{\text{est}} - \overline{D})^2}{\sum (D - \overline{D})^2}$$

where D_{est} is the estimated density from mathematical equations.

For Eq. (1), we can get the value of $r = 0.9935$ and $R^2 = 0.9873$ which indicate very good linear relationship between density and biodiesel percentage as shown in Fig. 1.

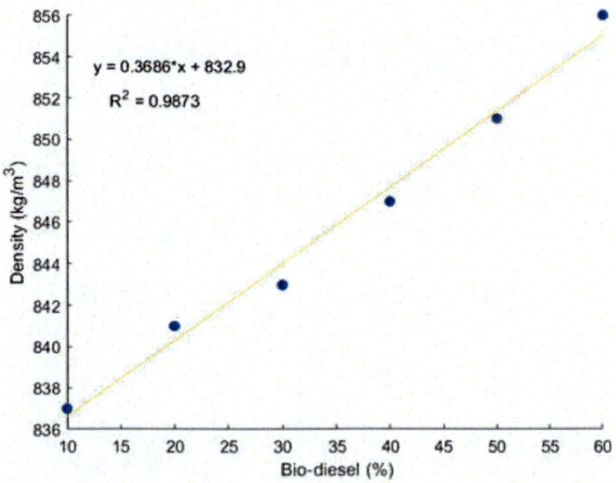

Fig. 1 Variation of density with biodiesel percentage in blended fuel

Table 2 Predicted value versus experimental value for density of blended fuel

Fuel	Predicted density @ 15 °C kg/m^3	Experimental density @ 15 °C kg/m^3
B 70	859	860
B 80	862	863
B 90	866	867
B 100	870	872

It has been observed from Fig. 1 that the density increases linearly with an increase in the biodiesel percentage in the biodiesel-diesel blends (B10 to B60) as the density of biodiesel is higher than diesel. Using Eq. (1) the predicted value of density for B70 to B100 can be calculated and obtained as in Table 2. A good correlation has been observed using regression analysis method for the identification of the density of blended fuel.

3.1 Variation of Kinematic Viscosity with Blending Ratio

Engine efficiency depends on the fluidity of fuel which is affected by viscosity. The kinematic viscosity influences the operation of a fuel injection system, particularly at low temperature conditions. For higher viscosity of fuel, atomization would be difficult which results in delayed combustion along with soot formation and carbon deposits. So for the identification of the relationship between kinematic viscosity and biodiesel percentage in blended fuel, a mathematical approach has been applied and the least square regression line of kinematic viscosity (V) versus biodiesel percentage (X) may be represented as

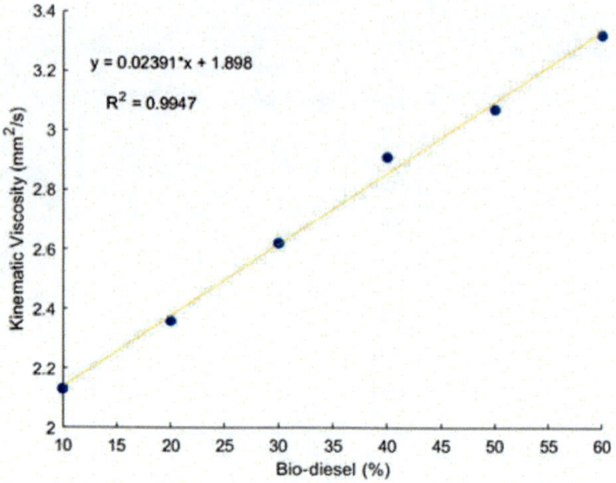

Fig. 2 Variation of kinematic viscosity with biodiesel percentage in blended fuel

Table 3 Predicted value versus experimental value for kinematic viscosity of blended fuel

Fuel	Predicted kinematic viscosity @ 15 °C kg/m^3	Experimental kinematic viscosity @ 15 °C kg/m^3
B 70	3.571	3.65
B 80	3.81	3.9
B 90	4.049	4.1
B 100	4.288	4.35

$$V = 0.0239\,X + 1.898 \tag{2}$$

For Eq. (2), we can get the value of $r = 0.9979$ and $R_2 = 0.9947$ which indicate good correlation between kinematic viscosity and blended fuel as shown in Fig. 2.

It has been observed from Fig. 2 that similar to density, the kinematic viscosity of biodiesel-diesel blends shows a linear increase with increase in volumetric percentage of biodiesel (B10 to B 60). Using Eq. (2) the predicted value of kinematic viscosity for B70 to B100 can be calculated and obtained as in Table 3. A good correlation has been observed using regression analysis method for the identification of kinematic viscosity of blended fuel.

3.2 *Variation of Cloud Point with Blending Ratio*

Cloud point is an important property of fuel for low temperature applications. The cloud point is the temperature at which a cloud or wax crystal first appears when fuel is cooled. So very low cloud point for any fuel is not desirable especially for cold countries. So for the identification of the relationship between cloud point and biodiesel percentage in blended fuel, a mathematical approach has been applied and the least square regression line of cloud point (C) versus biodiesel percentage (X) may be represented as

$$C = 0.1229\,X + 0.8667 \tag{3}$$

For the Eq. (3), we can get the value of $r = 0.9922$ and $R_2 = 0.9835$ which indcate good correlation between cloud point and blended fuel as shown in Fig. 3.

It has been observed from Fig. 3 that like density and kinematic viscosity, this property also show an increase in value with an increase in blending ratio (B10 to B60). Using Eq. (3) the predicted value of cloud point for B70 to B100 can be calculated and obtained as in Table 4. It shows that predicted value and experimental value are very close to each other which proves the correctness of the identified regression model.

Fig. 3 Variation of cloud point with biodiesel percentage in blended fuel

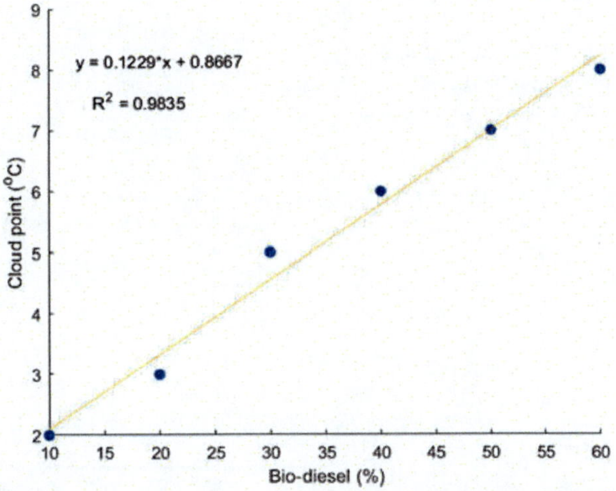

Table 4 Predicted value versus experimental value for cloud point of blended fuel

Fuel	Predicted cloud point (°C)	Experimental cloud point (°C)
B 70	9.46	9.9
B 80	10.69	11.2
B 90	11.92	12.3
B 100	13.15	13.5

3.3 Variation of Flash Point with Blending Ratio

The flash point of a fuel is important for safety purposes. It is the minimum temperature at which a fuel emits enough vapour to get ignited. Higher flash point indicates low volatility of fuel and greater safety. So from mathematical understanding, the relationship between flash point (F) and biodiesel percentage (X) in blended fuel least square regression line can be represented as

$$F = 1.321\,X + 47.8 \tag{4}$$

For Eq. (4), we can get the value of $r = 0.9958$ and $R_2 = 0.9916$ which indicate good correlation between cloud point and blended fuel as shown in Fig. 4

It has been observed from Fig. 4 that flash point shows an increasing tendency with increase in blending ratio as the flash point is higher for JCO biodiesel (B10 to B60). Using Eq. (4) the predicted value of flash point for B70 to B100 can be calculated and obtained as in Table 5. A good correlation has been observed.

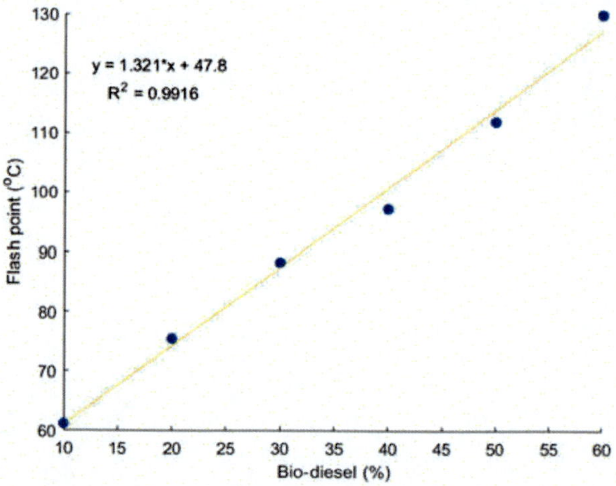

Fig. 4 Variation of flash point with biodiesel percentage in blended fuel

Table 5 Predicted value versus experimental value for flash point of blended fuel

Fuel	Predicted flash point (°C)	Experimental flash point (°C)
B 70	140.27	143
B 80	153.48	157
B 90	166.69	169
B 100	179.9	182

4 Conclusion

In the present research investigation, the suitability of blended fuel prepared from JCO biodiesel has been analyzed and it has been observed that the different physicochemical properties of blended fuel satisfy the standard specifications. With regard to kinematic viscosity and density, blended fuel show some higher values than diesel fuel. The different linear mathematical equations have been identified from the experimental observations to predict the blended biodiesel properties at different blending ratios. The developed mathematical models recognized a higher coefficient of determination (R^2) between biodiesel properties and blending ratio which shows excellent prediction of blended fuel properties. In future, there is a plan to study the engine performance and emission characteristics by using blended fuel from JCO biodiesel and diesel at a suitable blending ratio for best engine performance and emission characteristics which would be helpful for a sustainable environment.

References

1. Nurul HCH, Nozieana K, Bazlul MS, Mohd AH (2020) Potential of Jatropha curcas L. as Biodiesel Feedstoc, in Malaysia: a concise review. Processes 2020 8(7), 786
2. Reddy ANR, Saleh AA, Islam MS, Hamdan S, Rahman MR, Masjuki HH (2018) Experimental evaluation of fatty acid composition influence on Jatropha biodiesel physicochemical properties. J Renew Sustain Energy 10:013103
3. Abedin MJ, Masjuki HH, Kalam MA, Sanjid A, Rahman SMA, Fattah IMR (2014) Performance, emissions, and heat losses of palm and jatropha biodiesel blends in a diesel engine. Ind Crops Prod 59:96–104
4. Moniruzzaman M, Yaakob Z, Khatun R (2016) Biotechnology for Jatropha improvement: a worthy exploration. Renew Sustain Energy Rev 54:1262–1277
5. Hulwanm DB, Joshi SV (2011) Performance, emission and combustion characteristic of a multicylinder DI diesel engine running on diesel–ethanol–biodiesel blends of high ethanol content. Appl Energy 8(12):5042–5055
6. Jindal S, Nandwana BP, Rathore NS, Vashistha V (2010) Experimental investigation of the effect of compression ratio and injection pressure in a direct injection diesel engine running on Jatropha methyl ester. Appl Therm Eng 30:442–448
7. Ganapathy T, Gakkhar RP, Murugesan K (2011) Influence of injection timing on performance, combustion and emission characteristics of Jatropha biodiesel engine. Appl Energ 88(12):4376–4386
8. Anand K, Sharma RP, Mehta PS (2011) Experimental investigations on combustion, performance and emissions characteristics of neat karanji biodiesel and its methanol blend in a diesel engine. Biomass Bioenerg 35(1):533–541
9. Chen LY, Chen YH, Hung YS, Chiang TH, Tsai CH (2013) Fuel properties and combustion characteristics of Jatropha oil biodiesel-diesel blends. J Taiwan Inst Chem Eng 44:214–220
10. Mofijur M, Masjuki HH, Kalam MA, Atabani AE (2013) Evaluation of biodiesel blending, engine performance and emission characteristics of Jatropha Curas Methyl ester: Malaysian perspective. Energy 55:879–887
11. Nandi S, Bhattacharyya R (2016) Studies on enzymatic production of biodiesel from Jatropha Curcas Oil by varying alcohols. J Chem Biol Phy Sci 6(1):311–320
12. Nandi S, Bhattacharyya R (2015) Production of biodiesel from Jatropha Curcas oil with recycling of enzyme. Int J Appl Civ Environ Engg 1(1):01–05

13. Nandi S, Bhattacharyya R (2015) Biodiesel from Jatropha Curcas Oil: a comparative study between chemical and biocatalytic transesterification. Res J Rec Sci 4:44–50
14. Roy PK, Datta S, Nandi S, Basir FA (2014) Effect of mass transfer kinetics for the production of biodiesel from Jatropha Curcas oil: a mathematical approach. Fuel 134:39–44
15. Berber A (2016) Mathematical model for fuel flow performance of diesel engine. Int J Auto Engg Technol 5(1):17–24
16. El-Galad MI, El- Khatib KM, Abdelkader E, El-Araby R, El-Diwani G, Hawash SI (2018) Empirical equations and economical study for blending biofuel with petroleum jet fuel. J Adv Res 9:43–50
17. Acharya N, Nanda P, Panda S, Acharya S (2017) Analysis of properties and estimation of optimum blending ratio of blended mahua biodiesel. Engg Sci Technol Int J 20:511–517
18. Gulum M, Bilgin A (2019) Measurement and prediction of density and viscosity of different diesel-vegetable oil binary blends. Environ Cli Technol 23(1):214–228
19. Fassinou WF, Sako A, Fofana A, Koua KB, Toure S (2010) Fatty acids composition as a means to estimate the high heating value (HHV) of vegetable oils and biodiesel fuels. Energy 35(12):4949–4954
20. Gülüm M, Bilgin A (2017) Measurements and empirical correlations in predicting biodiesel-diesel blends' viscosity and density. Fuel 199:567–577
21. Gülüm M, Bilgin A (2016) Two-term power models for estimating kinematic viscosities of different biodiesel-diesel fuel blends. Fuel Pro Technol 149:121–130
22. Pratas MJ, Freitas SVD, Oliveira MB, Monteiro SC, Lima§ ÁS, Coutinho JAP (2011) Biodiesel density: experimental measurements and prediction models. Energy Fuels 25(5):2333–2340
23. Fahd MEA, Lee PS, Chou SK, Wenming Y, Yap C (2014) Experimental study and empirical correlation development of fuel properties of waste cooking palm biodiesel and its diesel blends at elevated temperatures. Renew Energy 68:282–288

Path Minimization Planning and Cost Estimation of Passive Optical Network Using Algorithm for Sub-optimal Deployment of Optical Fiber Cable

S. K. Biswas and Amitava Podder

Abstract Passive optical network (PON) is an ultimate solution for recent communication technology which accentuates on faster, less expensive and dependable communication system used as access network for optical fiber communication. The broadband carrier suppliers throughout the world are involved to innovate technique to pull down the complexity of network systems, required time of installation, and necessary skill set for installation which finally downsize the overall deployment expenses of the entire network structure. PON imparts one optical fiber to many end users by utilizing power splitters associated with various optical network unit (ONU) situated to customer's premises. One of the fundamental criteria of PON network planning is to design the path of optical link of a point-to-multipoint network which interfaces each end user through the central offices (COs) and power splitter(s), within a lower cost of deployment. This criterion needs to be executed considering the practical restrictions, such as conceivable fiber path, the splitting ratio of optical splitters, position of splitters, and presence of any obstacle on the deployment path. So, here, an algorithm is proposed to decrease the deployment cost by maintaining strategy of lower distance limit with avoiding the obstacle in the fiber path. So, ultimately, by this proposition, the network set-up cost will decrease, and the approximate deployment cost will be possible to compute.

Keywords PON · Optimization · ONU · CO · Remote node (RN) · Power splitter · Passive splitter combiner (PSC) · Cost estimation · End user

S. K. Biswas
Department of Electronics and Telecommunication Engineering, JIS School of Polytechnic, Kalyani, Nadia 741235, India

A. Podder (✉)
Department of Computer Science and Technology, JIS School of Polytechnic, Kalyani, Nadia 741235, India

M. Mitra et al. (eds.), *Computational Advancement in Communication, Circuits and Systems*, Lecture Notes in Electrical Engineering 786,
https://doi.org/10.1007/978-981-16-4035-3_7

1 Introduction

Passive optical networks (PON) might be used in various types as Ethernet PONs, Gigabit- PONs (G-PON), or WDM PONs. to serve as access network which convey information between CO(s) to ONU(s) located at end user's premises. For recent years, much research on PON has focused on areas like dynamic bandwidth allocation (DBA) among ONUs, framework up-gradation to achieve higher data transfer speed of PON, but little focus has been pointed on the technique to design cost-effective PON set-up. So, here, the proposed technique will help to identify a sub-optimal point-to-multipoint network that may be utilized to interconnect each customer (ONU) to their respective CO through the power splitters with the total length of the optical fiber as less as possible. When the positions of the central offices (COs), end users, locations to put optical splitter are known, we may utilize this technique as key strategies which is capable to develop a network diagram close to the shortest route. Further, when the probable path is recognized, there might be constrain where the shortest connections become impractical because of the presence of specific hindrances on the way, which cannot be traversed by any means (non-traversable impediments). Then, those said hindrances should be taken into count accordingly. Further, by applying hindrance circumvent technique described here, we may avoid the impediment on the way of optical fiber and complete to find the probable feasible route through which the fiber might be set-up. Optical fiber might be connected utilizing different topology like ring topology, bus topology, and tree topology, but tree topology is mostly utilized in practice. So, in this paper, our discussion is mainly for tree topology to connect the ONUs with CO through power splitters.

2 Optimization Process for Optical Path

2.1 Network Diagram Using k-nearest Neighbor Algorithm

The k-nearest neighbor (KNN) algorithm is a powerful tool which is essential to the algorithmic study of geometric problems [1]. Suppose, that N-points ($p_1, p_2, \ldots, p_N$), which is randomly generated, in a two-dimensional plane to mark the ONU are located at remote node and if all the N-points are allocated in the same plane along with M-points which indicate location of splitters ($s_1, s_2, \ldots, s_M$), and the position of CO is predefined, such that each N will be connected with CO through their nearest M-point by single fiber path, and the total Euclidean distance among each and every N-points with CO becomes minimum, and the result is partition of the plane into M-areas (where $M < N$). As we use tree topology, so in most of the case, the position of the splitter will be located at almost the center of each region. In nearest neighbor techniques, variants for multi-label classification, regression, and semi-supervised learning settings allow its application to classification of remote nodes located at different distance over a two-dimensional geometric plane.

2.2 Introduction to Classification and KNN Classifier

Classification

This is the problem to predict discrete class labels for unlabeled pattern or different remote nodes based on observations. Let, $\{(x_1,y_1), \ldots, (x_N, y_N)\}$ be the set of observations of q-dimensional patterns $X = \{x\}_{i=1}{}^{N} \subset R_q$, and the corresponding set of labels or location of splitter is $Y = \{y_i\}_{i=1}{}^{N} \subset \mathrm{R}_d$. The goal of classification is to calculate a functional model f that allows a reasonable prediction of class label y' for an unknown pattern or distribution of remote nodesx'. Remote nodes without labels should be assigned to labels.

KNN Classifier

Nearest neighbor classification which is also known as k-nearest neighbors (KNN) is based on the idea that the nearest patterns of remote node to a target pattern or distribution of remote node x', for which we seek the label, delivers useful label information. KNN assigns the class label of the majority of the k-nearest remote node in data. For this sake, we have to be able to define a similarity measure in data space. In R_q, it is reasonable to employ the Minkowski metric (p-norm)

$$\left\|x' - x_j\right\|^p = \left(\sum_{i=1}^{q} \left|(x_i)' - (x_i)_j\right|^p\right)^{1/p} \tag{1}$$

The Euclidean distance for p is 2.In the case of binary classification, the label set $\mathrm{Y} = \{1,-1\}$ is employed, and KNN is defined as:

$$f_{\mathrm{KNN}}(x') = +1 \text{ if } \sum_{i \in Nk(x')} y_i \geq 0; \text{ otherwise } -$$
$$f_{\mathrm{KNN}}(x') = -1 \text{ if } \sum_{i \in Nk(x')} y_i \leq 0 \tag{2}$$

With neighborhood size K and with the set of indices $Nk(x')$ of the k-nearest patterns the choice of K defines the locality of KNN.

2.3 The Proposed Algorithm

Most of the time, for two end points of each path where the optical fiber link might be set are confined by PSC position, existing network resources, obstructions (both traversable and non-traversable obstacles like streets and green-field territories). Holding this limitation viable, the network planner needs to design the PON route with the objective that the total fiber length to be as short as possible.

This issue can be displayed as an insignificant network designing issue in a given graph $G(N, M)$. In this graph, G is the collection of nodes N, and focus points of

Table 1 Input parameters to the proposed algorithm

Parameter	Description
$G(N, M)$	Set of ONU points and set of probable location of splitters
K	This number decides how many neighbors (where neighbors are defined based on the distance metric) influence the classification. This is usually an odd number if the number of classes is 2. If $k = 1$, then the algorithm is simply called the nearest neighbor algorithm
$N_{splitter}$	The splitting ratio of the optical splitter
L_{max}	The maximum allowable length of an optical fiber/aerial lead in line

splitter M are utilized to imitate the areas of ONU and splitters individually. Also, the network organizer needs to consider the parting proportion of the optical N splitter and the extreme admissible length of an aerial lead in line L_{max} that associates the optical splitter and ONU. We are considering L_{max} because good BER figure for better communication depends on length of the optical fiber [4]. Therefore, these parameters are also taken into count. The input parameters of the proposed algorithm are listed in Table 1.

The main challenge here is to avoid the obstacle (if any) situated in the path of the optical fiber deployment, which can be performed by following the obstacle avoidance technique described as follow.

2.4 *Obstacle Avoidance*

A larger portion of the operational network installation needs to be carried through densely populated area where the shortest route connections may not be feasible because of the presence of specific obstacles which cannot be traversed or may be traversed at a cost which is higher in comparison with a standard set-up cost of another optical fiber link. The said impediments must to be considered when searching for the shortest connecting route associated with two self-assertive nodes on a two-dimensional map. To start with, we have to construct a networking map containing various nodes. Obstacles may exist there between the shortest path of any two points or more in number on the map. Identify the obstacles from the map (Fig. 1). Next, we design convex hull structure using Graham scan process consisting with vertices of the hindrances we selected to bypass and the points we want to connect (Fig. 2). At this point, we need to pick the shorter route of the ways along the convex hull-shaped structure formed between the two nodal points (Fig. 3).

This technique is applicable for searching any path between CO to splitters and further splitters to ONU. So, to acquire a sub-optimal point-to-point (P2P) network which interconnects every customer's ONU to their respective CO with the total sum of optical fiber cable length close to the shortest path is identified. Next, we need to compute the approximate cost of the path.

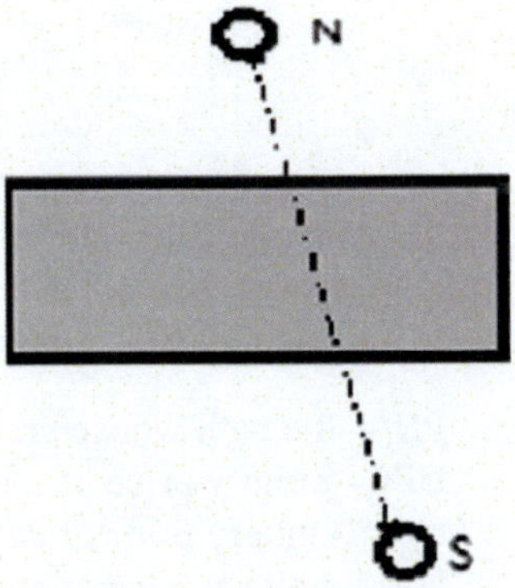

Fig. 1 Identify the obstacle located on the shortest path between two connecting nodes

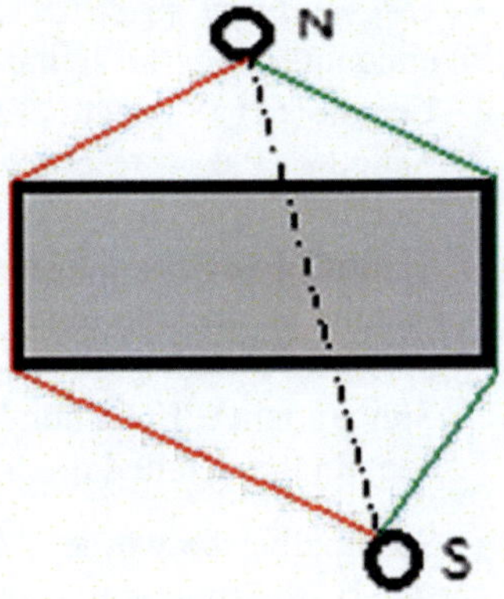

Fig. 2 Design convex hull structure using Graham scan process consisting with vertices of the hindrances

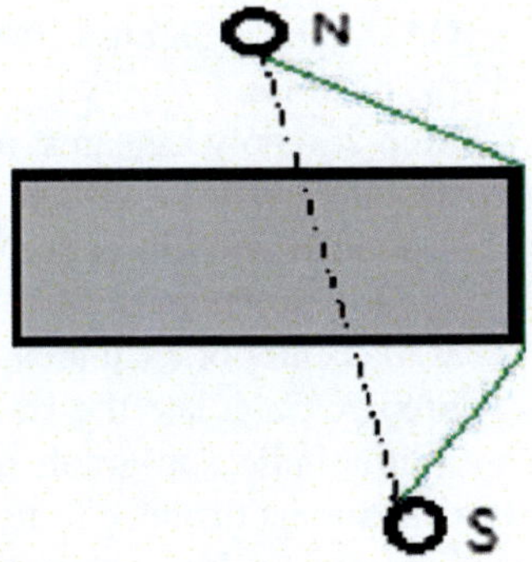

Fig. 3 Choose the shortest possible path by avoiding the obstacle

2.5 Approximate Cost Calculation of the Optimized Path

We can calculate the path price using following formula:

$$\begin{aligned}\text{Cost}(\text{PH}_1) = {} & \text{UP}_1.\text{Length}(\text{PH}_1) + \text{UP}_1.\text{Length}_{\text{R}}(\text{PH}_1) \\ & + (\text{UP}_2 + \text{UP}_3).\text{Length}_{\text{NR}}(\text{PH}_1) \\ & + \text{UP}_4.\text{Splice Count}(\text{PH}_1) \\ & + (\text{UP}_2 + \text{UP}_3 + \text{UP}_5).\text{Length}_{\odot}(\text{PH}_1)\end{aligned}$$

where

PH_l—the path whose price we are calculating,
UP_1—unitary price of the fiber cable,
UP_2—unitary price of deploying fiber in an already existing conduit,
UP_3—unitary price of creating a new cable conduit,
UP_4—price of realizing a fiber splice between two fiber sections,
UP_5—unitary price of deploying a fiber cable traversing a given obstacle encountered in the signal path.
Length (PH_l)—length of the path PH_l,
Length_R (PH_l)—resource length of a path PH_l defined as the total of lengths of individual links in the said path, which reuse an already existing resource,
Length_NR (PH_l)—non-resource length of a path PH_l defined as the sum of the lengths of all individual links in the said path which cannot reuse an existing resource Length $_\text{O}$ (PH_l)—length of the path PH_l that passes through an obstacle, Splice count (PH_l)—number of fiber splices in the path. The following subsections contain the detailed description.

So, finally, the whole optimization process along with cost estimation may be carried by the following steps as

Step 1: Construct a network diagram, with the assumption of the locations of ONU, they are arbitrarily distributed and that for the splitter and COs are fixed on the diagram.
Step 2: Apply k-nearest neighbor algorithm where the shortest route distance of the ONU with the splitter is considered. In this process, whole area of the diagram is divided into sub-area(s), where each sub-area consists with set of remote nodes and location points of one splitter. In most of the cases, splitters are located almost at the center of each area.
Step 3: Calculate the shortest distance route between the CO and each optical splitter in the sub-graph, and decide the path to every splitter. Similarly, calculate the shortest distance route from splitters to ONU.
Step 4: Check the connected shortest way from CO to ONU, and if any obstacle is there on the path, the route may be adjusted by the procedure described in the obstacle avoidance section, after which the feasible shortest path is possible to determine.
Step 5: Check the communication route from CO to ONU, and if the route length exceeds the maximum limit, then reject the route and search for new option.

Step 6: If the optimized path is determined, apply the approximate cost estimation technique on the optimized path to calculate the approximated cost of the possible route.

3 Simulation Results

For the purpose to evaluate the feasibility of the proposed algorithm, here, we conducted a numerical simulation. In this evaluation, we used a graph which is constructed with 30 randomly deployed remote nodes to approximate a network consisting of power splitters. Here, in the simulation, we used four locations for power splitters. As splitters are available in the ratio up to 1:32 even more, so the number of ONU connections from each splitter do not exceed the higher limit of splitting ratio. The graph is assumed to represent a set of paths where the optical fiber could be deployed. The input parameters used in the numerical simulation are listed here in Table 2.

Now the simulation results are shown and explained as follow:

Figure 4 First, we arbitrarily distribute ONUs or nodes in the network. It is presumed that the positions of the ONUs are at customer's premises around each fixed CO, and the areas of the splitters were put around the CO. Two obstacles are located on the same map. These are all the plots on the map of two-dimensional planes. The conditions are shown in Fig. 4.

Figure 5 After that, we have applied the k-nearest neighbor algorithm to the power splitters of network, expecting each power splitter has customers around it. For this reason, main graph is divided in sub-graph which is shown in Fig. 5.

Figure 6 Every sub-graph is a collection of ONUs and single splitter. ONUs in the network are connected with optical splitters which is shown in Fig. 6.

Now, by applying Step 3, the shortest path between the CO and each optical splitter which are located in the sub-graph is identified and related with the CO. Here, the map with four different locations of optical splitters causes the division of map to four different sub-regions which are marked by different colors. Next, we execute the obstacle avoidance process which is shown in Fig. 7.

In Fig. 8, we may observe that the obstacle in the route is bypassed, and the sub-optimal route is established from CO to ONU. For the cases, ONU to CO distance

Table 2 Input parameters to the proposed algorithm

Parameter	Value that is counted in the evaluation
$G(N, M)$	Graph of 30 randomly deployed nodes as ONU and 04 points indicating location of splitters in a geometric plane of 60 km^2
K	2
N_{splitter}	4
$L_{\max}$	25 km

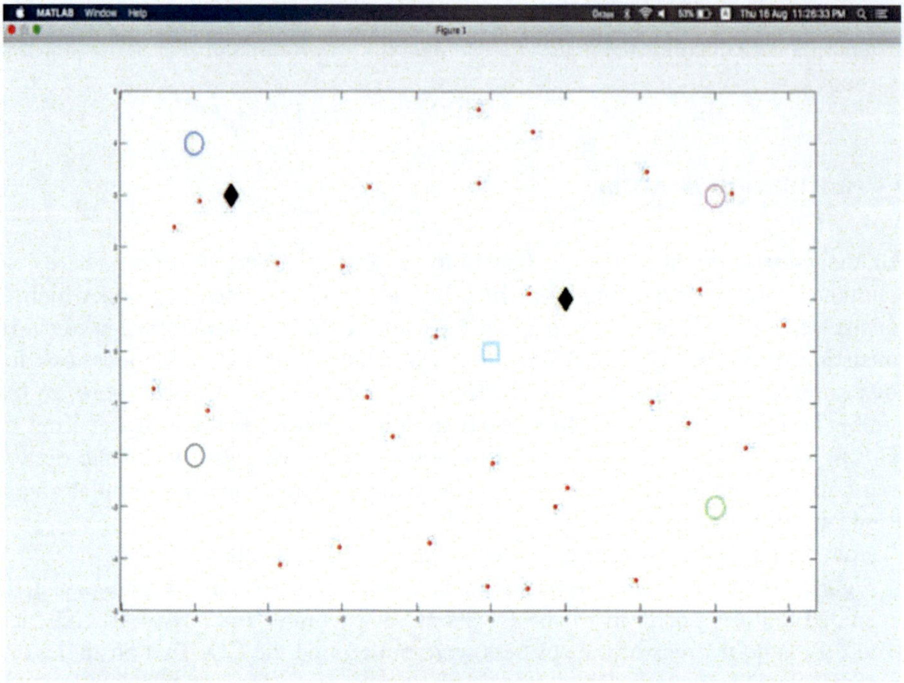

Fig. 4 Simulation Result 1: The CO, ONU, splitter and obstacles are plotted on the graph

crosses maximum limit $L_{\max}$, then ONUs are not marked under any splitter. By this technique, optical fiber path is determined through a sub-optimal path between the CO and ONUs.

4 Conclusion

By this algorithm, sub-optimal point-to-multipoint network is possible to generate. This point-to-multipoint network is capable to connect each ONU to the CO(s) with the help of power splitters. The splitting ratio of the optical splitter and the maximum allowable length of an optical drop cable that connects the optical splitter and subscriber are also considered under this calculation. From results of the simulations, we become confirmed that the algorithm can design the sub-optimal PON system in terms of total optical fiber length, and this is also helpful to calculate the deployment cost of the PON system which will be the optimized cost of network deployment.

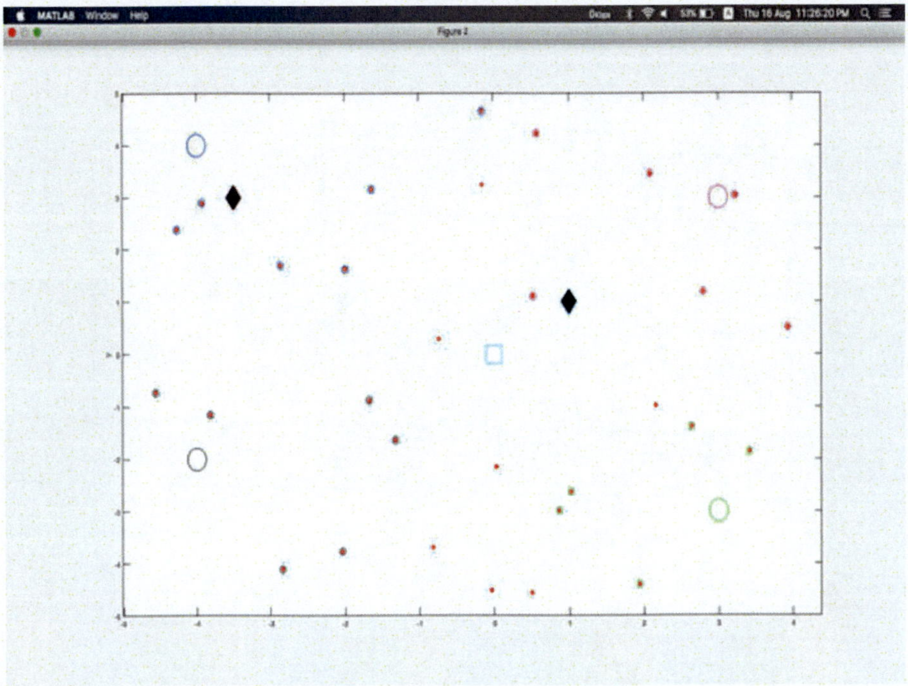

Fig. 5 Simulation Result 2: The graph is sub divided by application of K-th nearest neighbour algorithm

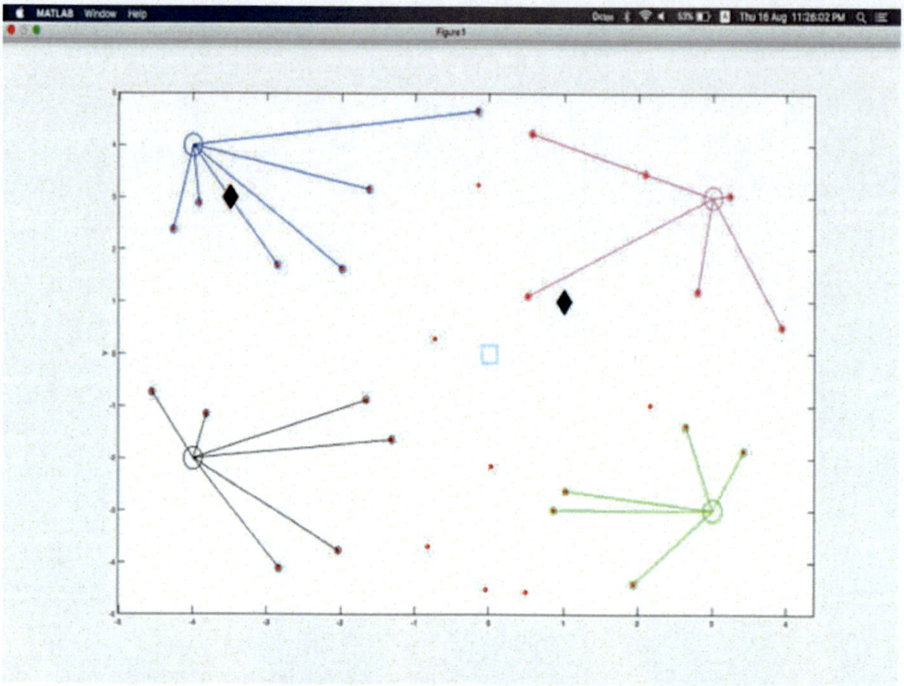

Fig. 6 Simulation Result 3: The route with shortest length is connected between splitter to ONUs

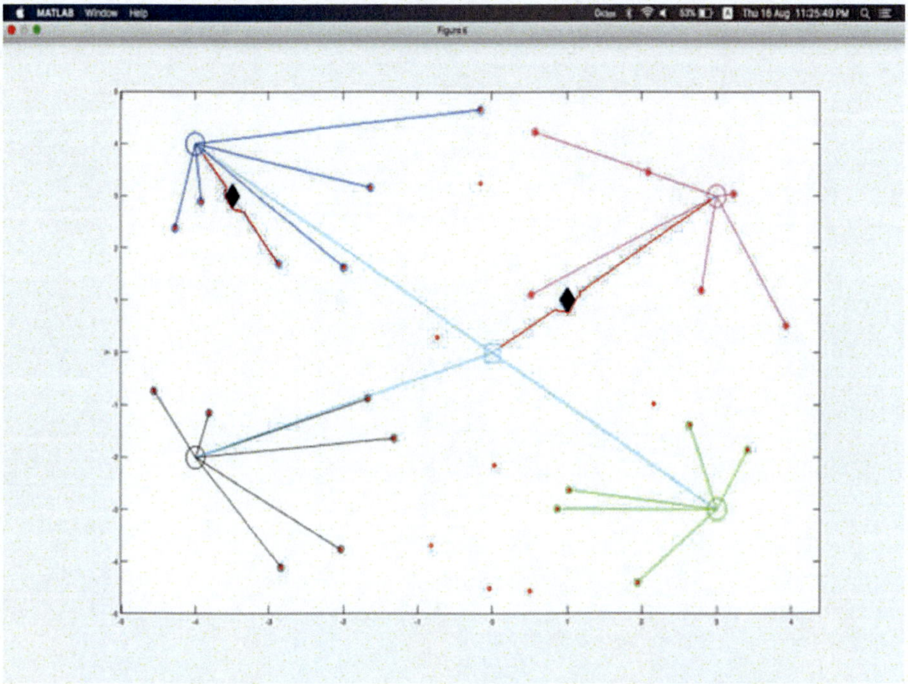

Fig. 7 The route with shortest length are connected between CO to splitters with application of obstacle avoiding techniques

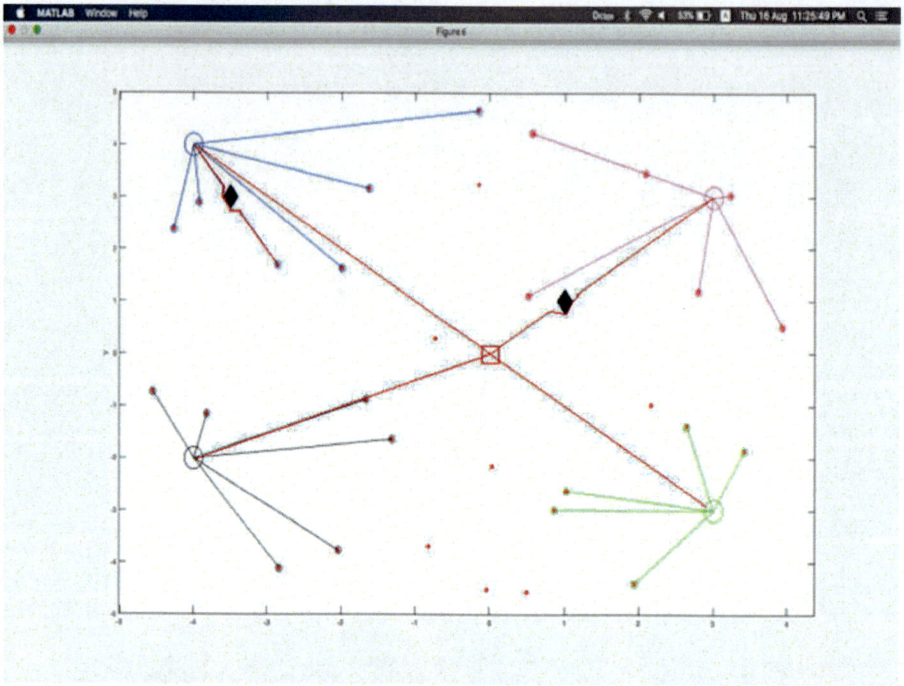

Fig. 8 The path with shortest length between CO to ONU is selected and the approximate cost of the path is calculated

References

1. Agata A, Horiuchi Y (2009) PON network designing algorithm for suboptimal deployment of optical fiber cables. In: Proceedings of SPIE-OSA-IEEE Asia communications and photonics, SPIE, vol 7633, 76330L, pp1–6
2. Lakic B, Hajduczenia M, da Silva H, Monteiro1 PP (2008) Using adapted visibility graphs for network planning. In: Proceedings of IEEE symposium on computers and communications 2008
3. Aurenhammer F (1991) Voronoi diagrams—a survey of a fundamental geometric data structure. ACM Comput Surv 23(3):345–405
4. Pandit J, Biswas SK (2017) Path optimization and cost estimation using algorithm for sub optimal deployment of passive optical network. In: Presented at ICCECE international conference, Kolkata
5. Biswas SK, Podder A (2020) Path minimization planning of passive optical network using algorithm for sub-optimal deployment of passive optical network. In: Presented at IRJET, vol. 7(9), pp 3484–3488 (September 2020)
6. Pandit J, Biswas SK (2016) Digital image transmission performance analysis through passive optical network. In: Presented at ICCECE international conference, Kolkata
7. Dijkstra EW (1959) A note on two problems in connection with graphs. NumerischeMathematik 1:269–271
8. Garg S, Jain RC (2006) Variation of k-mean algorithm: a study for high-dimensional large data set. Inf Technol J 5:1132–1135

9. Rattigan MJ, Maier M, Jensen D (2007) Graph clustering with network structure indices. In: Proceedings of 24th annual international conference on machine learning (ICML'07)
10. Shewchuk JR (1998) Tetrahedral mesh generation by Delaunay refinement. In: Proceedings of the 14th annual ACM symposium on computational geometry, pp 86–95
11. Yoshimoto N (2005) NTT's deployment of FTTH services and future developments. In: Presented at FTTH council conference, Las Vegas, NV, USA
12. Abrams M, Becker PC, Fujimoto Y, Byrne VO, Piehler D (2005) FTTP deployments in the United States and Japan—equipment choices and service provider imperatives. IEEE J Lightwave Technol 23:236–246
13. Podder A, Chanda PB (2017) Comparative performance study and analysis on different edge based image segmentation techniques of thermal images. In: Presented at IJRITCC, vol 5(5), pp 632–636
14. Nakyama Y, Yasunaga R (2020) ITU TWDM-PON module for ns-3. Wirel Netw 4(5), 532–533
15. Ji Y, Zhang J, Wang X, Yu H (2018) Towards converged, collaborative and co-automatic (3C). Opt Netw Sci 61:101–119
16. Satyanarayana K (2014) Recent trends in future proof fiber access passive networks: GPON and WDM PON. In: 2014 international conference on recent trends in information technology, 10–12 April 2014

Implementing Data Security in Delay Tolerant Network in Post-disaster Management

Chandrima Chakrabarti and Samir Pramanick

Abstract Disaster causes severe destruction to physical infrastructures. As a result, communication infrastructure has been getting disrupted for weeks. Wireless ad-hoc networks use mobile devices to deliver services. In any critical situation, ad-hoc network acts as delay-tolerant network (DTN). DTN is resource-constrained network, where nodes are required to cooperate with each other to relay messages in store-carry-forward feature. These messages are re-addressed to other nodes based on prearranged criteria and finally are conveyed to a destination node via multiple hops. Meanwhile, during the transmission of message from sender node to receiver node, the privileged message may be disclosed to the other node except sender node and receiver node. So, message should be encrypted by the sender and decrypted by the receiver to maintain proper data security in a DTN network; there may be periodic disruptions or long delays in the connection between the network devices. Opportunistic network environment (ONE) simulator is used for performance evaluation and comparison with other state-of-the-art schemes.

Keywords Ad-hoc network · Delay-tolerant network · Store-carry-forward · Multiple hops · Disaster

1 Introduction

In the post-disaster scenario, loss of communications will lead us to less connectivity, so transfer of messages is not an easy job in that scenario. We unreservedly depend on mobile communication for such cases. At that time, message transfer is a very important thing for rescue operation on that location; thus, the proper security of messages needs to be maintained. Delay-tolerant networks were introduced in 2003, for fighting the enormous delays involved in deep space communications (in the order of minutes, hours, or even days). Such retard cannot be manipulated

C. Chakrabarti (✉) · S. Pramanick
Computer Science and Engineering Department, Narula Institute of Technology, Agarpara, Kolkata, W.B 700109, India
e-mail: chandrima.chakrabarti@nit.ac.in

M. Mitra et al. (eds.), *Computational Advancement in Communication, Circuits and Systems*, Lecture Notes in Electrical Engineering 786,
https://doi.org/10.1007/978-981-16-4035-3_8

with subsist (terrestrial) networking technologies and, thus, must be manipulated at the application level. Extending the idea of delay tolerance, it may be doable with the same design to manipulate long-term disruptions as well. Disruptions (intentional or unintentional) can contemplate as unintended retard. Depending on climate and requisition scenario, a DTN may be a "delay-tolerant," a "disruption-tolerant" or even a "delay-tolerant and disruption-tolerant" network. Delay-tolerant networking (DTN) considers how to provide communication in contexts where it is unreasonable to assume end-to-end connectivity. We propose a security model that ensures encrypted and secured delivery of messages from source to destination. The model includes the impact of the disaster on the transportation network and that model's population and relief vehicle movement. Opportunistic network environment (ONE) simulator is used for simulation and performance analysis.

2 Related Work

Almost every year, the world is stricken by numerous catastrophic natural disasters [1–3], such as earthquake, hurricane, typhoon, and tsunami. When stricken by catastrophic natural disaster, such as Hud-Hud cyclone in Andhra Pradesh, flash floods in Himachal Pradesh, and landslides in Himachal Pradesh, emergency aid operation is very critical to numerous lives. Many people trapped in the disaster zone under collapsed buildings or landslides may have a large chance to survive if they are rescued in 72 h, referred as "golden 72 h." People evacuated from their home, jammed in highways, or dome shelters need to communicate, teach other for various reasons such as allocation of rescue and relief resource as well as reunion of family members. However, communication systems, fixed or mobile, were usually down due to various reasons. Rescue teams in each stricken area consist of few trained professional squads, army, police, fire fighters, and hundreds of thousands of disorganized volunteers. The loss of communication systems made the rescue operation extremely difficult. Although, establishing a temporary communication network to support emergency communications and networking is one of the most urgent tasks in the disastrous rescue mission. We propose to use Wi-Fi-ready mobile devices owned by rescue volunteers themselves to construct a DTN to support such a need. Because the popularity of Wi-Fi-ready mobile devices are very high nowadays, this solution would be highly feasible in many countries [1–3].

Researchers around the world are working on this issue. The preferred mode of securing data is through encryption. Secure multicast communication is a significant requirement in emerging applications in ad-hoc environments like scenarios of post-disaster management. DTNs (mobile ad-hoc network) are well-known for their flexibility and ease of communication [4]. The communication is purely based on trust, without any need of authentication. This often leads to insecure communication, causing information tampering. The traditional means of security are not sufficient to safeguard against the inherent dangers of DTN. Due to special features like hop-by-hop communications, wireless media, open border and easy to set up, DTN became popular for handling malicious nodes [5–11].

Hsu et al., in [12], studies six existing security models of PEKS/SCF-PEKS schemes and analyzes their efficiency and performance. Shikfa et al. in [12], uses PEKS to allow intermediate nodes to discover partial matches between their profile and the destination profile and uses policy-based encryption to enforce confidentiality of the payload. This scheme suits opportunistic networks well because it has a low storage and computation overhead, and it relies on an offline TTP only. Ramanathan et al., in [13], presented a novel mechanism called prioritized epidemic (PREP) that expires time instruction and topology awareness to decide which bundles to delete or hold back when faced with a resource (buffer, bandwidth) crunch. Joe et al., in [14], proposes a DTN message priority routing suitable for emergency situations. Mashhad et al., in [15], proposes an approach for priority-scheduling in participatory DTNs, whereby messages are being forwarded based on a combination of the likelihood of future encounters. Finally, John Burgess et al., in [16], proposes MaxProp, a protocol for effective routing of DTN messages, based on prioritizing both the schedule of packets transmitted to other peers and the schedule of packets to be dropped. Durst, in [17], has proposed an infrastructure security model for DTN based on public key technology. The work uses Pretty Good Privacy (PGP) systems, Public Key Infrastructure (PKI) systems, and Kerberos systems. Karim El Defrawy et al., in [18], takes advantage of well-known affiliated entities (AE) to which a destination may be linked. The scheme uses AEs' public key, and with as few as two usual entities, a source can send confidential messages to a destination. In all these works, existence of centralized security infrastructure is assumed for the providing authentication and certainty. In all these works, existence of centralized security infrastructure is assumed for the providing authentication and security. However, the availability of such infrastructure in a post-disaster scenario cannot be assumed. An onion routing technique for anonymous communication over a computer network where messages are repeatedly encrypted and then sent through several network nodes called onion routers has been proposed by Goldschlag et al. in [19]. Each onion router removes a layer of encryption to uncover routing instructions and sends the message to the next router where this is repeated. This prevents these intermediary nodes from knowing the contents of the message. In their work on secure group communication, Edelman et al., in [20], creates a group-oriented security solution for DTN where group keys are used at the application layer or may be integrated into a larger system encrypting all traffic. Zhengyi Le et al., in [21], proposes privacy-enhanced opportunistic networking (PEON) a framework that applies the principles of anonymous communications to the specific challenges of opportunistic networking. Durst, in [22], has proposed an infrastructure security model for DTN based on public key technology. The work uses Pretty Good Privacy (PGP) systems, Public Key Infrastructure (PKI) systems, and Kerberos systems. Asokan et al., in [23], has considered the applicability of identity-based cryptography as a solution for delay-tolerant networking security. Karim El Defrawy et al., in [24], takes advantage of well-known affiliated entities (AE) to which a destination may be linked. In all these works, existence of centralized security infrastructure is assumed for providing authentication and security. However, the availability of such infrastructure in a post-disaster scenario cannot be assumed [25–28]. Therefore, we, in this paper, try to eliminate the use of such centralized entities at the runtime and handle authentication and other security issues in a distributed and decentralized manner. Due to limited battery resource, it is important to reduce power consumption for realizing ad-hoc networks among handheld terminals.

3 System Model

When a disaster occurs, it becomes difficult to provide services in the affected areas due to the lack of transmission. All modes of transmission system get damaged, and it takes a lot of time to initiate an infrastructure network. To overcome this, the infrastructure less DTN is entrenched using Wi-Fi-enabled devices for effective and efficient rescue operations.

Managing disasters and emergency situations requires accurate and precise message delivery, on which sensitive decision-making in such scenarios is dependent. Since most of the infrastructure of any communication is eradicated by the strike of the disaster, this can affect rescuing missions and recovery which are time critical. The alternative should be planned and set for managing the post-disaster situations and emergency issues. A common issue to be solved in communications for disaster scenarios is security. During a disaster, information plays an important role, but the exchange of data must be secure and reliable. So the objective is in order to prevent various security attacks, some encryption and decryption algorithms need to be implemented.

Figure 1 shows secured communication in post-disaster situation. Here, control station, shelter, volunteer, and victim nodes are responsible to carry out the secure communication.

DTN is an infrastructure less network, in which collection of autonomous mobile nodes forming an interim (Ad-hoc) network without using any centralized assistance and base where all nodes in the network behave as routers and take part in the detection and maintenance of routes to other nodes in the network. The availability of routes at an instant can increase or decrease due to mobility; hence, availability of paths can vary in an ad-hoc network. This type of network is commonly set up on a temporary basis to facilitate communication in rough conditions and under specific situations. In DTN, all nodes are wireless, mobile, and generated through a battery. The participating nodes arrange themselves automatically and can be a standalone network or attached to a large network, including the Internet. The disaster rescue teams are required to communicate in an efficient manner to save people from their critical situation. The efficient communication needs different techniques and algorithms. Unlike some existing networking technologies such as Internet Protocol (IP) networks or cellular systems, wireless ad-hoc networks have the advantage that deployment of networks does not require preexisting infrastructures. Ad-hoc networks have neither fixed topology nor centralized servers; it is assumed

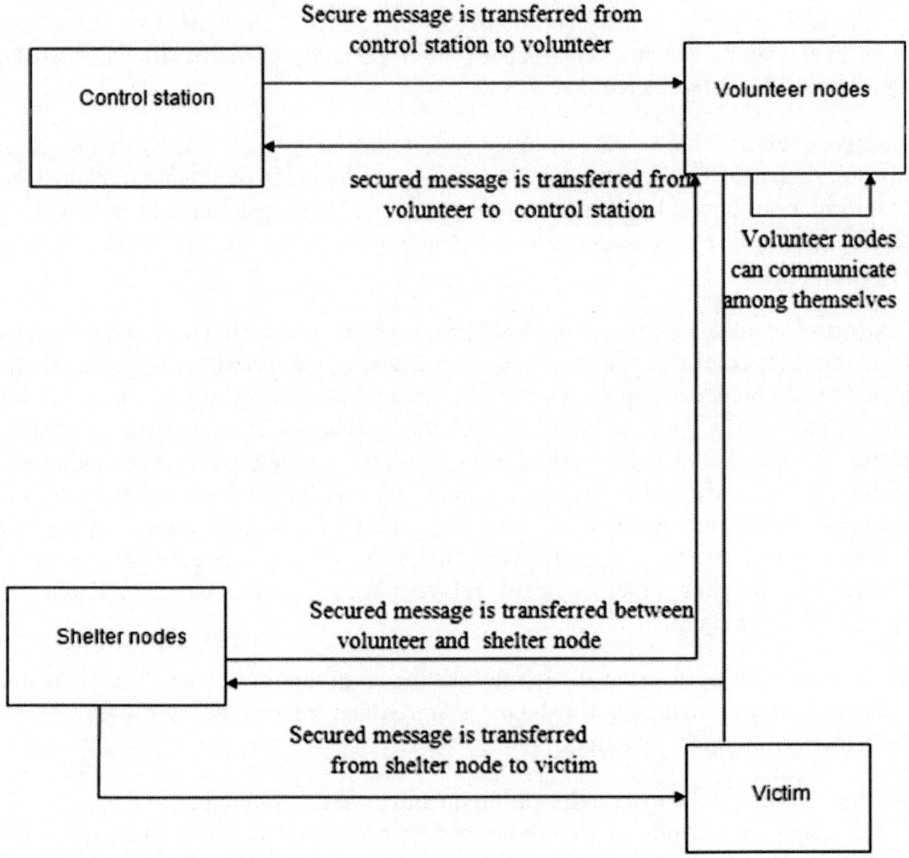

Fig. 1 Secured communication in post-disaster situation

that, once deployed, the network nodes would self-configure to provide affinity and form a communications network. Communications are vital for effective management and execution of disaster response and recovery efforts, emergency response personnel must be able to exchange information with each other from anywhere, at any time, to successfully fulfill their missions. DTN is a kind of ad-Hoc networks with special characteristics like an open network boundary, dynamic topology, distributed network, fast and quick implementation, and hop-by-hop communications. These characteristics of DTN made it popular, especially in military and disaster management applications. Due to the special features, wide-spread of DTN faced lots of challenges like peer-to-peer applications. A post-disaster situation demands an efficient transmission and coordination among rescue teams. The exchange of real-time information among responders and emergency management centers is crucial to saving lives. In such scenario, DTNs are suitable for providing communication mechanism, as they are easy to deploy and do not require elaborate infrastructure.

To evolve worthy security solutions for such habitat, we must first understand how DTNs can be striking. This paper provides a comprehensive study of a strike against mobile ad-hoc networks.

Control station This is an extremity actions center (control room) in the post-disaster scenario. The venture of the control room includes group and transmission of information concerning natural calamity and relief. They communicate with volunteer nodes to convey messages to shelter and to manage the situation of the disaster prone area.

Volunteer Volunteers are a common feature of disasters. They convey messages from shelter to control station and vice versa. There generally exists a group of volunteer. The volunteer can supply relief to the victim. Volunteers can also communicate among themselves. Despite concerted efforts for relaying crucial situational information, disaster relief volunteers experience significant communication challenges owing to failures of critical infrastructure and longstanding power outages in disaster-affected areas. Volunteers who are carrying smart phones are forwarder nodes (FN) that move across the disaster stricken area and forward-categorized shelter messages, to the corresponding relief camp by relaying them through forwarders that are approaching the camp.

Shelter node After disaster, all victims take shelter in certain safe site. That is shelter node. All relevant data about a shelter is stored and transmitted by a shelter node (SN) (e.g., a laptop in a shelter).

Victim The people who actually suffered from disaster is a victim.

Each and every node of the above needs to transmit between themselves for probing and rescuing the victims and to provide them proper relief.

For certain data transmission in the post-disaster scenario, we suggest encode and decode of messages. The algorithms are as follows.

Algorithm to generate cipher:(Encryption Algorithm)

```
1.Message is taken in the string message[ ]
2. Key=Generate random number and share this key to all other receiver
3. i=0
4. Encrypted message will be stored in Encrypted Message[ ]
5. Repeat steps from 6 to 24 until i <message[ ].length
6. ch= message[i]
7.   If ch>='a' and ch<='z'
8.        Then ch= ch+key
9.                 If ch>'z'
10.                    Then ch=ch - 'z'+'a' -1
11.                End if
12.       Encrypted Message[ ]=ch
13.  End if
14.  Else If ch>='A' and ch<='Z'
15.       Then ch= ch+key
16.                If ch>'Z'
17.                    Then ch=ch - 'Z'+'A' -1
18.                End if
19.       Encrypted Message[ ]=ch
20.  End of Else if
21.  Else
22.           Encrypted Message[ ]=ch
23. End of Else
24. i=i+1
25.End
```

4 Explanation of Encryption Algorithm

The above algorithm ensures the encoding of messages from sender node to receiver node. In the above algorithm, message is taken as a string. The receiver generates a random key or a cipher and shares the key to all other recipients to decode the message. None other than the sender and receiver knows the key. This message becomes secure enough. Encoded message is stored in an array of string, i.e., encrypted message. The work of this code is to restore each plain text letter with a distinct one random number of places down the alphabet. For example, if the random key is 3, then we have to replace the character by another character that is 3 positions down to it. Finally, the message can be encrypted and can be sent to receiver node. In this way, the sender can securely send the message to the receiver.

Algorithm to decrypt cipher:(Decryption Algorithm)

```
1.Encrypted Message is taken in the string message[ ]
2. Key is received from the sender
3. i=0
4. Decrypted message will be stored in DecryptedMessage[ ]
5. Repeat steps from 6 to 24 until i <message[ ].length
6. ch= message[i]
7.  If ch>='a' and ch<='z'
8.      Then ch= ch+key
9.          If ch>'a'
10.             Then ch=ch + 'z'-'a' +1
11.         Endif
12.     DecryptedMessage[ ]=ch
13. Endif
14. ElseIfch>='A' and ch<='Z'
15.     Then ch= ch+key
16.         If ch>'A'
17.             Then ch=ch + 'Z'-'A' +1
18.         Endif
19.     DecryptedMessage[ ]=ch
20. End of Elseif
21. Else
22.         DecryptedMessage[ ]=ch
23. End of Else
24. i=i+1
25.End
```

5 Explanation of Decryption Algorithm

The above algorithm ensures the decode of messages at receiver node. In the above algorithm, encoded message is taken as a string. The receiver receives the key from the sender. None other than the sender and receiver knows the key. This message becomes secure enough. Decoded message is stored in an array of string, i.e., decrypted message. As recipient knows the key of sender, receiver uses the same key of sender to decode the original message. Finally, the message can be decoded, and the receiver gets the original message.

6 Result Evaluation

In order to have a secure communication between nodes, encryption and decryption of messages are done using the proposed encryption and decryption algorithms.

Table 1 Simulation parameter

Simulation time	43,200 s
Routing	Spray and wait, epidemic, PRoPHET, MaxProp, energy aware router
Movement model	Shortest path map-based movement, map-based movement, random waypoint
No of Control station	1
No of shelters	3
No of volunteers	3 groups, each group has 10 volunteers
No of victims	3 groups, each group has 20 victims
Buffer size for shelter	100 Mb
Buffer size for Volunteer	80 Mb
Buffer size for control station	100 Mb
Buffer size for victim	50 Mb
Speed of volunteer	7–10 m/s
Speed of victim	2–5 m/s

Delivery probability, latency average, and overhead ratio of the secure message is also analyzed. Energy level report is also generated from different nodes and movement model. Finally, we can assure secure message is transferred between all nodes. In simulation scenario, one control station, three shelters, three groups of volunteer, and three groups of victim are taken. Here, the post-disaster management has been done by this group of nodes. We can now evaluate the result of a simulation using opportunistic network environment (ONE) simulator [29]. The simulation is done in 43,200 s.

Here the following simulation parameters are used in Table 1.

6.1 *Delivery Probability*

These are the probability successfully delivering of messages that to the destination from the source.

6.2 *Latency Average*

It is the average of messages delayed from creation to delivery. Recess is a very important parameter as it constitutes to the network speed. The term recess refers to any kind of detain typically incurred in a transform of network data. A low recess network link is that which normally experiences small delay, whereas high recess usually experiences long delays. In DTN, recess is high due to its network value.

6.3 *Overhead Ratio*

It is the negation of the number of messages relayed to the number of messages carry. Low value of the overhead means less processing required while conveying the relayed messages. The objective of the algorithm is to minimize the value of overhead.

7 Delivery Probability, Overhead Ratio, Latency Average Before Encryption or Without Security

7.1 *Delivery Probability*

In Fig. 2, *X*-axis is taken as time and *Y*-axis is taken as delivery probability, for shortest path map-based movement, spray and wait router has the highest delivery probability.

The highest delivery probability ensures good and effective communication among nodes. So, in order to get higher delivery probability, the combination of shortest path map-based movement and spray and wait router is recommended.

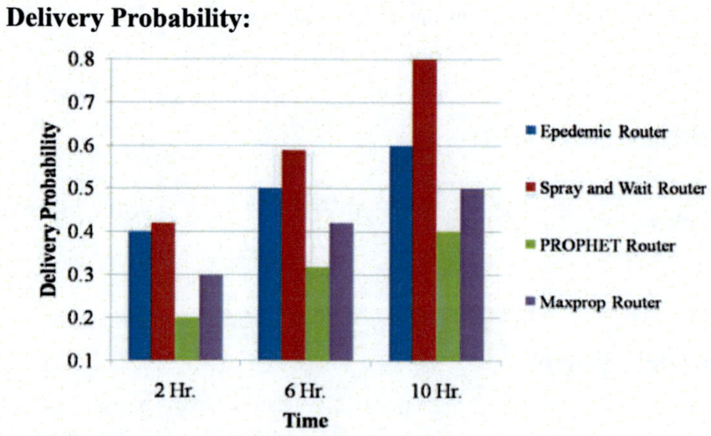

Fig. 2 Delivery Probability using different routers

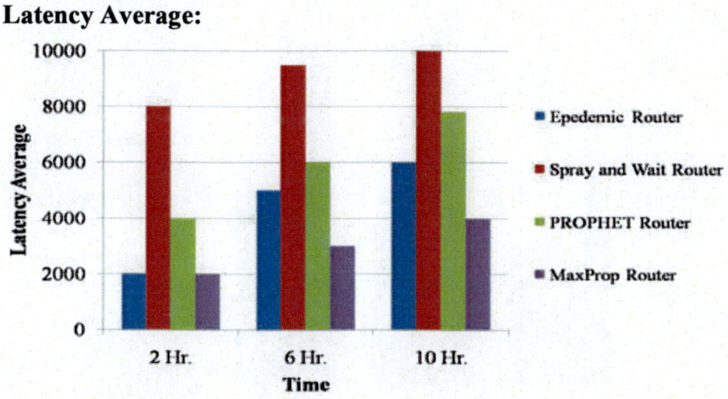

Fig. 3 Average latency using different routers

7.2 Latency Average

In Fig. 3, *X*-axis is taken as time and *Y*-axis is taken as latency average. Shortest path map-based movement and spray and wait router have the excessive latency average. The lowest latency average ensures good and effective communication among nodes. So, in order to get the lowest latency, the combination of shortest path map-based movement and MaxProp router is recommended.

7.3 Overhead Ratio

In Fig. 4, *X*-axis is taken as time and *Y*-axis is taken as overhead ratio. Shortest path map-based movement and epidemic router have the highest overhead ratio. The highest overhead ratio ensures less efficient communication among nodes, and the lowest overhead ratio ensures more good and effective communication. So, in order to get the lowest overhead ratio, the combination of shortest path map-based movement and spray and wait router is recommended.

Overhead Ratio:

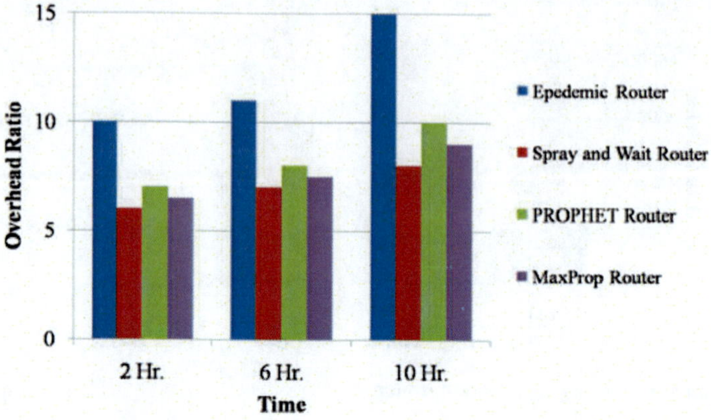

Fig. 4 Overhead ratio using different routers

8 Delivery Probability, Overhead Ratio, Latency Average After Encryption or with Data Security

8.1 Delivery Probability

Delivery probability can also be examined after encryption or with security. In Fig. 5, X-axis is taken as time and Y-axis is taken as delivery probability. Shortest path map-based movement using spray and wait router has the highest delivery probability.

Delivery Probability, Overhead ratio, Latency average after encryption or with data security:

Delivery Probability:

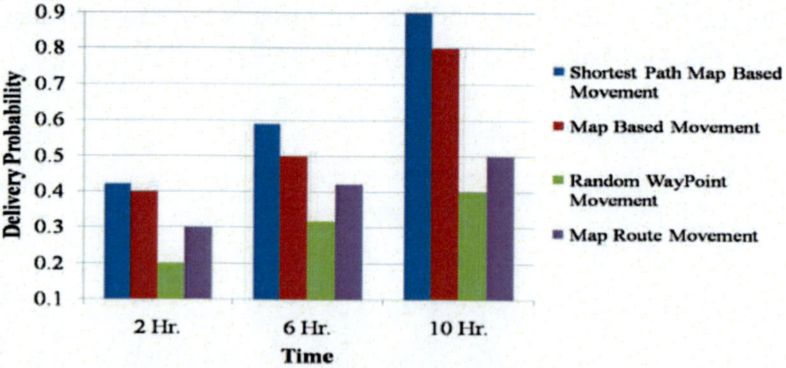

Fig. 5 Delivery probability using different movement models

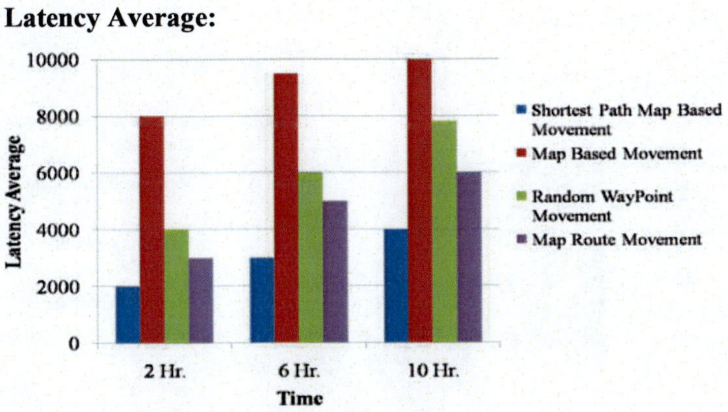

Fig. 6 Latency average using different movement models

The highest delivery probability ensures good and effective communication among nodes. So, in order to get higher delivery probability, the combination of shortest path map-based movement and spray and wait router is recommended. The same combination of router is recommended as before encryption or without security.

8.2 *Latency Average*

Latency average can also be analyzed after encryption or with security. In Fig. 6, *X*-axis is taken as time and *Y*-axis is taken as latency average. Shortest path map-based movement and spray and wait router have the lowest latency average. The lowest latency average ensures good and effective communication among nodes. So, in order to get the lowest delivery probability, the combination of shortest path map-based movement and spray and wait router is recommended.

8.3 *Overhead Ratio*

Overhead ratio can also be analyzed after encryption or with security.

In Fig. 7, shortest path map-based movement and spray and wait router have the lowest overhead ratio. The highest overhead ratio ensures less efficient communication among nodes, and the lowest overhead ratio ensures more good and effective communication. So, in order to get the lowest overhead ratio, the combination of shortest path map-based movement and spray and wait router is recommended. The same combination of router is recommended as before encryption or without security.

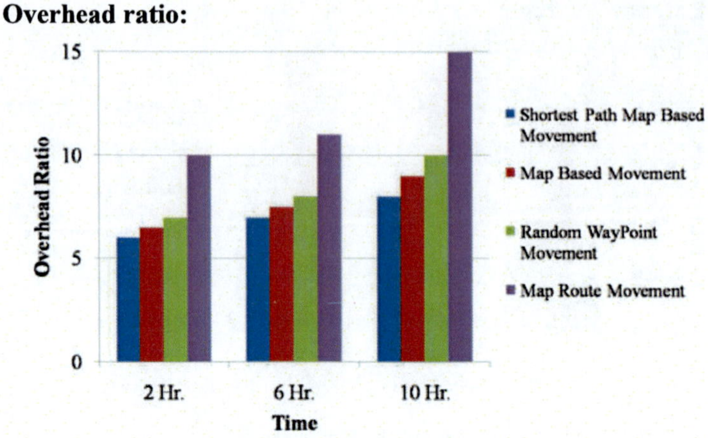

Fig. 7 Overhead ratio using different movement models

9 Conclusions

In this paper, the ONE simulator is presented as an opportunistic networking evaluation system that offers a variety of tools to create complex mobility scenarios that come closer to reality than many other synthetic mobility models. Through this paper, the researcher had tried to understand what DTNs are, their traits, and uses. The various criteria, upon which the safety of the network is evaluated, are also realized. Crucially, the various vulnerabilities in DTNs and also the possible attacks that can occur are studied. The study of the above makes well-equipped on the understanding of the possible problems in DTNs. It helps in deciding an effective technique to solve the presented problem. The final study on the security measures shows the possible solutions to all given problems. Moreover, the study of the above would help researchers to understand the underlying methods and the shortcomings of the existing systems and gives them a clear idea on the direction the research should proceed to develop a better system with enhanced features. This project focuses on how to establish a secure key management service in an ad-hoc networking environment. This paper has discussed different features and properties of delay-tolerant network. DTN enables people to take a planned approach in disaster-affected areas where network infrastructure can vary from bad to worst while coping under harsh conditions. It can also span across multiple networks.

References

1. Hao Y, Haiyun L, Fan Y, Songwu L, Lixia Z (2004) Security on mobile ad hoc networks: challenges and solutions. 1536-1284/04/IEEE Wirel Commun
2. Zhou L, Haas JZ (1999) Securing ad hoc networks. IEEE Netw Mag 13(6):24–30

3. Nishu G, Mahapatra RP (2009) DTN security issues. IJCSNS Int J Comput Sci Netw Sec 9(8)
4. Kumar BA, Siddhartha G, Kumar SY, Anil G Security in adhoc networks. Computer Science Department, University of Kentucky
5. Tushar Er G (2010) Overview of security threats in mobile ad-hoc network. J High Perform Commun Syst Netw 2(1–2):1–10
6. Burg A (2003) Ad hoc network specific attacks. In: Seminar ad hoc networking: concepts, applications and security. Technische University Munchen
7. Bouam S, Othman BJ (2003) Data security in ad hoc networks using multipath routing. In: Proceedings of the 14th IEEE PIMRC, pp 1331–1335, 7–10 September
8. Schafer G (2002) Research challenge in security for next generation mobile networks. In: Position papers PAMPAS' 02-workshop on requirements for mobile privacy and security, 16–17 September
9. Wu B, Chen J, Wu J, Cardei M A survey of attacks and countermeasures in mobile ad hoc networks. Department of Computer Science and Engineering, Florida Atlantic University, http://student.fau.edu/jchen8/web/papers/SurveyBookchapter.pdf
10. Stallings W (2002) Wireless communication and networks. Pearson Education
11. Borisov N, Goldberg I, Wagner D (2001) Interception Mobile communications: the insecurity of 802.11. In: Conference of mobile computing and networking
12. Hsu ST, Yang CC, Hwang MS (2013) A study of public key encryption with keyword search. Int J Netw Sec 15(2)
13. Ramanathan R, Hansen R, Basu P, Hain RR, Krishnan R (2007) Prioritized epidemic routing for opportunistic networks. In: MobiOpp 2007, New York, USA
14. Joe I, Kim SB (2010) A message priority routing protocol for delay tolerant networks in disaster areas. Fut Gen Inform Technol 727–737
15. Mashhadi A, Capra L (2011) Priority scheduling for participatory DTN. In: Proceedings of WoWMoM
16. Burgess J, Gallagher, Jensen BD, Levine BN (2006) MaxProp: routing for vehicle-based disruption-tolerant networks. In: Proceedings of IEEE Infocom
17. Durst CR (2002) An infrastructure security m networks. www.dtnrg.org/papers/dtn-sec-wp-v5.pdf, July
18. Defrawy EK, Solis J, Tsudi G (2009) Contacts for message confidentiality in COMPSAC. (1) 271–279
19. Goldschlag MD, Reed GM, Syverson FP (1996) Hiding routing information. In: The workshop on information hiding (1996)
20. Edelman TP, Donahoo JM, Sturgill BD (2010) Secure group communications for DTN. In: ICITST
21. Le Z, Vakde G, Wright M (2009) PEON: privacy enhanced opportunistic networks with applications in assistive environments. In: Proceedings of PETRA 2009
22. Durst CR (2002) An infrastructure security model for delay tolerant networks. dtnrg.org/papers/dtn-sec-wp-v5.pdf
23. Asokan N, Kostiainen K, Ginzboorg P, Ott J, Luo C (2007) Towards securing disruption-tolerant networking. Nokia Research Centre, NRC-TR-2007-007 (2007)
24. Defrawy EK, Solis J, Tsudik G (2009) Leveraging social contacts for message confidentiality in delay tolerant networks. In: COMPSAC, (1), pp 271–279
25. Chakrabarti C, Roy S (2015) Adapting mobility of observers for quick reputation assignment in a sparse post-disaster communication network. In: IEEE proceedings of AIMoC 2015, Kolkata, India, pp 29–35
26. Chakrabarti C, Roy S, Basu S (2018) Intention aware misbehavior detection for post-disaster opportunistic communication over peer-to-peer DTN. In: Peer-to-peer networking and applications. Springer, pp 1–19
27. Chakrabarti C, Basu S (2019) A blockchain based incentive scheme for post-disaster opportunistic communication over DTN. In: Proceedings of ICDCN 2019, ACM, New York, NY, USA, pp 385–388

28. Chakrabarti C, Roy S (2019) iSecure: imperceptible and secure peer-to-peer communication of post-disaster situational data over opportunistic DTN. In: Proceedings of ICDCN 2019, ACM, New York, NY, USA, pp 465–468
29. Keranen A, Ott J, Karkkainen T (2009) The one simulator for DTN protocol evaluation. In: Proceedings of SIMUTools 2009, Article No. 55

Face Detection and Extraction Using Viola–Jones Algorithm

Mayukh Ghosh, Tathagata Sarkar, Darshan Chokhani, and Anilesh Dey

Abstract In the current times, face detection by computer system has become a major field of interest. Face detection technology can be applied to various fields-including security, biometrics, law enforcement, entertainment, and personal safety—to provide surveillance and tracking of people in real time. Face detection applications use algorithms to find only the human faces within larger images. Face detection algorithms typically start by searching for human eyes, one of the easiest features to detect. The algorithm might then attempt to detect eyebrows, mouth, nose, nostrils and the iris. Once the algorithm concludes that it has found a facial region, it applies additional tests to confirm that it has, in fact, detected a face. In this report, we propose a human face detection method for colored as well as gray images. Also, we cropped the particular detected facial image and extracting and showing the individual cropped image if the input image contains many faces.

Keywords Image processing · Face detection · Viola–Jones algorithm · Cascade object detector · Human faces · Gray images

1 Introduction

Image processing is the tool which is used to process digital images through an algorithm.

Image processing is a multidisciplinary field, with contributions from different branches of science including mathematics, physics, optical, and electrical engineering. Moreover, it overlaps with other areas such as pattern recognition, machine learning, artificial intelligence, and human vision research. Different steps involved in image processing include importing the image with an optical scanner or from a digital camera, analyzing and manipulating the image (data compression, image enhancement, and filtering), and generating the desired output image [1].

M. Ghosh · T. Sarkar · D. Chokhani · A. Dey (✉)
Narula Institute of Technology, Agarpara, India
e-mail: anileshdey@ieee.org

M. Mitra et al. (eds.), *Computational Advancement in Communication, Circuits and Systems*, Lecture Notes in Electrical Engineering 786,
https://doi.org/10.1007/978-981-16-4035-3_9

The generation and development of digital image processing are mainly affected by three factors: first, the development of computers; second, the development of mathematics (especially, the creation and improvement of discrete mathematics theory); third, the demand for a wide range of applications in environment, agriculture, military, industry, and medical science [2–4]. In medicine, it is used for diagnostic imaging modalities such as digital radiography, positron emission tomography (PET), computerized axial tomography (CAT), magnetic resonance imaging (MRI), and functional magnetic resonance imaging (fMRI). Industrial applications include manufacturing systems such as safety systems, quality control, and automated guided vehicle control [12]. Face detection, also known as facial detection, is an artificial intelligence (AI)-based computer technology which is used to find and identify human faces in digital images. Many attempts have been made to solve human face detection problem. The previous trials are aimed for gray-level images only, and image pyramid schemes are necessary to scale with unknown face sizes.

The aim of this project is to successfully detect all the human faces present in a picture and to count the number of faces present in that picture. This algorithm is effective in screening all the non-human faces (dogs, cats, etc.). Such an algorithm which does not involve too many parameters, which is very simple to understand and code, yet giving the desired results with greater efficiency adds to the success of this project.

The scope of this project in the near future is detecting human faces

- in harsh lighting conditions without missing out even a single face,
- even if all the parameters of face detection may not be present (example—a side facing person where one half portion of the face is visible),
- even if the image is blurred,
- if the camera is far away from the subject.

2 Materials and Methods

In this research work, an image containing one or more human faces is given by the user. The code analyzes the image and detects all the human faces present in that image, only when the faces satisfy all the requirements of face detection. In this work, cascade object detector is used which follows Viola–Jones algorithm. Once the detection is completed, all the detected human faces are cropped out from the image and are shown in another output screen. The code also calculates the number of human faces detected in the input image and does not consider any other faces except humans even if they satisfy all the basic requirements of face detection [7].

2.1 Data Used for Image Insertion

The user inputs an image and the MATALB reads it using imread() function. The image can be of jpg, png, jpeg, or any other format. The code also calculates the size of the image using size() function and also resize if width is greater than "320" using imresize() function.

2.2 Data Used for Human Face Detection

For detecting human face, an object called vision.CascadeObjectDetector(_) has been used. This helps us to detect only the human faces with 95% accuracy. Now a function called step() has been used to return the location of the detected regions. For detecting the number of detected region, we use size() function. Then numbers are converted into string using num2str() function. Finally to concatenate the strings, the function strcat() has been used. Now in the output window along with the detected faces, the number of faces detected will also be shown.

Viola–Jones Algorithm Developed in 2001 by Paul Viola and Michael Jones, the Viola–Jones algorithm is an object-recognition framework that allows the detection of image features in real time. Despite being an outdated framework, Viola–Jones is quite powerful, and its application has proven to be exceptionally notable in real-time face detection [5]. Viola–Jones algorithm is used within cascade object detector to detect people's faces, noses, eyes, mouth, or upper body [6]. The Viola–Jones algorithm is a widely used mechanism for object detection. The main property of this algorithm is that training is slow, but detection is fast. This algorithm uses Haar basis feature filters, so it does not use multiplications [12].

The problem to be solved is detection of faces in an image. A human can do this easily, but a computer needs precise instructions and constraints. To make the task more manageable, Viola–Jones requires full view frontal upright faces. Thus, in order to be detected, the entire face must point toward the camera and should not be tilted to either side. While it seems, these constraints could diminish the algorithm's utility somewhat, because the detection step is most often followed by a recognition step, in practice these limits on pose are quite acceptable [11].

The characteristics of Viola–Jones algorithm which make it a good detection algorithm are: Robust—very high detection rate (true-positive rate) and very low false-positive rate always. Real time—For practical applications, at least 2 frames per second must be processed.

Face detection only (not recognition)—The goal is to distinguish faces from non-faces (detection is the first step in the recognition process).

The algorithm has four stages:

1. Haar feature selection
2. Creating an integral image

3. AdaBoost training
4. Cascading classifiers.

Haar-Like Features Often in computer vision, features are extracted from input images rather than using their intensities (RGB values, etc.) directly. Haar-like features are one example. Other examples include histogram of oriented gradients (HOG), local binary patterns (LBP), etc. A Haar-like feature consists of dark regions and light regions. It produces a single value by taking the sum of the intensities of the light regions and subtract that by the sum of the intensities of dark regions. There are many different types of Haar-like features, but the Viola–Jones object detection framework only uses the ones in Fig. 1. The different types of Haar-like features let us extract useful information from an image such as edges, straight lines, and diagonal lines that we can use to identify an object (i.e., the human face) [9] (Fig. 2).

Creating an Integral Image An integral image is an intermediate representation of an image where the value for location (x, y) on the integral image equals the sum of the pixels above and to the left (inclusive) of the (x, y) location on the original image [12]. This intermediate representation is essential because it allows for fast calculation of rectangular region. To illustrate, figures shows that the sum of the red region D can be calculated in constant time instead of having to loop through all the pixels in that region. Since the process of extracting Haar-like features involves

Fig. 1 Working of Viola–Jones algorithm

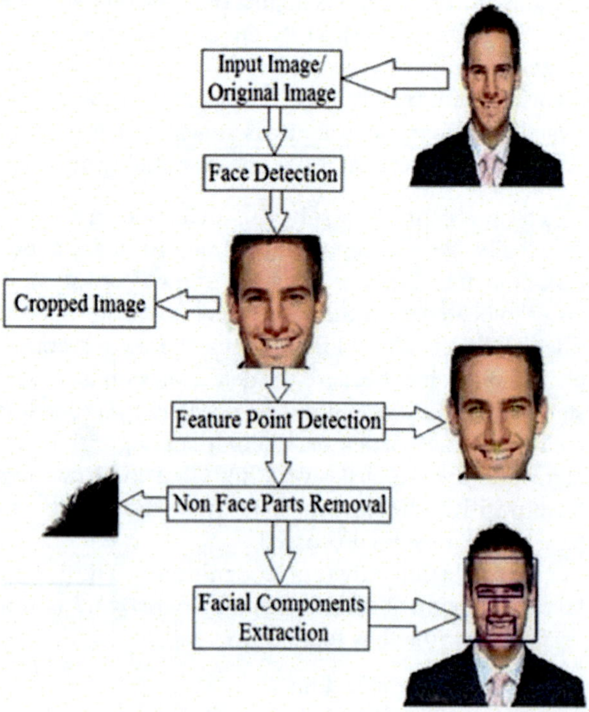

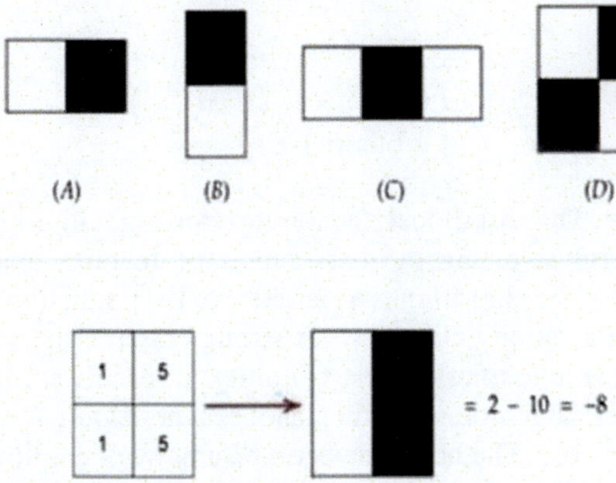

Fig. 2 Haar-like features (top) and how to calculate them (bottom)

calculating the sum of dark/light rectangular regions, the introduction of integral images greatly cuts down the time needed to complete this task [10] (Fig. 3).

$$\theta_j, s_j = \arg\min_{\theta,s} \sum_{i=1}^{N} w_j^i \varepsilon_j^i \tag{1}$$

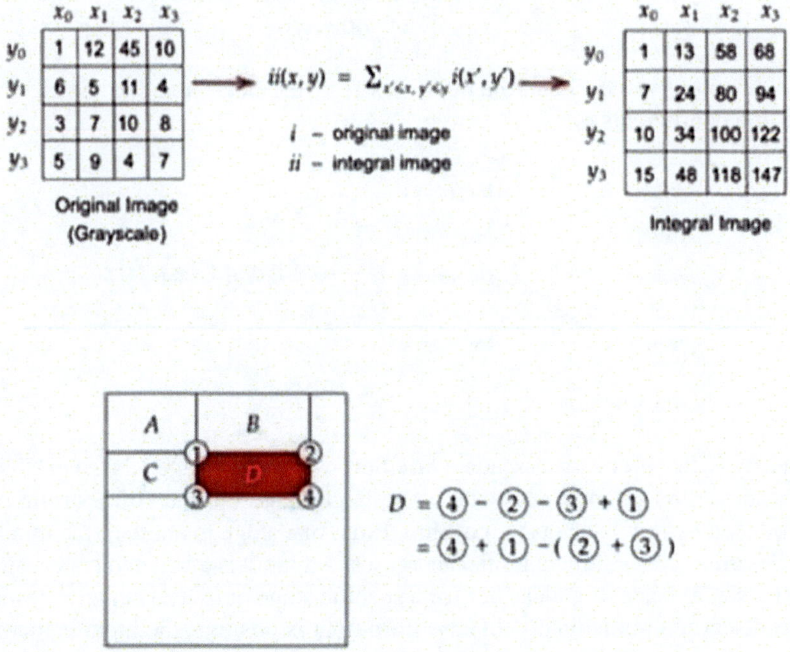

Fig. 3 Conversion of original image into integral image

where,

$$\varepsilon_j^i = \begin{cases} 0 \text{ if } y^i = h_j\left(\mathbf{x}^i, \theta_j, s_j\right) \\ 1 \text{ otherwise} \end{cases} \tag{2}$$

Adaboost Training The AdaBoost (adaptive boosting) algorithm is a machine learning algorithm for selecting the best subset of features among all available features. The output of the algorithm is a classifier (a.k.a prediction function, hypothesis function) called a "strong classifier." A strong classifier is made up of a linear combination of "weak classifiers" (best features). In each iteration, the algorithm finds the error rate for all features and then chooses the feature with the lowest error rate for that iteration [10]. The object detection framework employs a variant of the learning algorithm AdaBoost to both select the best features and to train classifiers that use them. This algorithm constructs a "strong" classifier as a linear combination of weighted simple "weak" classifiers.

$$h(\mathbf{x}) = \operatorname{sgn}\left(\sum_{j=1}^{M} \alpha_j h_j(\mathbf{x})\right) \tag{3}$$

Each weak classifier is a threshold function based on the feature f_j.

$$h_j(\mathbf{x}) = \begin{cases} -s_j & \text{if } f_j < \theta_j \\ s_j & \text{otherwise} \end{cases} \tag{4}$$

The threshold value θ_j and the polarities $s_j \in \pm 1$ determined in the training, as well as the coefficients a_j.

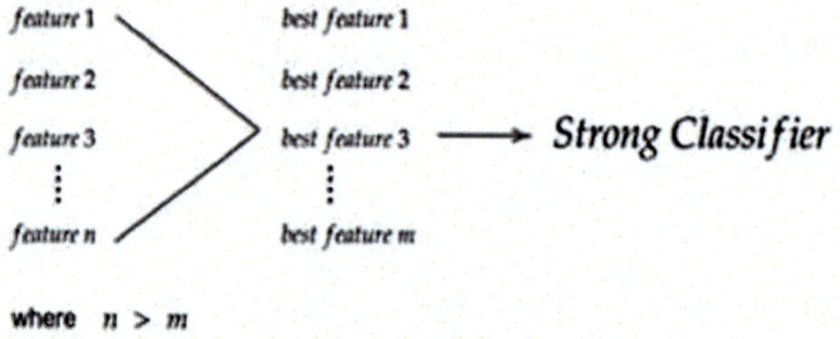

Cascading Classifiers A cascade classifier is a multi-stage classifier that can perform detection quickly and accurately. Each stage consists of a strong classifier produced by the AdaBoost algorithm. From one stage to another, the number of weak classifiers in a strong classifier increases. An input is evaluated on a sequential (stage by stage) basis. If a classifier for a specific stage outputs a negative result, the input is discarded immediately. In case the output is positive, the input is forwarded onto the next stage [5] (Fig. 4).

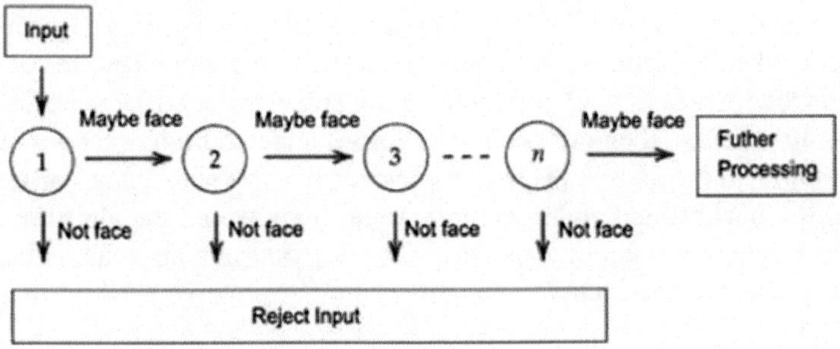

Fig. 4 Working of cascading classifiers

Cascading Basis Cascading is another sort of "hack" to boost the speed and accuracy of our model. So, we start by taking a sub-window, and within this sub-window, we take our most important or best feature and see if it is present in the image within the sub-window. If it is not in the sub-window, then we do not even look at the sub-window, we just discard it. Then if it is present, we look at the second feature in the sub-window. If it is not present, then we reject the sub-window. We go on for the number of features have, and reject the sub-windows without the feature. Evaluations may take split seconds; but since you have to do it for each feature, it could take a lot of time. Cascading speeds up this process a lot, and the machine is able to deliver better results much faster than ever before.

2.3 Datasets

Datasets	Face detection accuracy (%)	Distant image (%)	Non-human face (%)	Blurred image (%)	Side image (%)
[1]	100	0	0	0	0
[2]	87.5	12.5	0	0	0
[3]	0	0	0	100	0
[4]	100	0	50	0	0
[5]	100	0	50	0	0
[6]	100	0	75	0	0
[7]	0	50	50	0	100
[8]	0	0	0	0	100
[9]	66.6	0	0	33.3	0
[10]	100	0	100	0	0
Mean	65.41	6.25	32.5	13.33	20

The following datasets were taken in order to measure the accuracy of the algorithm. A total number of ten datasets were taken out of which five of them gave 100%

accuracy, and the mean of all the accuracies resulted to be 65.41%. The datasets which were taken in random included pictures having only people, people with animal, blurred images as well as people facing sideward out of which if in an image contains only animals, then our accuracy of non-detected human face is 100%.

For example, in the dataset [4] which consists of a boy with a dog lying on a green grassy field in background and harsh light conditions where the algorithm properly extracts the boy's face without considering the dog face and also calculates that only one face is present in the image.

2.4 Algorithm

- Read an image using "imread()" function and store in a variable "the_image." (.png /.jpg / jpeg / etc.)
- Store the size of image in a matrix form of order "width × height" using "size()" function.
- Resize the image to the width of "320" using "imresize()" function if width of input image is greater than "320."

- Detect the face using "vision.CascadeObjectDetector()" function and return the location of the detected region of face using "step()" function and store in a variable "location_of_face."
- Calculate the number of extracted faces using "size(location_of_face,1)" function and store in a variable "*n*" and convert it into string using "num2str(n)".
- Insert a shape over the detected faces using "insertShape()" function.
- Crop the detected image using "imcrop()" function.
- Now extract the individual faces from the detected image using a for loop which runs from "1 to *n.*"
- Calculate the location of the individual extracted faces using "location_of_face(i,:)" and crop the respective extracted faces using "imcrop(the_image,predicted_loc)."
- Plot the original image, the face detected image and also show the total number of faces detected.

2.5 Flowchart

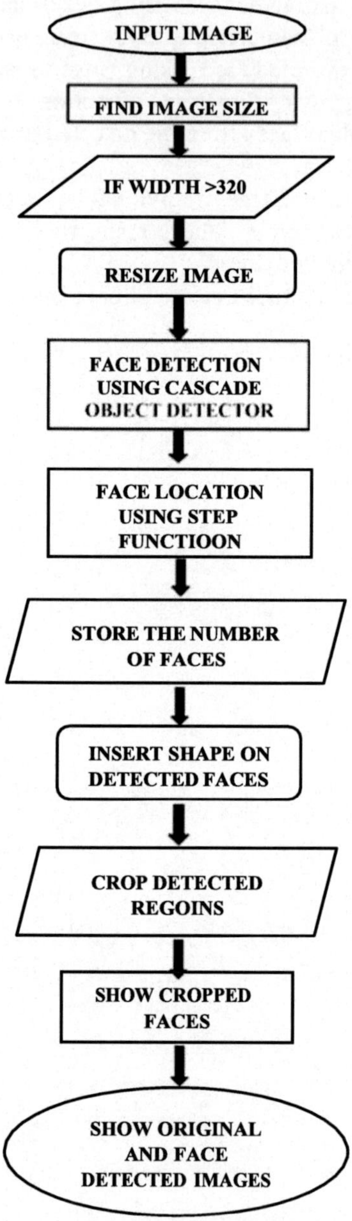

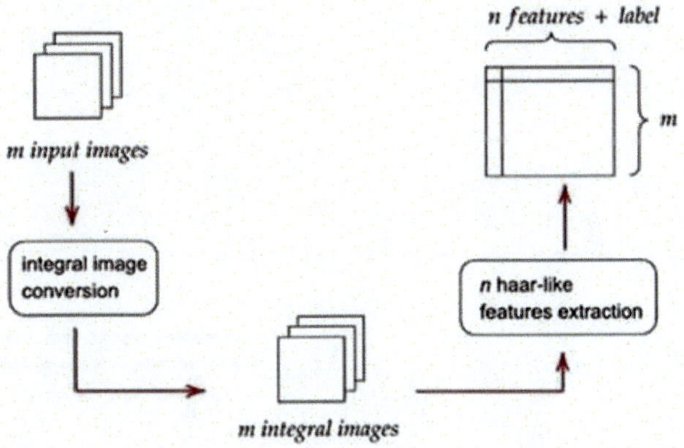

Fig. 5 Data preparation

3 Results and Discussion

3.1 Data Preparation

Assuming that you already have a training set consisting of positive samples (faces) and negative samples (non-faces), the first step is to extract features from those sample images. Viola and Jones [12] recommend the images to be 24 × 24. Since each type of Haar-like features can have different sizes and positions in a 24 × 24 window, over 160,000 Haar-like features can be extracted. Nonetheless, in this stage, all 160,000+ Haar-like features need to be calculated. Fortunately, the introduction of integral images helps speed up this process [10] (Fig. 5).

3.2 Constructing a Cascade Classifier with a Modified AdaBoost Algorithm

As you can imagine, using all 160,000+ features directly are computationally inefficient. Viola and Jones [12] proposed two solutions that can solve this. First, reduce the number of features to only a handful of useful features with the AdaBoost algorithm. Second, split the remaining features into stages and evaluate each input in a stage by stage (cascading) fashion. Viola and Jones [12] devised a modified version of the AdaBoost algorithm to be able to train a cascade classifier [10] (Fig. 6).

Fig. 6 AdaBoost algorithm

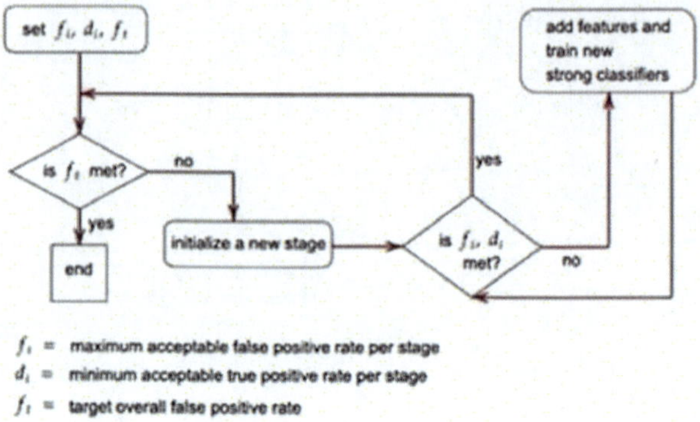

3.3 Statistical Analysis

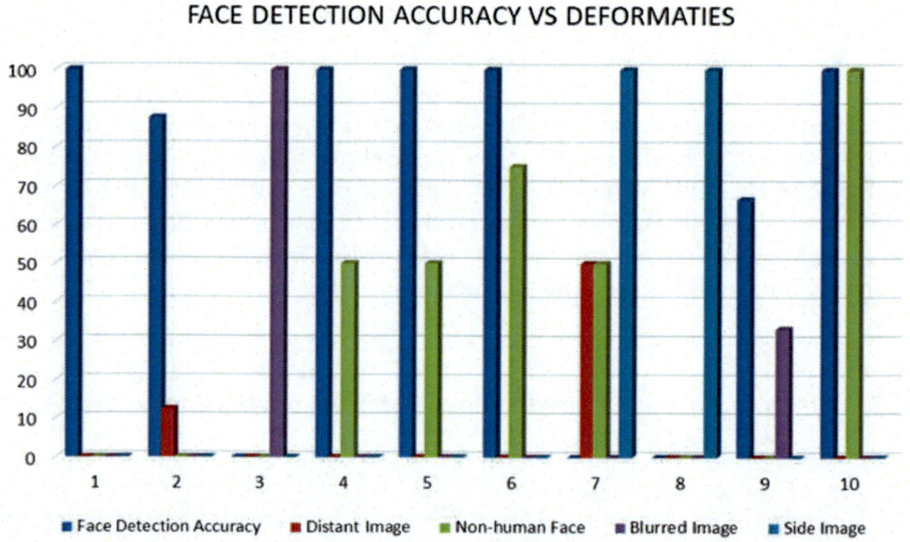

The graph is based on certain parameters which includes face detection accuracy, distant images, non-human faces, blurred images, and side view images on the x-axis, and for the y-axis, it contains the percentage. The percentage here is calculated on the basis of the number of elements present in picture as well as the algorithm which is opting here.

For example, the dataset [10] tells us that non-human entity that means either any kind of animal or bird is present in the picture; that is why it is a hundred percent on that aspect as well as the accuracy of the detected face is hundred percent as the algorithm is designed so that it can detect no other face than human faces.

Again, in case of dataset [1], the accuracy of face detected is hundred percent where all the other parameters are zero. This means that it is a proper image of a person or a group of people having no deformities such as blurred image, non-human face, distant image, or side viewing angle.

In case of dataset [8], the percentage of sideward image is hundred percent, whereas the face detection accuracy is zero percent. For example, consider a picture of a woman holding her baby, both are facing sideward.

For the dataset [4] the percentage of non-human image is fifty percent as only one dog is present, whereas the accuracy of face detection is hundred percent because only one human face is detected. For example, consider a picture of a boy and a dog lying side by side on green field.

3.4 Application

Imagine that we need to detect faces in the above image. Viola and Jones [12] use a sliding window approach where window of various scales is slid across the entire image. The scale factor and the shifting step size are parameters for you to decide upon. So, for the above image, there are *m* sub-windows to evaluate. For a sub-window "*I*," the framework resizes the image of that sub-window to a base size of 24 × 24 (to match training data), convert it into an integral image, and feed it through a cascade classifier produced during the training phase. A face is detected if a sub-window passes through all the stages in the cascade classifier (Figs. 7 and 8).

Now, we are also extracting the particular facial images with their respective identification numbers (Fig. 9).

Fig. 7 Input image

Fig. 8 Detected image

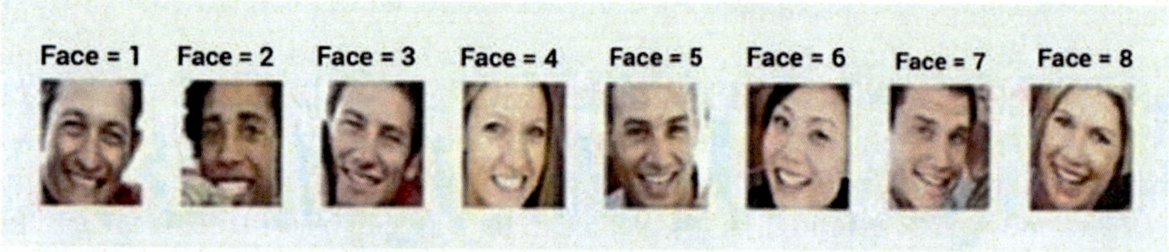

Fig. 9 Individually extracted images

4 Conclusion

To conclude, one learns about the Viola–Jones object detection framework and its application for face detection. Many technologies today benefited from Paul Viola and Michael Jones's work. By understanding how the framework works, one can confidently implement their own version of their work or used an open-source implementation like the one provided by OpenCV. It is hoped that the explanation moves someone forward in that direction and compels him/her to create amazing technologies that uses this awesome framework. This research work has various applications in the field of forensic, security, biometrics, law enforcement, entertainment, and personal safety to provide surveillance and tracking of people in real time. Image processing is also used in cameras for detecting the faces for bokeh mode and many such modes. This can also be used in the CCTV cameras present in the school and college classrooms for tracking the number of students present in the class. This is an efficient weapon against bank robberies.

So just to add some close remarks about the Viola–Jones Algorithm:

- The algorithm was developed in 2001 by Paul Viola and Michael Jones, the first of its kind, and was primarily used for facial detection applications.

- There are two steps to the algorithms: there is training with facial and non-facial images, and then there is the actual detection.
- There are two steps for training: training the classifiers and AdaBoost.
- There are two steps for detection: detecting the Haar-like features and creating the integral image.
- Viola–Jones is one of the most powerful algorithms of its time, and even though there are better models out there today, Viola–Jones set the foundation for it in the field of facial detection.

References

1. Rosenfeld A (1969) Picture processing by computer. Academic Press, New York
2. Gonzalez RC (2008) Digital image processing. In: Woods RE (Richard Eugene), 1954, 3rd ed. Prentice Hall, Upper Saddle River, N.J., pp 23–28
3. Chakravorty P (2018) What is a signal? [Lecture Notes]. IEEE Sig Process Mag
4. Gonzalez R (2018) Digital image processing. Pearson, New York, NY
5. Papageorgiou C, Oren M, Poggio T (1998) A general framework for object detection. In: International conference on computer vision
6. Rathore D, Das GK, Saxena A (2020) Concept of face recognition: a review. Int J Technol Res Eng 7(5)
7. Blake C, Keogh E, Merz C (1999) UCI repository of machine learning databases. Department of Information and Computer Science, University of California at Irvine, Irvine, CA
8. Freund Y, Schapire RE (1996) Experiments with a new boosting algorithm. In: Machine learning: proceedings of the thirteenth international conference. Morgan Kauman, San Francisco, pp 148–156
9. Mordvintsev A, Abid K (2013) Face detection using Haar cascades. OpenCV-Python Tutorial
10. Jensen OH (2008) Implementing the Viola-Jones face detection algorithm. PhD thesis, Technical University of Denmark, DTU
11. Ramsri (2012, September 16) Viola-Jones face detection and tracking explained. In: Cen K (2016) Study of Viola-Jones real time face detector
12. Viola P, Jones M (2001) Rapid object detection using a boosted cascade of simple features. In: Proceedings of the 2001 IEEE computer society conference on computer vision and pattern recognition, vol 1, pp I-511–I-518. https://doi.org/10.1109/CVPR.2001.990517

FPGA-Based Efficient Implementation of CBNS Computational Circuits: A Modular Approach

Madhumita Mukherjee and Salil Kumar Sanyal

Abstract Complex binary number system (CBNS) finds extensive applications in the faster computation of various digital signal processing (DSP) algorithms. In this paper, an attempt has been undertaken to develop various computational circuits based on CBNS for implementation in Spartan XC3S700A FPGA platform. The circuits have been designed following a modular approach. The designed modules involve simple logic gates leading ultimately to efficient implementation on FPGA. The codes for the modules have been developed using verilog hardware description language (HDL). Structural-level designs of nibble size CBNS adder, multiplier, and subtractor have been exclusively accomplished involving these modules. In the design of multiplier and subtractor, a new concept of sub-block has been introduced to efficiently utilize the limited input capability of the designed modules. The proposed design involves less hardware complexity, silicon area, and path delay compared to existing works. Simulation results and performance metrics for all the three CBNS circuits have been included.

Keywords CBNS · FPGA · Modular approach · Nibble size · Conceptual approach

1 Introduction

The arithmetic computational units like adder, subtractor, multiplier form the backbone for the hardware design of any digital systems. These arithmetic units are the core element of modern digital signal processors (DSP), multimedia processor and almost every data processing systems. Demand for high-speed area-efficient and low-power processors required exploration of new architecture for these arithmetic units. To design, these optimized computational units for the data processing applications

M. Mukherjee (✉)
Department of ETCE, Jadavpur University, Kolkata 700032, India

S. K. Sanyal
Department of ECE, Narula Institute of Technology, Agarpara, Kolkata 700109, India

M. Mitra et al. (eds.), *Computational Advancement in Communication, Circuits and Systems*, Lecture Notes in Electrical Engineering 786,
https://doi.org/10.1007/978-981-16-4035-3_10

attempt have been undertaken by the researchers to deviate from conventional number system to some other number system such as Residue number system (RNS)[1] and complex binary number system (CBNS) [2]

With RNS, we can design a reliable communication system as in this number system arithmetic operations can be carried out independently and concurrently in several residue channels [3]. Thus, carry-free arithmetic and lack of ordered significance among the residues are the key attractive features of the RNS. These properties motivate application of RNS in CDMA known as RNS-CDMA [4].

While computing signal processing and image processing algorithms, complex number plays a unique and vital role as it locates a point within two dimensions of Cartesian coordinate system. As a result complex number is required in determining filtering operations, frequency domain transformation, orthogonal transformation, image compression, and restoration. Thus, it is important to emphasize on efficient representation and manipulation of complex numbers. According to the conventional number system, the complex number arithmetic operation still involve the 'divide-and-conquer' technique [5]. In such technique, a complex number is broken into real and imaginary parts and then arithmetic operations are carried out on each part as if it is a real arithmetic. Finally, the overall result of the arithmetic operations is obtained by the accumulation of individual results. This process of separate treatment of complex numbers has resulted in defining a binary number system with base other than 2, which will provide a single unit representation of complex number for the significant reduction of computational complexity.

In 1960, Kunth [6] described a 'quarter—imaginary' number system with base 2j and analyzed the arithmetic operations of numbers using this imaginary base. However, the state of the art is restricted in providing not only the division process but also the algorithm for the circuit-level implementation involving the process of single unit representation. Walter Penney [7], in 1965 attempted to define a complex number system, first by using a negative base of (−4) and then by using a complex number $(-1 + j)$ as the base. The change of base is adopted in Penney's number system targeted the efficient computation of complex arithmetic operations such as complex addition, subtraction, multiplication, and division. But he could not provide the detailed algorithm for the implementation of these arithmetic circuits involving single unit representation. Jamil et al. [8], in 2000, revisited Penney's number system with base $(-1 + j)$ and presented a detailed analysis of the different processes behind the complex arithmetic operations. He also renamed the $(-1 + j)$ base complex number system as complex binary number system (CBNS) [8, 9]. Jamil [10] proposed FPGA-based architectures for performing complex arithmetic involving digital circuits like decoders and multiple input OR gates. He also proposed, an architecture for ripple carry adder using addition unit and some specially designed combinational units like the zero detection unit, the extended carry generation unit, and the output generation unit.

Upon critical investigation of Jamil's work, it has been observed that there is ample scope to revisit the design at structural level to explore the possibility of achieving optimized implementation in terms of area and path delay of various complex arithmetic circuits using CBNS on FPGA platform. Thus, the objective behind this work is the efficient implementation of CBNS-based computational circuits on FPGA platform occupying lesser silicon area and having reduced path delay when compared with existing works.

A new functional block diagram for CBNS adder has been incorporated based on the modular approach. The CBNS adder designed by Jamil involves one 8 × 256 decoder, and several multiple input OR gates occupying not only large silicon area but also exhibiting larger path delay. This drawback motivates us to design CBNS-based computational circuits using modular approach leading to efficient implementation on FPGA platform. The modules have been designed using simple logic gates having limited input–output terminals. Three categories of modules, namely module 0, module 1, and module 2, have been designed using simple verilog codes to develop CBNS-based adder, multiplier, and subtractor. All such computational circuits have been developed for hierarchical level design approach involving the simple modules. A novel conceptual approach has also been developed for the required interconnection of these modules associated with various computational circuits to be culminated as netlist. The approach has been followed in CBNS adder, multiplier, and subtractor. However, in case of multiplier and subtractor there is a possibility of exceeding the limited number of input–output terminals of the modules. This problem has uniquely been solved by introducing another novel concept of sub-block-based computation. The novel and unique concept of modular approach followed by the concept of sub-block has culminated in the overall satisfaction of our objective.

The remainder of this paper proceeds as follows: Sect. 2 presents the design of modules. These are the fundamental building block of a CBNS arithmetic computational units. Section 3 presents the design of nibble size adder, multiplier, and subtractor. Experimental results of the implemented circuits on re-configurable (FPGA) processor have been presented in Sect. 4 with a comparative performance evaluation of Jamil's design approach. At last, Sect. 5 encompasses the conclusion.

2 Design of Modules

The CBNS can be expressed involving the power series as shown below:

$$a_{n-1}(-1+j)^{n-1} + a_{n-2}(-1+j)^{n-2} + \cdots + a_1(-1+j)^1 + a_0(-1+j)^0$$

where the coefficients $a_{n-1}, a_{n-2}, \ldots, a_1, a_0$ are binary. Conversion process of integers [2] and fractions [11] has been illustrated [12]. The procedure of CBNS arithmetic computation rule [12] has also been illustrated [13].

A module is the basic unit of the CBNS arithmetic operations. The interconnection of these core components leads us to the design of a VLSI system having regular and simple architectures for performing CBNS computations. Thus, depending upon the processing of input and generation of output these building blocks are divided into three parts.

2.1 Design of Module 0

This module consists of 2-input and 4-output terminals as shown in Fig. 1. To design this fundamental unit, we have described the behavioral-level representation written in verilog code. According to this HDL code 'A' is the 2 bit CBNS input nomenclatured as $A[0]$ and $A[1]$, sum is a register type single-bit CBNS output and C is 3-bit register type CBNS carries. The code reveals that depending upon the four combinations of the CBNS input terminals, the output bit patterns are set to logic high or logic low.

Algorithm 1 Verilog code for module 0

```
Inputs [1 : 0] A;
Output C,Sum;
reg [2 : 0] C;
reg Sum;

always @ (A)
begin
case (A)
2'b00:begin
Sum=1'b0;C= 3'b000; end
2'b01,2'b10 :begin
Sum=1'b1;C= 3'b000;end
2'b11 :begin
Sum=1'b0;C= 3'b110;end
endcase
end
```

The code for module 0 reveals that the processing of this circuit is similar to the binary half adder except in the last case where two CBNS logic high input combinations generate the sum as logic low state and three carries $C[0]$ as logic low state, $C[1]$ and $C[2]$ as logic high state. The requirement of these three carries is mainly to express ($2_{\text{base}10}$)equivalent to $1100_{\text{base}(-1+j)}$. Thus, the output terminals provide the functionality of CBNS-based half adder.

The FPGA device utilization summary shows that to implement the module 0 we require only one slice or two 4-input LUTs, whereas the timing analysis indicates 7.054 ns as the maximum combinational path delay. The Xilinx Xpower Analyzer shows that this design requires 0.032 W quiescent power.

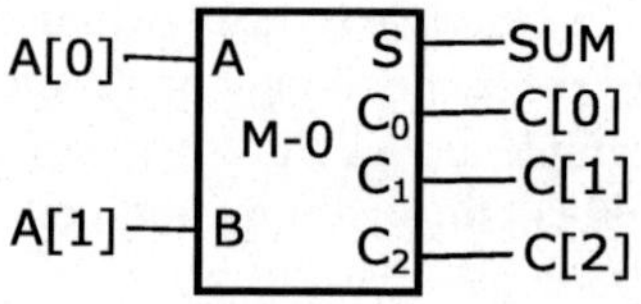

Fig. 1 Block diagram of module 0

2.2 Design of Module 1

Module 1 operates as the combination of 3-input and 4-output terminals as shown in Fig. 2. The behavioral-level description of this component is shown using binary decision diagram (BDD). BDD is the representation method of Boolean function. According to this flow diagram, 'A', 'B', and 'C' are the three single-bit CBNS input and S, C_0,C_1, and C_2 are the single-bit CBNS output. The position of each carry bit has been indicated in their suffix.

The BDD for sum (S) expression shown in Fig. 3a reveals that the node A can have logic low state as indicated by dotted lines in the branch whereas logic high state has been shown by solid lines. Depending upon the condition of input 'A', the node 'B' sets the logic state. Finally, the sum is set to logic high condition on the decision of the input 'C'. Similarly, BDD for the two propagating carries C_1 and C_2 is shown in Fig. 3b. The BDD shows that the functions C_1 and C_2 are set to logic high state depending upon three specific paths AB, $A\overline{B}C$, $\overline{A}BC$. After the synthesis of the behavioral-level description of the module 1, we obtain the combinational logic flow of input–output. Using K map the gate-level description shows that C_0 = logic 0

$$S = A \oplus B \oplus C \tag{1}$$

$$C_1 = C_2 = A \cdot (B + C) + (B \cdot C) \tag{2}$$

As the functionality of this circuit is a CBNS-based full adder, thus in last condition the output combination will express 3_{base10} equivalent to $1101_{\text{base}(-1+j)}$. The FPGA-

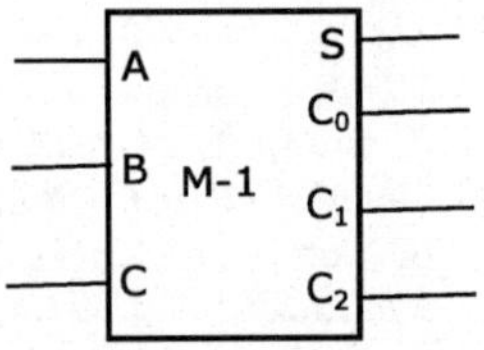

Fig. 2 Block diagram of module 1

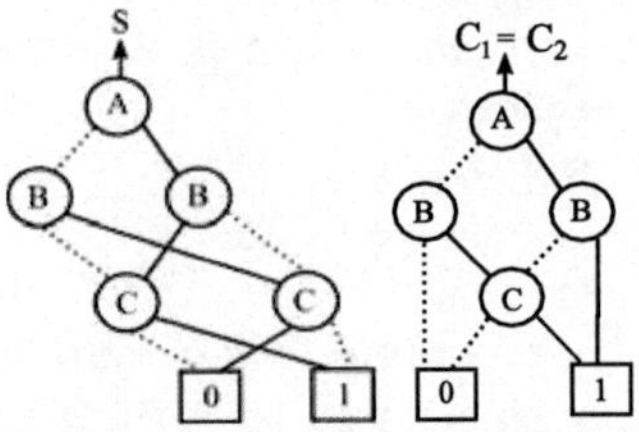

(a) BDD for sum

(b) BDD for carry

Fig. 3 BDD for module 1

level device utilization shows that to design the circuit of module 1 we require two 4-input LUTs, one for implementing the 'S' output and another for the two carries 'C_1' and 'C_2'. The timing analysis revels that critical path of this combinational circuit is 7.054 ns, whereas the Xilinx Xpower Analyzer indicates that this combinational circuit requires 0.032 W quiescent power.

2.3 Design of Module 2

Module 2 performs the CBNS computation on the various combination of 4-input terminals to provide output at the 9-output terminals as shown in Fig. 4. In order to describe the design architecture, we have first presented the behavioral style of modeling using the HDL code.

Algorithm 2 Verilog code for module 2

```
Inputs [2 : 0] A;
Inputs D;
Output C,Sum;;
reg [7 : 0] C;
reg Sum;

parameter CBNS 0 =9'b000000000, CBNS 1 =9'b000000001 ,CBNS 2 =9'b000001100 ,
CBNS 3 =9'b000001101 , CBNS 4=9'b111010000;

always @ (A or D)
begin
case (A)
3'b000:begin
{C,Sum}= D ? CBNS 1 : CBNS 0; end
3'b001,3'b010,3'b100 :begin
{C,Sum}= D ? CBNS 2 : CBNS 1; end
3'b011,3'b101,3'b110 :begin
{C,Sum}= D ? CBNS 3 : CBNS 2; end
3'b111 :begin
{C,Sum}= D ? CBNS 4 : CBNS 3; end
endcase
end
```

According to this verilog code, we have described 'A' as a 3-bit CBNS input nomenclatured as $A[0]$, $A[1]$, $A[2]$, and 'D' as a single-bit CBNS input. Sum is a register type single-bit output, and Cs are also register type 8-bit carry terminals. Depending upon the condition of the input terminal 'A', we are having four case statements. In each case statement, the conditional operation on the single-bit CBNS input 'D' is being checked. The concatenated output sum (S) and C are fed with the constant parameter values depending upon the logic functionality of the circuit.

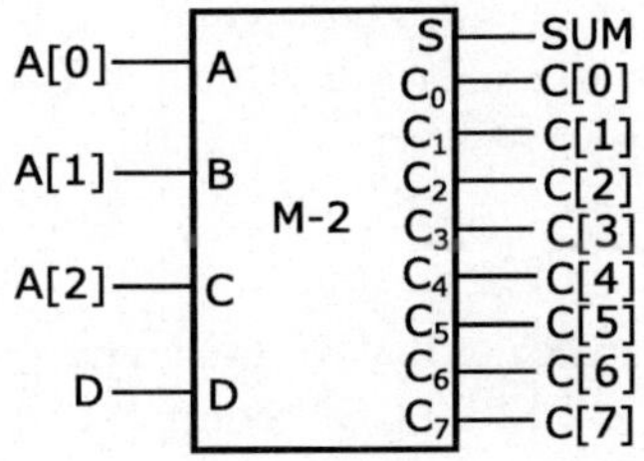

Fig. 4 Block diagram of module 2

According to the last case statement when the CBNS input 'A' is set to logic high, then depending upon the condition of the input 'D' the output Sum (S) and carries are set to the constants $CBNS_i$; $i = 3$ or 4. In this statement, only if CBNS input 'D' is set to logic high, then the Sum (S) and C's are set with the 'CBNS 4' being logically equivalent to $111010000_{\text{base}(-1+j)}$ $(4_{\text{base}10})$.

After the FPGA-level synthesis of the behavioral HDL code, we have obtained that the module 2 consumes three 4-input LUTs. The requirement of one LUT is to implement the sum expression, another one LUT is for implementing the $C[1]$ and $C[2]$, whereas the third LUT will design the $C[3]$, $C[5]$, $C[6]$and$C[7]$. The critical path delay analysis shows that module 2 circuit exhibits a maximum combinational path delay of 7.278ns. As there is an increase in the number of logic gates in this circuit so we have obtained an increase in the critical combinational path in comparison with module 1 and module 0. The power analysis based on Xpower Analyzer shows requirement of the quiescent power of 0.032W.

3 Design of Nibble Size Computation Units

In this section, the design issues of CBNS-based computational units have been illustrated. These computational architectures perform the arithmetic operations on nibble size data. To design the parallel arithmetic architectures, we have adopted a hierarchical design approach. In this structural-level design method, we first describe the architecture of fundamental modules. After the module-level abstraction, we then introduce the detail description of the interconnection of these modules to design the top level architecture of several parallel arithmetic computational units.

3.1 Structural-Level Design of a Nibble Size Parallel CBNS Adder

Adder is the backbone of various arithmetic circuits such as subtractor and multiplier. Thus emphasizing on this fundamental element, we have designed the nibble size parallel CBNS adder circuit. The schematic of the top level block as shown in Fig. 5

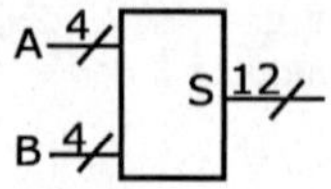

Fig. 5 Block diagram for nibble size parallel CBNS addition

indicates that we are having 4-bit CBNS inputs 'A' and 'B' as ($A_3, \ldots A_0, B_3, \ldots B_0$) and 12-bit CBNS output port as ($S_{11}, \cdots, S_0$). The positions of the input and output bits are indicated in their indices. In order to illustrate the structural-level design of this nibble size CBNS adder circuit, we have to begin with the introduction of the conceptual approach as shown in Fig. 6. In this conceptual approach, the traditional bit-by-bit addition of the input combinations has been illustrated. The first stage (St0) performs the CBNS addition of the two single-bit CBNS input A_0 and B_0. So this stage specifies the requirement of module 0 as the hardware component. The output of this stage is a single-bit CBNS Sum indicated as S_0 and the propagating carries C_{ij} where i (0, 1, ..., 11) represents the particular stage and j (0, 1, ..., 7) represents the particular carry output relevant to the specific module. The carries C_{ij} propagate in their neighboring positions. Thus, C_{01} signifies that this is the carry bit generated from (St0) and will be placed in the second adjacent position with respect to St0. Since for module 0 implementation C_0 is always zero, thus carry bit C_{00} is indicated as zero with a pointed arrow mark. The second stage (St1) is similar to St0 except we are feeding the CBNS input as A_1 and B_1. The output of this stage is the sum S_1 with three carry bit indicated as C_{10}, C_{11}, and C_{12}.

In the third stage (St2), we require the CBNS addition of the input A_2 and B_2 along with the carry bit C_{01}. This specification demands for the module 1 to function like a CBNS-based full adder circuit for St2. The output of this stage is the sum S_2 with three carries C_{20}, C_{21}, and C_{22}. The BDD of module 1 shows that C_{20} is always set to a logically active low condition. The fourth stage (St3) implements the CBNS addition of each single-bit CBNS input A_3 and B_3 with the two carry bits C_{02} and C_{11}. The carry C_{02} is the third positional bit of first stage (St0), whereas the carry bit C_{11} is the second positional bit of the second stage (St1). This stage specifies that for the hardware design we require module 2 as the basic component. This module generates the sum S_3 as a single-bit output with 8-bit propagating carries ($C_{37}, \ldots, C_{30}$) in their

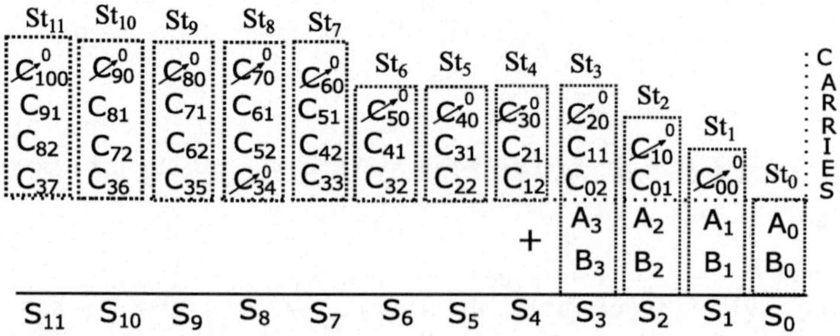

Fig. 6 Conceptual approach for nibble size parallel CBNS addition

Table 1 Nibble size CBNS addition of input test vectors

Decimal			CBNS		
Augend	Addend	Sum	Augend	Addend	Sum
2	3	5	1100	1101	000111010001
$-1-j$	$-2j$	$-1-3j$	0110	0100	000000110010
1/2	$1/2j$	$1/2+1/2j$	01.11	00.01	0000001110.10

adjoining positions as shown in Fig. 6. According to the behavioral- and gate-level descriptions of the module 2, we have logically set the two carry bits C_{30} and C_{34} to active low state. The CBNS addition of the propagating carries is performed in the subsequent stages. We can implement these stages using the basic modules. The selection of the particular module mainly depends on the specific requirement of the input port for implementing the CBNS bit-by-bit addition operation.

Based on the conceptual approach for the nibble size parallel CBNS adder, we can now proceed toward the structural-level design involving the various modules. The interconnections among the different modules are being dictated by the conceptual approach of Fig. 6 leading to the netlist as shown in Fig. 7. The verilog code for the netlist is written in ISE design suite 13.4. This HDL code has been synthesized in the same software platform. For the reliable evaluation of the netlist under reference, simulation has been carried out with arbitrary test input vectors. The simulation result for checking the functionality of the nibble size parallel CBNS adder is shown in Fig. 8 using the test bench waveform and Table 1.

Next we have implemented the designed netlist on the FPGA-level platform, XC3S700A. The synthesis result for the netlist along with the comparative study of Jamil's design metrics has been discussed in the result section. In the simulation waveform of Fig. 8, a[3:0] represents a_3, a_2, a_1, and a_0 as the 4-bit CBNS input A, while b[3:0] corresponds to b_3, b_2, b_1, and b_0 of B and S[11:0] is the 12-bit CBNS output of Fig. 7. Between $2\mu s$ and $3\mu s$ time interval, the CBNS addition corresponding to the second row of Table 1 has been demonstrated.

3.2 *Structural-Level Design of Nibble Size CBNS Multiplier*

Multiplier forms the fundamental building element in signal processing and various arithmetic operations. Therefore, greater emphasis has been placed in designing high-speed multiplier at structural level. This section deals with the structural-level design of nibble size CBNS multiplier. We first discuss the conventional approach related to the nibble size CBNS multiplication algorithm. Figure 9a shows the bitwise logical AND operation between each multiplier bit and the whole multiplicand generating the partial product (PP) terms. According to the conventional multiplication rule, the partial product array (PPA) has been shifted one bit to the left and then these PPAs are

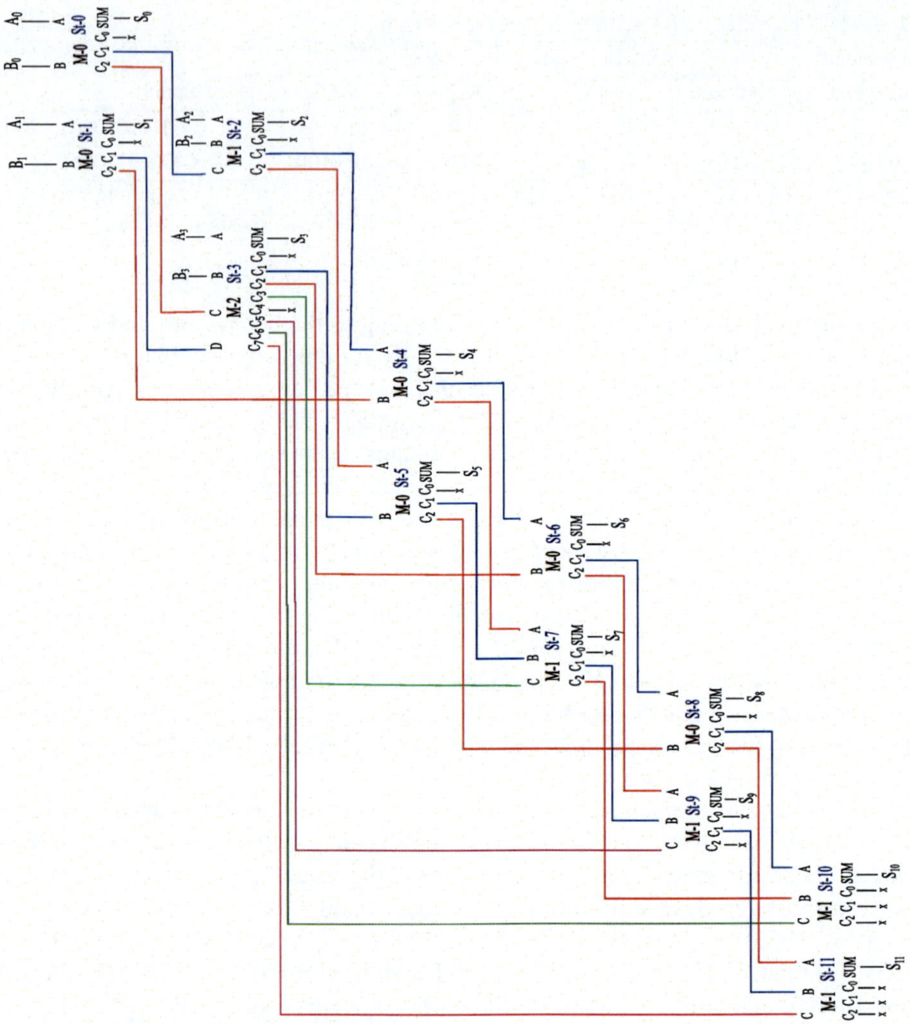

Fig. 7 Netlist for nibble size parallel CBNS adder

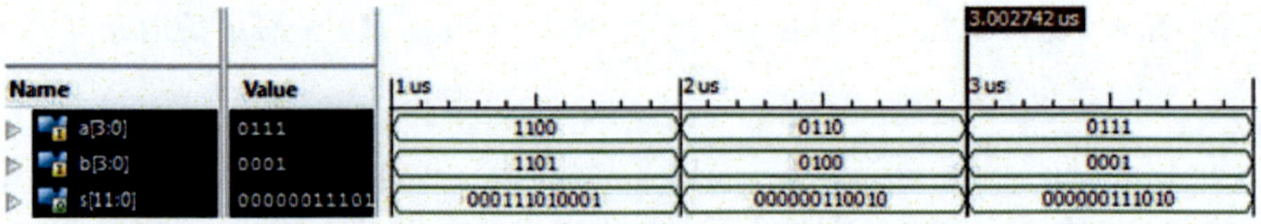

Fig. 8 Simulation waveform of nibble size parallel CBNS addition

added according to the CBNS addition algorithm. The design of the CBNS multiplier circuit is to be implemented with the basic modules already discussed. Since module 2 can handle a maximum of 4-input, there is a possibility of handling more than 4 inputs during the process of multiplication. Accordingly to implement the CBNS multiplier with the basic modules having limited inputs, a new concept of separation into two sub-blocks has been incorporated in the structural-level design. The sub-block-1 performs the CBNS addition of PPA_2, PPA_1, and PPA_0 generating the 12-bit ($E_{11} \ldots E_0$) intermediate results (IR) terms. Module 0 is required to get E_1 and E_6, Module 1 is required to get E_2, E_5, E_7, E_8, and E_9, and Module 2 is required to get E_3, E_4, E_{10}, and E_{11}. Figure 9b shows the conceptual approach followed in sub-block-1. The architecture of sub-block-1 is cascaded with sub-block-2. The sub-block-2 performs the CBNS addition of the PPA_3 with IR from sub-block-1 to produce the final 12-bit ($P_{11} \cdots P_0$) product terms. The conceptual approach of the sub-block-2 is shown in Fig. 9c. The product term P_2, P_1 and P_0 of the nibble size CBNS multiplier is equal to the E_2,E_1 and E_0 IR of sub-block-1. To generate the product terms P_3 and P_4, we require module 0 as the hardware component. Similarly, P_5, P_7, P_8, P_9, and P_{11} require module 1, whereas P_6 and P_{10} require module 2. The cascading of sub-block-1 and sub-block-2 has been done according to the rule of interconnection of IR.

The HDL code is synthesized, and the simulation has been done in this software platform with arbitrary input test vectors for the evaluation of the functionality of the circuit as shown in Table 2.

3.3 *Structural-Level Design of Parallel Nibble Size CBNS Subtractor Circuit*

Subtraction operation is also an integral part of the hardware circuit used in digital signal processing. The first step of the structural-level design of parallel nibble size CBNS subtractor is to multiply the nibble size subtrahend ($A_3 \ldots A_0$) with $11101_{\text{base}(-1+j)}$ maintaining the CBNS multiplication rule. This multiplication process generates an 11-bit intermediate results (IR-1) denoted as ($D_{10} \ldots D_0$). In this step-1, four partial product arrays(PPA_s) have been generated. These PPA_s are added according to CBNS addition rule. We have incorporated two sub-blocks to design Step-1. In the first sub-block PPA_2, PPA_1 and PPA_0 are added. The result of this CBNS addition is the intermediate results (IR-2) denoted as ($F_{10} \ldots F_0$). The conceptual approach of the sub-block-1 has been shown in Fig. 10. In sub-block-1, the terms F_0 and F_1 are equal to subtrahend A_0 and A_1. Module 0 is required to generate F_2, F_3, and F_7, module 1 generates the F_6, F_8, F_9, and F_9, whereas F_4 and F_5 have been generated using module 2. In the next sub-block, we have added the IR-2 with PPA_3 to produce IR-1. Sub-block-2 shows that the term D_0, D_1, D_2, and D_4 is equal to F_0, F_1, F_2, and F_4. Module 0 is required to get D_3, whereas module 1 generates D_5, D_6, D_7, D_8, D_9, and D_{10}. After designing the two sub-blocks using

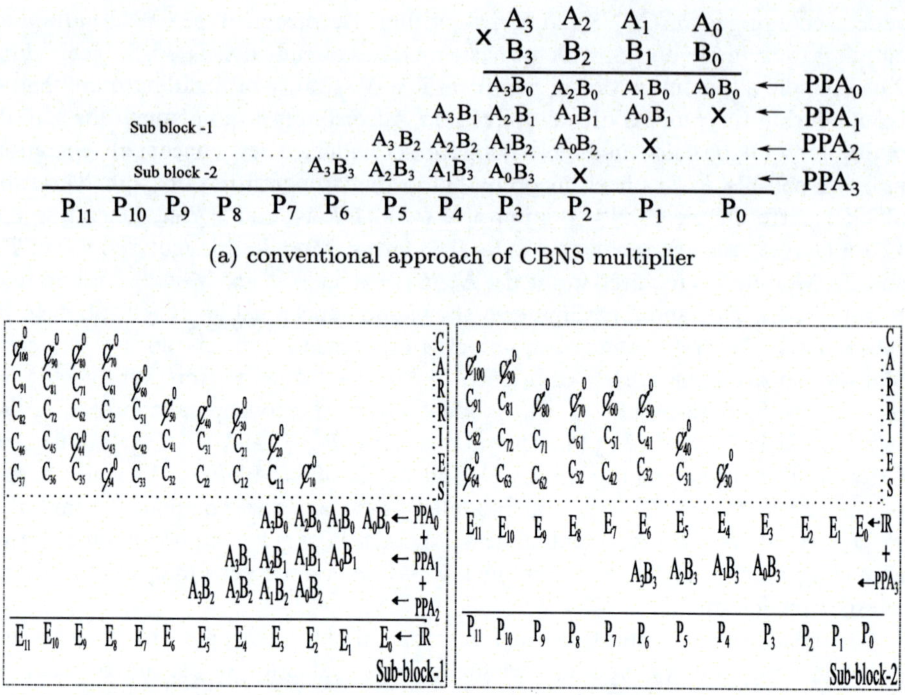

(a) conventional approach of CBNS multiplier

(b) Sub-block -1

(c) Sub-block -2

Fig. 9 Nibble size CBNS multiplication rule

Table 2 Nibble size CBNS multiplication of some numbers

Decimal			CBNS		
Multiplicand	Multiplier	Product	Multiplicand	Multiplier	Product
3	3	9	1101	1101	000111000001
$2+j$	$-2j$	$2-4j$	1111	0100	000000111100
$-1+j$	1/2	$-1/2+1/2j$	0010	01.11	0000000011.10

the appropriate basic hardware components, we have to connect these modules maintaining the rules of CBNS addition as defined in the conceptual approach in Fig. 10. After cascading the two sub-blocks, the IR-1 is fed as the input to the next step. In Step-2, IR-1 is added with the nibble size ($B_3 \cdots B_0$) minuend. This CBNS addition generates the 11-bit result ($E_{10} \ldots E_0$). The basic hardware component module 0 is required to get E_0 and E_1, module 1 provides E_2, E_4, E_5, E_6, and E_8, whereas module 2 generates E_3, E_7, E_9, and E_{10}. The verilog code for the nibble size CBNS subtractor has been developed in ISE design suite 13.4. After the synthesis of the HDL code, simulation has been done in this software platform with arbitrary input test vectors for the evaluation of the functionality of the circuit as shown in Table 3.

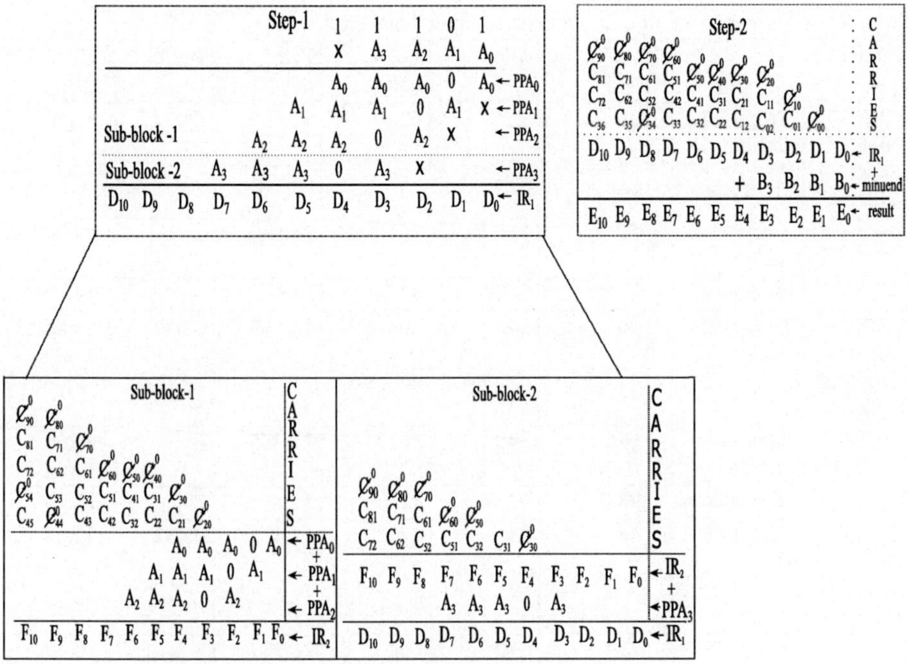

Fig. 10 Conceptual approach of CBNS nibble size subtractor

Table 3 Nibble size CBNS subtraction of some input test vectors

Decimal			CBNS		
Minuend	Subtrahend	Difference	Minuend	Subtrahend	Difference
3	1	2	1101	0001	000000001100
$-1-j$	j	$-1-2j$	0110	0011	000011101001
$-1/2$	$-1/4j$	$-1/2-1/4j$	0.1100	0.0111	00000000.1111

4 FPGA Implementation and Performance Evaluation

In order to model the proposed computational architecture, we have selected a standard computing processor with Windows 7 OS. The hardware code is written in verilog on ISE design suite 13.4 software platform. The implementation of HDL code have been done on Spartan 3 A family XC3S700A FPGA device. In the FPGA platform, we have provided the input bit pattern for the functional verification of the circuit through the sliding switches. We have observed the output bit pattern through the LED arrays. The re-configurable FPGA processor provides the flexibility to carry

Table 4 Performance metric of arithmetic computational circuits

Parameter		LUTs	Slice	Logic gates	Path delay (ns)	Power (W)	Net delay (ns)
CBNS Nibble Size adder	Proposed design	16	9	62	12.065	0.032	3.456
	Jamil's design	59	31	354	24.839	–	4.024
	% reduction	72.8	70.96	82.5	51.4	–	14.12
CBNS Nibble Size Multiplier	Proposed design	45	24	204	16.119	0.032	5.468
	Jamil's design	676	340	4617	25.085	–	6.268
	% reduction	93.3	92.94	95.58	35.7	–	12.76
CBNS Nibble Size Subtractor	Proposed design	46	24	222	13.853	0.032	4.362
	Jamil's design	734	370	5040	26.289	–	10.494
	% reduction	93.19	92.94	95.59	47.3	–	58.43

out testing of the circuit with arbitrary test input vectors. The design issues of proposed modular architecture of computational units are compared with Jamil's [10] minterms arithmetic design approach. The evaluation of device utilization for both the design approaches can be comprehended easily with the help of result in Table 4. The analysis reveals that in case of CBNS adder circuit proposed design reduces the number of LUTs by 72.8% and number of slices by 70.96%. However for CBNS subtractor circuit, there is 93.19% reduction in number of LUTs and 92.94% reduction in number of slice. These implementation statistics reveal the fact that the proposed design approach provides a better device utilization. For CBNS adder circuit, the proposed architecture provides 51.4% reduction in the path delay as compared to the existing design circuit. However for the CBNS subtraction circuit, there is 47.3% reduction in path delay. It is now clearly visible that our proposed method is highly efficient in terms of Si area, path delay, and power consumption.

5 Conclusion

The design of various CBNS-based computational circuits has been incorporated in this work involving simple modular approach. The necessary background for CBNS computation including the conversion process has been developed. Nibble size adder, multiplier, and subtractor have been designed following hierarchical approach based on the modules. Verilog codes have also been developed for such design. The concept of sub-block eliminates the limited number of input–output terminals of the modules for large size circuits. Extensive simulation confirms the terminal behavior of the various circuits. The design approach ultimately leads to efficient implementation on FPGA platform with significant reduction of hardware requirement and combinational path delay.

References

1. Miguens MP, Chaves R, Sousa L (2015) Arithmetic-based binary-to-RNS converter modulo $2^n pmk$ for jn -bit dynamic range. IEEE Trans VLSI Syst 23(3):603–607
2. Jamil T (2011) An introduction to complex binary number system. In: Processing of international conference on information and computing (ICIC), pp 229–232
3. Sengupta A, Natarajan B (2013) Performance of systematic RRNS based space-time block codes with probability-aware adaptive demapping. IEEE Trans Wirel Comm 12(5):2458–2469
4. Zhang S, Zhang Y, Yang LL (2011) Redundant residue number system based multicarrier DS-CDMA for dynamic multiple-access in cognitive radios. In: Processing of IEEE vehicular technology conference (VTC Spring), pp 1–5
5. Parhi M, Yingjie L, Parhi KK (2014) Canonic real-valued FFT structures. In: Processing of Asilomar conference on signals, systems and computers, pp 1261–1265
6. Knuth DE (1960) A imaginary number system, commun. ACM 3:245–247
7. Penney W (1965) A "binary", systm for complex numbers. JACM 12(2):247–248
8. Jamil T, Holmes N, Blest D (2000) Towards implementation of a binary number system for complex numbers. In: Processing of IEEE southeastcon, pp 268–274
9. Jamil T (2002) The complex binary number system: basic arithmetic made simple. IEEE Potentials 20(5):39–41
10. Jamil T (2011) Design of arithmetic circuits for complex binary number system. J Am Inst Phys IAENG Trans Eng Technol 1373(1):83–97
11. Jamil T (2013) Complex binary number system algorithms and circuits. Springer, India, pp 5–6
12. Mukherjee M, Sanyal SK (2017) Design of CBNS nibble size adder using pass transistor logic circuit: extension to FPGA implementation. In: Communication and networking technologies. 8th international conference on computing. Delhi, India
13. Mukherjee M, Sanyal SK (2017) Design of a high-speed reconfigurable fast hartley transform processor using CBNS. In: 14th IEEE India council international conference. Roorkee, India

Checking and Coloring Graphs Through Quantum Circuits: An IBM Quantum Experience

Asmita Banerjee, Bikash K. Behera, Kunal Das, and Prasanta K. Panigrahi

Abstract Checking the graph state and coloring it with possible least number of colors (or chromatic number) has many useful applications such as coloring of maps, solving Sudoku, making schedules to name a few. Though classical algorithms are used to solve this problem, it is believed that with quantum approach, the space and time complexity can be reduced. Thus, here we present quantum algorithms and design quantum circuits to check the graph states having two- and three-vertices with given edges and color them accordingly. We also propose a quantum algorithm for coloring of trees. For each of the cases, we discuss the quantum costs and complexities and compare it with the classical approach.

Keywords IBM quantum experience · Graph coloring · Quantum circuits · Quantum cost analysis

---

A. Banerjee (✉)
Department of Computer Science and Engineering, Narula Institute of Technology, Nilgunj Road, Agarpara 700109, West Bengal, India

K. Das
Department of Computer Science, Acharya Prafulla Chandra College, New Barrackpur, Kolkata 700131, West Bengal, India
e-mail: kunal@apccollege.ac.in

B. K. Behera · P. K. Panigrahi
Department of Physical Sciences, Indian Institute of Science Education and Research Kolkata, Mohanpur 741246, West Bengal, India
e-mail: bikash@bikashsquantum.com

P. K. Panigrahi
e-mail: pprasanta@iiserkol.ac.in

M. Mitra et al. (eds.), *Computational Advancement in Communication, Circuits and Systems*, Lecture Notes in Electrical Engineering 786,
https://doi.org/10.1007/978-981-16-4035-3_11

1 Introduction

Computers are no more limited to work in bits of 0 and 1's. Through quantum mechanical phenomena like superposition and entanglement [1], quantum computers are designed to work with superposition state, where they encode information as quantum bits or qubits. Qubits represent quantum versions of bits and their respective control devices that work together to act as computer memory and a processor. Because a quantum computer can contain these multiple states simultaneously, it can easily crack algorithms like that of encryption [2], used today in very less time, whereas it takes billions of years by the best supercomputer available today. For example, a classical computer of n bits can have n states at a time but a quantum computer of n qubits can deal with 2^n states at a time. A processor that can use registers of qubits will be able to perform calculations using all the possible values of the input registers simultaneously. A quantum computer maintains a sequence of qubits, which can represent a 1, a 0, or any quantum superposition of those two qubit states. Superposition causes a phenomenon called quantum parallelism and is the motivating force behind the research being carried out in quantum computing. Graph algorithms are sure to benefit from parallelization, while on the other hand they contain a number of NP-complete problems [3] which may be liable for quantum boost or speedup.

Quantum algorithms are often probabilistic, in that they provide the correct solution only with a definite known probability. A quantum bit like electrons is with two spin states: "down" and "up" (typically written using Dirac's bra–ket notation $|0\rangle$ and $|1\rangle$). In general, the state of a quantum bit (or qubit) is described by: $\alpha|0\rangle + \beta|1\rangle$ where α and β are complex numbers, satisfying $|\alpha|^2 + |\beta|^2 = 1$. In quantum information theory, a quantum circuit is a model for quantum computation and is a sequence of quantum gates, which are reversible transformations on a quantum mechanical analog of an n-bit register. Here, various operations are performed using the following quantum gates (which are Pauli matrices), namely (i) X gate: It transforms $|0\rangle$ to $|1\rangle$ and vice-versa, (ii) Hadamard gate: It creates superposition states, (iii) CNOT gate: It performs not operation on the second qubit only if the first qubit is in $|1\rangle$ state, (iv) anti-CNOT gate: It performs not operation on the second qubit only if the first qubit is in $|0\rangle$ state.

Quantum algorithms are run on realistic model of quantum computers such as provided by IBM quantum experience platform (IBM QE) [4], D-wave quantum computers [5], trapped ion quantum computer (IonQ) [6] to name a few. IBM QE has recently gained popularity by providing the 5-qubit and 14-qubit quantum chips to the community and making easily accessible through QISKit. A number of quantum information processing tasks in the field of quantum simulation [7–10], quantum machine learning [11, 12], quantum error correction [13–16], quantum information theory [17–19], quantum cryptography [20], quantum optimization problems [21], designing quantum communication devices [22, 23] are tested on the real chips, and feasible results are obtained.

A graph G(V,E) consists of a set of objects called vertices V $\{v_1, v_2, \ldots\}$ and another set E $\{e_1, e_2, \ldots\}$, whose elements are edges such that edge e_k or e_{ij} is associated with vertices v_i and v_j. Graph coloring is a way of assigning colors to these vertices of a graph such that no two adjacent vertices are colored using the same color. Chromatic number is the smallest number of colors needed to color a graph. In the following text, we provide quantum algorithms for checking the connection between vertices and assigning colors according to the chromatic number of the graph. In the original work of Hondt [24], algorithms were proposed of 2-coloring and 3-coloring graphs based on qubits and qutrits, respectively. In the recent Ref. [25], quantum circuits were proposed to check and color the graphs for 2- and 3-coloring for two and three vertices system.

The organization of the paper is as follows. In Sect. 2, we discuss the 2-coloring of graphs with circuit implementation of appropriate quantum operators. Following which, 3-coloring is explained by mapping qutrits into qubits and developing quantum circuits to generate the graph states, to check and color the graph in Sect. 3. Finally, we conclude in Sect. 6.

2 Algorithm for 2 Vertices Coloring

2.1 *Algorithm and Implementation*

We start by generating a superposition of all possible colorings of this graph when edges are ignored, as follows $H^{\otimes n}|0\rangle^{\otimes n} = 1/2^n \sum_{i=0}^{2^n-1} |i\rangle$, where the i is written in binary. For example, for two vertices we would have $|00\rangle$, $|01\rangle$, $|10\rangle$ and $|11\rangle$ with normalization factor 1/2. Now for each back edge, e_{ij} considers the projective measurement M_{ij} with projection operators O_{ij} and $I - O_{ij}$, where $O_{ij} = |01\rangle_{ij}\langle 01| + |10\rangle_{ij}\langle 10|$ and $I - O_{ij} = |00\rangle_{ij}\langle 00| + |11\rangle_{ij}\langle 11|$.

Now we have two vertices 0 and 1 in the graph, where there may be an edge connection or not. We can show in forms of two qubits that there are four possible states of connection between the two vertices, namely $|00\rangle$, $|01\rangle$, $|10\rangle$ and $|11\rangle$ states. Here we can show that the superposition of $(|00\rangle + |11\rangle)/\sqrt{2}$ are for no edge connected and that of $(|01\rangle + |10\rangle)/\sqrt{2}$ for an edge connected (Fig. 1).

For the circuit in Fig. 2, we use the first two qubits for vertices (namely 0 and 1) and the edge joining these vertices are denoted by $|\psi_{01}\rangle$. For an edge connected, the user puts an X gate in the second qubit to specify the connection. The third qubit is the ancilla qubit which remains $|0\rangle$ for no edges connected and changes to $|1\rangle$ for an edge connected. The fourth and fifth qubit output the connection or graph state and show the superposition of $|00\rangle$, $|11\rangle$ or $|01\rangle$, $|10\rangle$ according to no or an edge connected. The sixth and seventh qubits denote the colors of the vertices, where $|0\rangle$ and $|1\rangle$ represent two different colors. It outputs $|00\rangle$ for no edges connected and $|01\rangle$ for an edge connected. The colors are shown hierarchically for each case (Table 1).

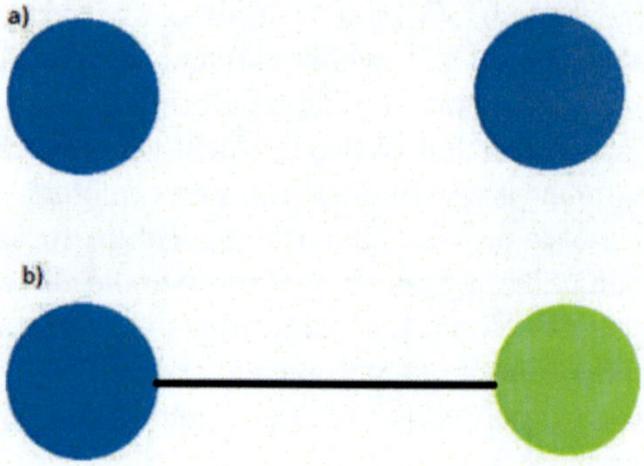

Fig. 1 2 vertices coloring: **a** No edge is connected. Thus, both the vertices are of same color, i.e., color 0 or blue. **b** Edge is connected. Thus, each of the vertices is of different colors, i.e., color 0 and 1 or blue and green

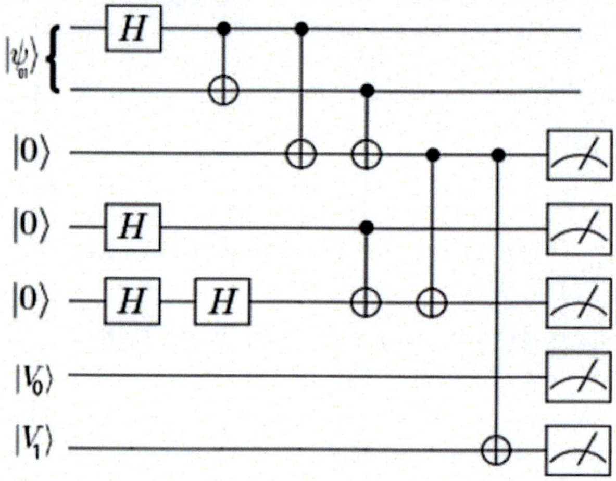

Fig. 2 Quantum circuit for 2 vertices coloring. The first two qubits represent the two vertices, the next is the ancilla qubit, the next two qubits show the graph state, and the last two depict the colors

Table 1 Quantum costs according to given figures

Figure no	Quantum cost
2	10

2.2 *Quantum Cost*

2.3 *Classical Complexity*

Coloring a graph with minimum number of colors is NP-complete problem. In classical computers, an adjacency matrix of space complexity O(n^2) is also assigned to store information of connected vertices. However in case of a graph with just two vertices, the problem can be solved in constant time and space (Fig. 3).

Fig. 3 **3 vertices coloring**: **a** When none of the edges are connected, all vertices will have the same color, i.e., color 0 or blue. **b** When any two of the vertices are connected they will have different colors, i.e., color 0 and 1 or blue and green; the third vertex will have color 0 or blue. **c** When all the vertices are connected by edges, the three vertices will have three different colors, i.e., color 0, 1, and 2 or blue, green, and red

3 Algorithm for 3 Vertices Coloring

3.1 Algorithm and Implementation

Here we have 3 vertices 0, 1, and 2 in a graph where there may be no edges or one to three edges are connected. In this case, we can show in forms of two qutrits that there are nine possible states of connection between any two vertices, namely $|00\rangle$, $|01\rangle$, $|02\rangle$, $|10\rangle$, $|20\rangle$, $|12\rangle$, $|21\rangle$, $|11\rangle$ and $|22\rangle$. Each qutrit represents a vertex, and again we start out by generating unrestricted coloring, $|\psi_1\rangle = H^{\otimes n}|0\rangle^{\otimes n} = 1/3^n \sum_{i=0}^{3^n-1} |i\rangle$. The measurement M_{ij} for each edge e_{ij} is a straightforward generalization of that from the previous section: We split the Hilbert space in two sub-spaces, associated with allowed and disallowed coloring of vertices i and j, respectively, and project onto these sub-spaces. Concretely, the projection operators read as follows; $O_{ij} = |01\rangle_{ij}\langle 01| + |10\rangle_{ij}\langle 10| + |02\rangle_{ij}\langle 02| + |20\rangle_{ij}\langle 20| + |12\rangle_{ij}\langle 12| + |21\rangle_{ij}\langle 21|$ and $I - O_{ij} = |00\rangle_{ij}\langle 00| + |11\rangle_{ij}\langle 11| + |22\rangle_{ij}\langle 22|$.

We can show that superposition of $(|00\rangle + |11\rangle + |22\rangle)/\sqrt{3}$ are for no edges connected; superposition of $(|01\rangle + |10\rangle)/\sqrt{2}$ for 0 and 1 connected and likewise for 0,2 and 1,2; superposition of $(|01\rangle + |10\rangle + |02\rangle + |20\rangle)/\sqrt{4}$ for edges between 0,1 and 0,2 and likewise for 1,2 and 1,0 and also 1,2 and 0,2. Lastly, we have superposition of the states $(|01\rangle + |10\rangle + |02\rangle + |20\rangle + |12\rangle) + |21\rangle)/\sqrt{6}$ for all the three edges connected between the three vertices. We now map the two-qutrit system

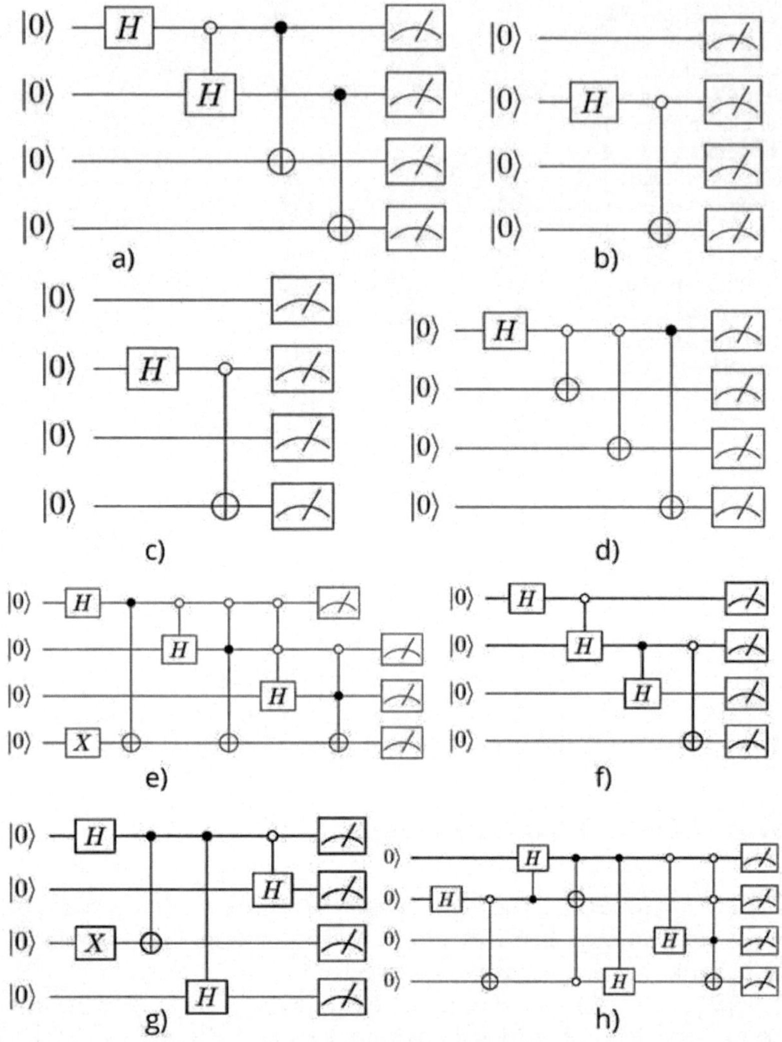

Fig. 4 **Quantum circuit for 3 vertices coloring**: Mapping qutrit to qubit system, **a** when there is no edge between any vertices, the qubits will show superposition of |0000⟩, |0101⟩ and |1010⟩ states, **b** when there is an edge between vertices 0 and 1, the qubits will show superposition of |0001⟩ and |0100⟩ states, **c** when there is an edge between vertices 0 and 2, the qubits will show superposition of |0010⟩ and |1000⟩ states, **d** when there is an edge between vertices 1 and 2, the qubits will show superposition of |0110⟩ and |1001⟩, **e** when there are edges between vertices 0,1 and 0,2,the qubits will show superposition of |0001⟩, |0100⟩, |0010⟩ and |1000⟩ states, **f** when there are edges between vertices 0,1 and 1,2, the qubits will show superposition of |0001⟩, |0100⟩, |0110⟩ and |1001⟩, **g** when there are edges between vertices 0,2 and 1,2, the qubits will show superposition of |0010⟩, |1000⟩, |0110⟩ and |1001⟩, **h** when there is an edge between every two vertices, the qubits will show superposition of |0001⟩, |0100⟩, |0010⟩, |1000⟩, |0110⟩, and |1001⟩

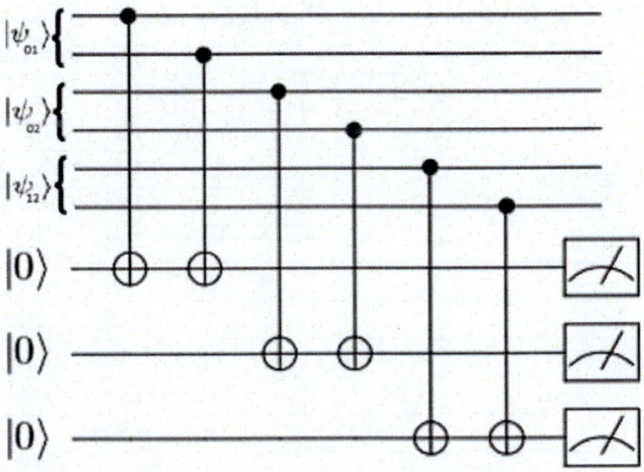

Fig. 5 Quantum circuit for 3-coloring: connection between ancilla qubits and vertices

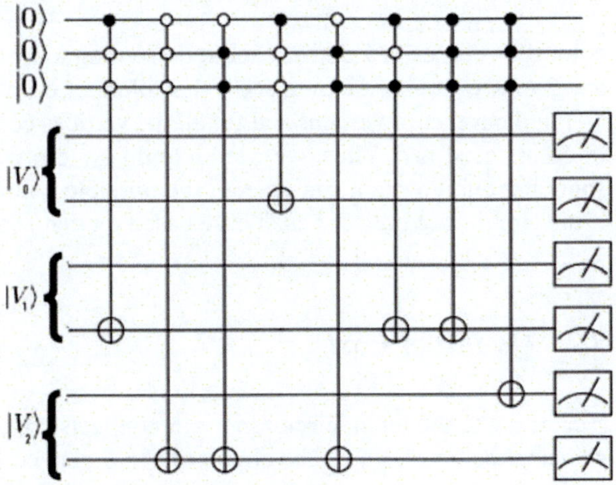

Fig. 6 Quantum circuit for 3 vertices coloring: connection between ancilla qubits and colors. The colors are mapped as $|00\rangle$, $|01\rangle$, and $|10\rangle$ for colors 0, 1, and 2 respectively

to four-qubit system changing $|00\rangle$ to $|0000\rangle$, $|01\rangle$ to $|0001\rangle$, etc., where the first two qubits denote each of the first qutrit and second two qubits denote the each of the second qutrit. This will help us in producing outputs in terms of binary and quantum computer readable language. In Fig. 4, there are eight circuits for each case where we output the vertex combinations and thus the graph state.

In Fig. 5, we show the connection between the vertices and the ancilla qubits. For the circuit, we use each of the first two qubits of the six for vertices (namely 0,1 and 0,2 and 1,2) edges joining these vertices are denoted by $|\psi_{01}\rangle$ and $|\psi_{02}\rangle$ and $|\psi_{12}\rangle$. For an edge connected, the user puts an X gate in the second qubit of the edges to specify the connection. The next three qubits are the ancilla qubits which remains $|0\rangle$ for no edges connected and changes to $|1\rangle$ for an edge connected. They denote the connection between given two vertices. The connection between the ancilla qubits and the vertex coloring is shown in Fig. 6. According to the given connection, we

Table 2 Quantum costs according to given figures

Figure no	Quantum cost
4a	12
4b	4
4c	4
4d	8
4e	85
4f	20
4g	19
4h	109
5	6
6	508

color the vertices 0, 1 and 2: each representing a different color. We again map 0, 1 and 2 to two-qubit system to show the colors of each vertex hierarchically.

For d number of vertices and for all the vertices connected, the maximum number of colors used is d. Thus, we need a qudit system which when mapped to a corresponding qubit system can give all the possible color combination for all the cases if the circuits are built for each case based on that showed for 3-coloring.

3.2 Quantum Cost

According to the input, circuit in Fig. 5 connects with one of the circuits from Fig. 4 showing the graph state. The circuit of Fig. 5 also connects with the circuit of Fig. 6 to show the respective coloring of the vertices. For example, for no edges connected, Fig. 5 is combined with Figs. 4a and 6; (Table 2).

3.3 Classical Complexity

The adjacency matrix with space complexity O(n^2) is required to denote connections between vertices. Actions like finding an edge between two given vertices or an edge removal can be done in constant time. The greedy algorithm requires O($V^2 + E$) time complexity in the worst case.

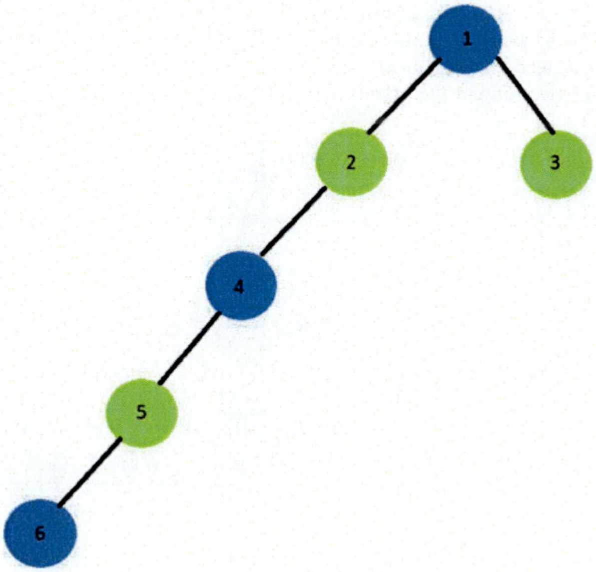

Fig. 7 Representation of a tree: a given tree

4 Trees

In the previous sections, we denoted graph states as superposition of vertices and showed corresponding graph state's vertices' coloring. We now extend our coloring algorithms to trees as 2 colorable bipartite graphs. In graph theory, trees have only one connection between any two vertices. Each vertex can have zero to many children vertices. We use this property of one connection between parent and child vertices as a logical qubit. The connection is changed to $|1\rangle$ via an X gate if a connection is present. The circuit is developed with the vertices initially and then their corresponding connections. The vertices output their colors. Since only 2 colors are applicable, we either have $|0\rangle$ or $|1\rangle$ as outputs. For our paper, we assume a given tree in Fig. 7 and show its corresponding quantum circuit in Fig. 10. This circuit can be used as a reference for a tree with n vertices with given connections (Fig. 8).

4.1 Quantum Cost

Counting each connection between each parent and child, let us assume we have c connections for a tree of n vertices. We can calculate the quantum cost as $18c + c + 1 = 19c + 1$. Also, we can calculate the cost as $18(n - 1) + 1 + c = 18n + c - 17$. For fig, our cost is 96.

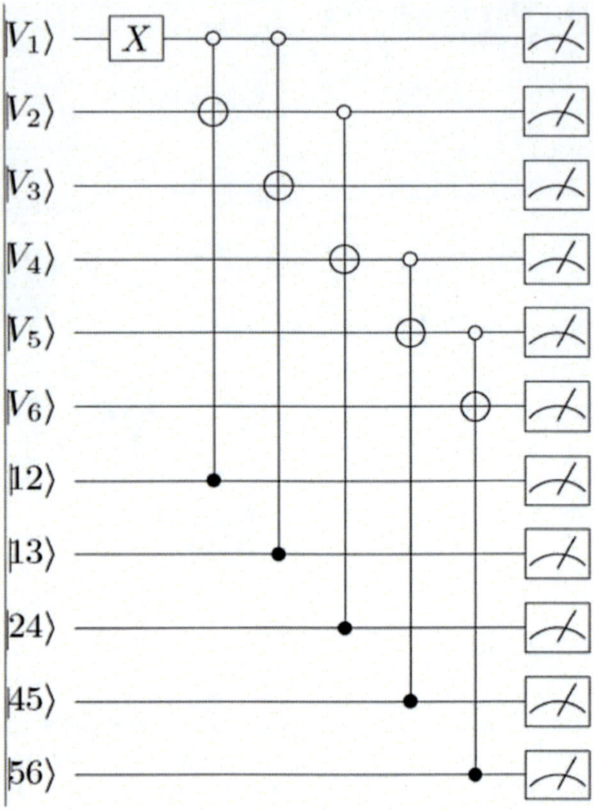

Fig. 8 Quantum circuit for the given tree: Vertices and connections are shown. The vertices output their colors

4.2 Classical Complexity

We determine a bipartite graph and color it with 2 colors by computing breadth first traversal (BFT) or depth first traversal (DFT). This is done in linear time complexity.

5 Binary Trees

Binary trees are a special form of trees where every parent vertex has exactly two children vertices. We now present a circuit specifically for binary trees. Here we do not denote connections as starting from the root vertex, every parent–children connection is previously known. In our paper, we assume a given binary tree shown in Fig. 9 and show its quantum circuit for outputting colors in Fig. 10. This reference circuit can be extended to n vertices.

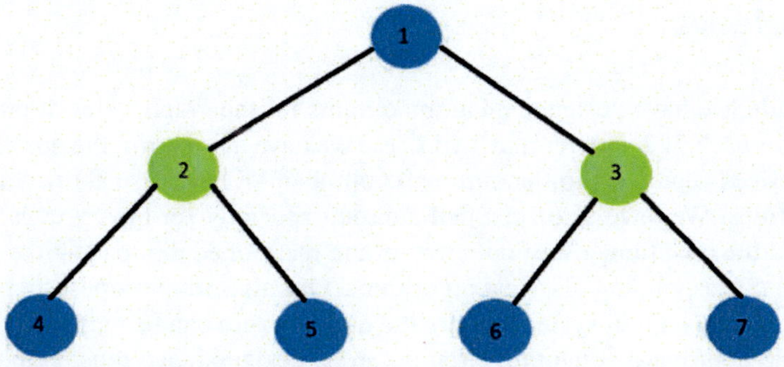

Fig. 9 **Representation of a binary tree**: a given binary tree

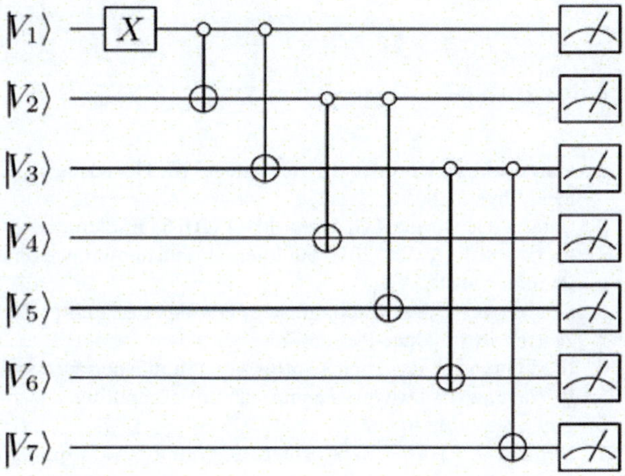

Fig. 10 **Quantum circuit for the given binary tree**: Vertices output their corresponding colors

5.1 Quantum Cost

For a binary tree of n vertices, the quantum cost corresponding to Fig. 10 can be calculated as $3n + 1$. For our circuit in Fig. 9 with 7 vertices, the quantum cost is 19.

5.2 Classical Complexity

Binary trees take O(n) to color its vertices.

6 Conclusion

To conclude, we have designed quantum circuits for the graph states through superposition for each of 2 vertices and 3 vertices. We have also shown the possible colors of the vertices according to the chromatic number. W have also shown circuits for coloring trees. We have also specified a circuit specially for binary trees. We have mentioned the quantum cost of our circuits and mentioned the space–time complexity for the corresponding classical algorithms. Our algorithm proposes a pattern for d-coloring using a qudit system, where the qudit system can be mapped into a qubit system and appropriate quantum circuits can be designed, and thus, graph coloring can be achieved. The approach used here can be used in applications such as solving Sudoku [26, 27], setting time table [28], and coloring of maps.

References

1. Horodecki R, Horodecki P, Horodecki M, Horodecki K (2009) Quantum entanglement. Rev Mod Phys 81:865
2. Cangea O, Oprina CS, Dima M-O (2016) Implementing quantum cryptography algorithms for data security. In: 2016 8th international conference on electronics, computers and artificial intelligence (ECAI)
3. Brucker P (1979) NP-complete operations research problems and approximation algorithms. Zeitschrift fur Oper-Res 23:73
4. IBM Quantum Experience. https://quantumexperience.ng.bluemix.net/qx/community
5. D-Wave. www.dwavesys.com/quantum-computing
6. IonQ. https://ionq.co/
7. Viyuela O, Rivas A, Gasparinetti S, Wallraff A, Filipp S, Martin-Delgado MA (2018) Observation of topological Uhlmann phases with superconducting qubits. NPJ Quant Inf 4:10
8. Aggarwal D, Raj S, Behera BK, Panigrahi PK, Compactly generated t-structures in the derived category of a commutative ring. arXiv:1806.00781
9. Zhukov AA, Remizov SV, Pogosov WV, Lozovik YE (2018) Algorithmic simulation of far-from-equilibrium dynamics using quantum computer. Quantum Inf Process 17:223
10. Malik GR, Singh RP, Behera BK, Panigrahi PK, First experimental demonstration of multi-particle quantum tunneling in IBM quantum computer. https://doi.org/10.13140/RG.2.2.27260.18569
11. Dutta S, Suau A, Dutta S, Roy S, Behera BK, Panigrahi PK, Demonstration of a quantum circuit design methodology for multiple regression. arXiv:1811.01726
12. Woerner S, Egger DJ, Quantum risk analysis. arXiv:1806.06893
13. Vuillot C (2018) Is error detection helpful on IBM 5Q chips. Quant Inf Comput 18:0949
14. Ghosh D, Agarwal P, Pandey P, Behera BK, Panigrahi PK (2018) Automated error correction in IBM quantum computer and explicit generalization. Quant Inf Process 17:153
15. Singh RK, Panda B, Behera BK, Panigrahi PK, Demonstration of a general fault-tolerant quantum error detection code for (2n+1)-qubit entangled state on IBM 16-qubit quantum computer. arXiv:1807.02883
16. Satyajit S, Srinivasan K, Behera BK, Panigrahi PK (2018) Nondestructive discrimination of a new family of highly entangled states in IBM quantum computer. Quant Inf Process 17:212
17. Alsina D, Latorre JI (2016) Experimental test of Mermin inequalities on a five-qubit quantum computer. Phys Rev A 94:012314

18. Kalra AR, Gupta N, Behera BK, Prakash S, Panigrahi PK (2019) Demonstration of the no-hiding theorem on the 5-Qubit IBM quantum computer in a category-theoretic framework. Quant Inf Process 18:170
19. Swain M, Rai A, Behera BK, Panigrahi PK (2019) Experimental demonstration of the violations of Mermin's and Svetlichny's inequalities for W and GHZ states. Quant Inf Process 18:218
20. Behera BK, Banerjee A, Panigrahi PK (2017) Experimental realization of quantum cheque using a five-qubit quantum computer. Quant Inf Process 16:312
21. Srinivasan K, Satyajit S, Behera BK, Panigrahi PK, Efficient quantum algorithm for solving travelling salesman problem: an IBM quantum experience. arXiv:1805.10928
22. Behera BK, Reza T, Gupta A, Panigrahi PK, Router designing quantum in IBM quantum computer. arXiv:1803.06530
23. Behera BK, Seth S, Das A, Panigrahi PK (2019) Demonstration of entanglement purification and swapping protocol to design quantum repeater in IBM quantum computer. Quant Inf Process 18:108
24. D'Hondt E (2009) Quantum approaches to graph colouring. Theor Comput Sci 410:302
25. Saha M, Behera BK, Panigrahi PK, Quantum algorithms for colouring of graphs. https://doi.org/10.13140/RG.2.2.19222.19523
26. Pal A, Chandra S, Mongia V, Behera BK, Solving Sudoku game using quantum computation. https://doi.org/10.13140/RG.2.2.19777.86885
27. Majumder A, Kumar A, Das N, Chakraborty N (2016) Recursive backtracking for solving 9*9 Sudoku puzzle. Bonfr Int J Data Min 6:07
28. Nanda A, Pai MP, Gole A (2012) An algorithm to automatically generate schedule for school lectures using a heuristic approach. Int J Mach Learn Comp 2:492

Design of FPGA-Based QPP Interleaver for LTE/LTE-Advanced Application

Bijoy Kumar Upadhyaya and Salil Kumar Sanyal

Abstract Modern wireless communication systems have witnessed increasing use of channel coding techniques to enhance the throughput and to reduce latency. Interleavers are playing an important role to make the communication systems more robust and resilient in such channel coding approaches. The Long-Term Evolution (LTE)/LTE-Advanced of the 3rd Generation Partnership Project (3GPP) uses Quadrature Permutation Polynomial (QPP) interleaver in its Turbo coding scheme. The address generator of the interleaver contains a quadratic expression having square and modulus function whose direct digital hardware is not yet available in the literature. A novel algorithm has now been proposed which can provide low complexity hardware solution to implement the interleaver address generator. This paper describes VHDL model and timing simulation of the proposed address generator using ModelSim XE-III software. Due to absence of implementation results in the literature, comparison of this work is made by implementing conventional LUT-based technique on the same FPGA. Such comparison shows better FPGA resource utilization by 71.16% and improved operating speed by 82.26% in favour of the novel proposed technique.

Keywords Long-Term Evolution · Quadrature Permutation Polynomial · Interleaver · FPGA · VHDL

B. K. Upadhyaya (✉)
Department of Electronics & Communication Engineering, Tripura Institute of Technology, Agartala, Tripura 799015, India

S. K. Sanyal
Department of Electronics & Communication Engineering, Narula Institute of Technology, Kolkata, West Bengal 700109, India

M. Mitra et al. (eds.), *Computational Advancement in Communication, Circuits and Systems*, Lecture Notes in Electrical Engineering 786,
https://doi.org/10.1007/978-981-16-4035-3_12

1 Introduction

The demand for ubiquitous mobile Internet services requires high bandwidth connectivity. To cater this demand, new technologies like the Long-Term Evolution (LTE) of 3GPP standards [1] have been developed. LTE has rapidly become the dominant global standard for fourth generation cellular networks [2]. It has brought together many technological innovations from different area of research such as digital signal processing, Internet protocols, network architecture and security, and is poised to dramatically change the way we use the world wide mobile network in the future. LTE-Advanced (LTE-A) [3] is the project name of the evolved version of LTE that is being developed by 3GPP [4]. LTE-A is set to meet or exceed the requirements of the International Telecommunication Union (ITU) for the fourth generation (4G) radio communication standard known as IMT-Advanced [5].

LTE/LTE-A uses Turbo coding as channel coding scheme. Turbo encoder and decoder is one of the major blocks in a LTE wireless transceiver. Turbo encoder/decoder employs interleaver to reduce the effect of burst error in the channel. Turbo decoders provide best performance but suffer from high decoding latency due to the iterative decoding process [6]. This is due to the forward–backward recursion in the maximum a posteriori (MAP) decoding algorithm and the interleaving/de-interleaving between iterations [7].

Literature survey shows limited number of works in connection with design of address generator for Quadrature Permutation Polynomial (QPP) interleaver used in LTE/LTE-A transceiver. A unified approach of Turbo decoder design suitable for mobile WiMAX and 3GPP-LTE applications has been proposed by Kim and Park [8]. This work includes design of QPP interleaver with support the radix-4 single-binary turbo decoding. The work is claimed to be implemented on 0.13 μm CMOS technology. However, implementation result of QPP interleaver is not available. Sun and Cavallaro [7] propose a low complexity QPP interleaver of LTE/LTE-A and implemented using 65 nm CMOS technology. The authors extended the work to design Turbo code decoder to be used in 3GPP LTE/LTE-A. Another work of designing QPP interleaver for high throughput HSPA/LTE multi-standard turbo decoder is proposed in [9] by the same group of researchers of [7]. The work shows improved LUT-based approach where inter-row and inter-column parameters are pre-computed and stored in memory. Based on the input parameter and the inputs received from various intermediate pre-processing units, desired interleaver addresses are generated. The work is reported to be implemented on 45 nm CMOS technology. FPGA implementation of cyclic redundancy code (CRC) implementation for LTE-Advanced has been described in [10]. The authors demonstrate the integration of three CRC circuits into one with aim to reduce the total area of the LTE-Advanced transceiver.

In this paper, we propose an efficient algorithm to model the interleaver address generator used in LTE/LTE-A. The address generator of QPP interleaver involves a quadratic expression having square and modulus function whose corresponding digital hardware is not available. In our approach, the said expression is divided into two parts. The first part generates raw addresses which is described by a novel algorithm. The second part computes modulus on the raw addresses and employs a modified technique over [11]. Both algorithms are transformed into digital circuits for efficient FPGA implementation. VHDL model has been prepared to implement the hardware in Xilinx Spartan 6 FPGA. Functionality of the interleaver address generator is verified through timing simulation using ModelSim XE-III software. In spite of best possible effort in literature survey, the authors could not find similar works implemented on FPGA platform having implementation results. Consequently, comparison could not be drawn with existing works implemented on CMOS and other platforms. As a result, for the sake of comparison, the authors implemented the conventional look up table (LUT)-based technique with improved memory modelling on the same FPGA. Comparative analysis of our proposed work with improved LUT-based technique in terms FPGA resource utilization (Block RAM) and operating speed shows betterment by 71.16% and 82.26%, respectively, at the negligible cost of FPGA slice requirement.

2 Interleaving in LTE/LTE-A

Turbo coding is employed in many standards for forward error correction techniques due to its impressive performance [5]. The interleavers are used in turbo coding/decoding to improve the error performance. During decoding of information through iterative process, extrinsic information is exchanged between sub blocks. Interleaving operation of this extrinsic information in parallel decoder suffers due to memory access contention [12]. The 3rd Generation Partnership Project (3GPP) LTE and LTE-A standard proposed to use the Quadratic Permutation Polynomial (QPP) interleaver [13]. Such interleaver poses contention free property and allows parallel decoders to decode one code word with improved throughput.

The interleaving operation in a QPP interleaver defined for LTE/LTE-A may be expressed as

$$\Pi(i) = \left(f_1 \times i + f_2 \times i^2\right) \bmod \mathrm{K} \tag{1}$$

where i stands for the original address, and $\Pi(\mathrm{x})$ is the interleaved address. The parameters f_1 and f_2 are related to the block size K and are defined in [1]. In the 3GPP LTE/LTE-A standard, there are 188 different block sizes ranging from 40 to 6144, and each size has its different interleaver parameters f_1 and f_2.

3 Proposed Algorithm for QPP Interleaver

This section describes proposed algorithm for address generator of interleaver used in LTE/LTE-A. Firstly, a MATLAB program has been developed using (1) to determine sequence of addresses to be generated against each value of K, f_1 and f_2 parameters. Table 1 shows certain portion of such address sequences obtained for three instances with (a) $K = 40$, (b) $K = 1008$ and (c) $K = 6144$.

Due to non-availability of corresponding hardware to implement (1), we propose a novel algorithm which leads to low complexity implementation the address generator. In our approach, (1) is divided into two subparts out of which the first part computes the raw addresses. These raw addresses pass through the mod function in the second part. Both parts are described by (2) and (3), respectively. Accordingly, previous MATLAB program is partially modified to generate the raw addresses without the mod function, i.e. implementation of (2). The raw addresses so generated for the same three cases of Table 1 are shown in Table 2(a)–(c).

$$y(i) = \left(f_1 \times i + f_2 \times i^2\right) \quad (2)$$

$$\Pi(i) = y(i)\text{mod}K \quad (3)$$

Table 1(a) Address sequences with $K = 40, f_1 = 3, f_2 = 10$

13	6	19	12	25	18	31	24	37	30	3	36
9	2	15	8	21	14	27	20	33	26	39	32
…											
21	14	27	20	33	26	39	32	5	38	11	4

Table 1(b) Address sequences with $K = 1008, f_1 = 171, f_2 = 204$

375	150	333	924	915	306	105	312	927	942	357	180
411	42	81	528	375	630	285	348	819	690	969	648
…											
423	438	861	684	915	546	585	24	879	126	789	852

Table 1(c) Address sequences with $K = 6144, f_1 = 263, f_2 = 480$

743	2446	5109	2588	1027	426	785	2104	4383	1478	5677	4692
4667	5602	1353	4208	1879	510	101	652	2163	4634	1921	168
…											
951	5726	5317	5868	1235	3706	993	5384	4591	4758	5885	1828

Table 2(a) Raw address sequences with $K = 40, f_1 = 3, f_2 = 10$

13	46	99	172	265	378	511	664	837	1030	1243	1476
1729	2002	2295	2608	2941	3294	3667	4060	4473	4906	5359	5832
…											
320,947	324,540	328,153	331,786	335,439	339,112	342,805	346,518	350,251	354,004	350,251	354,004

Table 2(b) Raw address sequences with $K = 1008, f_1 = 171, f_2 = 204$

375	1158	2349	3948	5955	8370	11,193	14,424	18,063	22,110	26,565	31,428
36,699	42,378	48,465	54,960	61,863	69,174	85,020	93,555	102,498	111,849	121,608	131,775
…											
6,421,383	6,493,974	6,566,973	6,640,380	6,714,195	6,788,418	6,863,049	6,938,088	7,013,535	7,089,390	7,165,653	7,242,324

Table 2(c) Raw address sequences with $K = 6144, f_1 = 263, f_2 = 480$

743	2446	5109	8732	13,315	18,858	25,361	32,824	41,247	50,630	60,973	72,276
84,539	97,762	111,945	127,088	143,191	160,254	178,277	197,260	217,203	238,106	259,969	282,792
…											
15,084,471	15,255,134	15,426,757	15,599,340	15,772,883	15,947,386	16,122,849	16,299,272	16,476,655	16,654,998	16,834,301	17,014,564

Careful examination of the sequence of raw addresses in Table 2(a)–(c) shows certain correlation between them which may be expressed by the following novel algorithm for address generator of LTE/LTE-A interleaver:

Define:
$F = f_1 + f_2$;
Fixed Increment: $INC = 2*f_2$
Initial condition: $PA = PI = F$
Subsequent address: Previous address (PA) + {previous increment (PI) + INC}
e.g. (a) for $K = 40, f_1 = 3, f_2 = 10$
Initial Conditions:
$PA = PI = F = 13$, $INC = 20$,
Subsequent addresses: 13 + (13+20) = 46; 46 + (33+20) = 99; 99 + (53+20) = 172 and so on.
(b) For $K = 1008, f_1 = 171, f_2 = 204$
Initial Conditions:
$PA = PI = F = 375$, INC = 408,
Subsequent addresses: 375 + (375+408) = 1158; 1158 + (783+408) = 99; 99 + (1191+20) = 1290 and so on.
(c) For $K = 6144, f_1 = 263, f_2 = 480$
Initial Conditions:
$PA = PI = F = 743$, $INC = 960$,
Subsequent addresses: 743+ (743+960) = 2446; 2446 + (1703 + 960) = 5109; 5109 + (2663+960) = 8732 and so on.

The proposed algorithm computes the raw addresses recursively without involving multiplier and squarer circuit thus ensuring low complexity implementation.

Implementation of (3), i.e. computation of modulus on raw addresses is done by suitably modifying the algorithm proposed by Butler and Sasao [11]. This algorithm computes $x \bmod z$ as a modulo reduction process, where at each stage, the magnitude of x is reduced, but the residue remains the same which is continued until only the residue remains. As the interleaver block size K (divisor) has 188 different values ranging from 40 to 6144, the second method of Butler and Sasao [11] where the divisor is an independent variable is adopted. In the proposed case, as shown in Table 2(c) I, the maximum value whose modulus is to be computed is 17014564 which in binary requires 25-bits representation. As a result value of $I = 25$ in computation of $\theta = z \times 2i$ where $i\text{-}1$ represents number of comparison stages, z represents the divisor (= K) and θ is defined to be the first value to be subtracted from the dividend, X. Subsequent values of θ are computed by dividing present value of θ by 2. In our work, division is accomplished by right shift which is more resource efficient than direct division technique.

4 Hardware Realization

In order to test the functionality of the proposed algorithm, corresponding digital hardware is designed. Top level view of the hardware is presented in Fig. 1. The first block is raw address generator, implementing (1), receives initiation pulse (*INIT*), clock signal (*CLK*) and a memory pointer (*I*) to retrieve corresponding values of *K*, f_1 and f_2 stored in an LUT [1]. The raw addresses (*X*) and *K* values are passed on to the MOD circuit which finally generates the desired interleaver addresses.

The raw address generator as shown in Fig. 2 comprises of two subsidiary units namely, an LUT (a), a *START* signal generator (b) along with the main unit (c). The LUT stores the values K, f_1 and f_2 and has 32-bit width. The organization of data in the LUT is described in Table 3. A desired combination of the block size K, f_1 and f_2 is read from the LUT by supplying appropriate address of LUT in *I*. An adder is used to generate $F = f_1 + f_2$. A *START* signal is generated by the hardware shown in Fig. 2(b) either at the beginning of the operation or after completion of one period of generating 188 interleaver addresses. *INIT* pulse is responsible for generating the *START* signal in the beginning whereas an active reset signal generates the subsequent

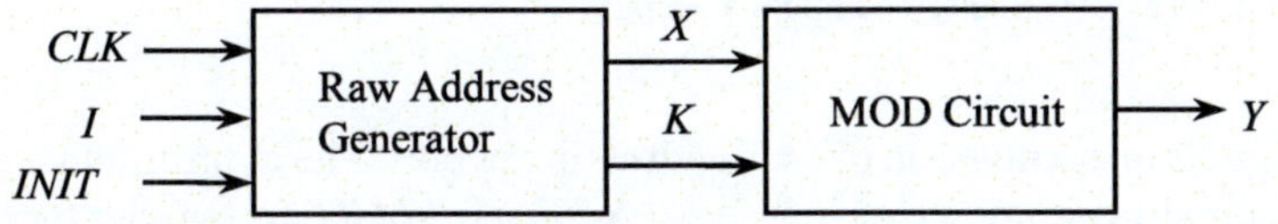

Fig. 1 Top level view of the QPP interleaver address generator

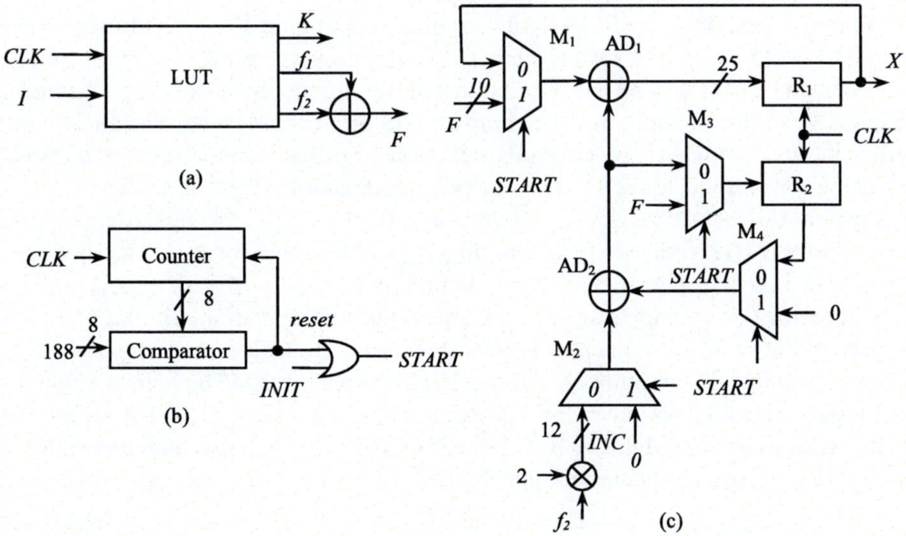

Fig. 2 Proposed hardware of raw address generator. (a) LUT. (b) START signal Generator. (c) MAIN Unit

Table 3 Organization of LUT

I (9 bit)	*K* (13 bit)	f_1 (9 bit)	f_2 (10 bit)
000H	01400C0AH		
001H	01801C0CH		
002H	01C04C2AH		
…	…		
05BH	01F80DC54H		
…	…		
0BBH	C0041DE0H		

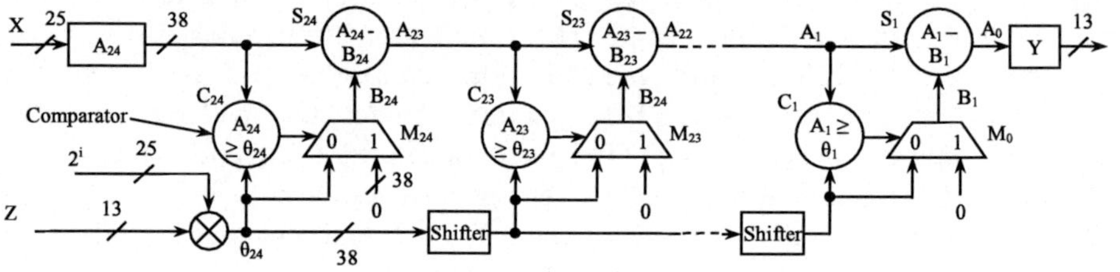

Fig. 3 Modified MOD circuit to compute *X mod Z*

START signals on completion of each addressing cycles. The main circuit responsible for raw address generation receives signal F, f_2 and *START* from other two ancillary units and the system *CLK*. Active *START* pulse causes $M_1 = F, M_2 = INC = 2 \times f_2, M_3 = F \, and \, M_4 = 0$ representing the initial condition of the circuit. Registers R_1 and R_2 store the output of A_1 and M_3, respectively.

Figure 3 describes modified hardware used to compute *X mod K*. In this application, $i = 25$ and z is a 13-bit number resulting in θ to be a 38-bit number. This consequents 25-bit *X* to be converted into a 38-bit number by appending 13 zeros in the MSBs so that *X* and θ may be compared. In order to make our design resource efficient, we computed θ by appending 25 number of zeros in the LSB of *Z* instead of using a multiplier directly. The left most comparator (C_{24}) as shown in Fig. 3 compares A_{24} (38-bit version of *X*) with θ_{24}. If $A_{24} \geq \theta_{24}$, the select input of the multiplexer (M_{24}) receives a 0, thus routing θ_{24} to the output ($= B_{24}$). The subtractor (S_{24}) performs first stage reduction by computing $A_{23} = A_{24} - B_{24}$. In case, $A_{24} \ngeq \theta_{24}$, M_{24} receives $B_{24} = 0$, the stage performs pass through operation. Similar operation is carried out in 23 subsequent stages as shown in Fig. 3 till A_0 is computed which is converted into 13-bit number, *Y* removing 25 zeros from MSBs before sending it as output of the address generator. Subsequent θ_s (i.e. $\theta_{23}, \theta_{22}, \ldots, \theta_1$) are computed from previous θ values dividing by 2. Instead of using a divisor, the authors employed shifter to perform $\div$ 2 operation, thereby making design more resource efficient.

5 Simulation Result

The simulation results in the form of timing diagram obtained using ModelSim Xilinx Edition-III for $i = 1 (K = 40)$, 91 $(K = 1008)$ and 188 $(K = 6144)$ are shown in Fig. 4(a)–(c), respectively. The captured portion shows the interleaver addresses generated for the first few cases and is identical with Table 1(a)–(c), respectively. The circuit fetches $K = 40, f_1 = 3$ and $f_2 = 10$ from the LUT of Fig. 2(a) when supplied with $i = 1$. As shown in Fig. 4(a)–(c), *init* = 1 in the beginning for one clock pulse to enable hardware of Fig. 2(b) to generate *START* pulse. On completion of an iteration of address generation, the *START* pulse for subsequent iterations is generated by the reset signal. Based on the values received against the signals *START*, *F* and f_2, the main unit of the address generator as shown in Fig. 2(c) computes the desired addresses and is made available at the output as y in Fig. 4(a)–(c). The authors have generated and verified addresses for all values of *i*; however, to avoid clumsiness other, results are not included.

6 FPGA Implementation Result and Analysis

The proposed hardware structure of LTE/LTE-advanced interleaver address generator is transformed into VHDL model using Xilinx Integrated Software Environment (ISE 12.1) and is implemented on Xilinx Spartan 6 FPGA. Table 4 shows the HDL synthesis report for the implementation. The adder of Fig. 2(a) is realized by the 10-bit adder. The 13-bit and 38-bit subtractors are used in the mod circuit to determine the Y and $A_x - 1 = A_x - B_x$. AD_1 and AD_2 adders of Fig. 2(c) are implemented through two 25-bit adders. The input applied through I is converted into LUT address with the help of the 9-bit subtractor to access the content of LUT described by Table 3. An 8-bit up-counter is used to implement the counter of Fig. 2(b). The 1-bit, 25-bit registers are used to implement reset signal, R_1 and R_2 register, respectively. Figure 2(b) uses the 8-bit greater than comparator to generate the reset pulse, whereas the 24 numbers of 38-bit less than equal to comparators are used in the mod circuit to implement C_{24}, C_{23}, C_{22} .....,C_1. Similarly, the mod circuit of Fig. 3 also uses 24 number of 38-bit 2-to-1 multiplexers (M_{24}, M_{23}, M_{22}, ..., M_1). The 25-bit 2-to-1 multiplexers implements M_1-M_4 of Fig. 2(c)

Direct comparison of our proposed work with the existing works in [7–9] is not possible due to dissimilarity in implementation platform and non-availability of implementation result for interleaver alone. However, in order to compare the implementation results, we have modelled and implemented the conventional LUT-based technique in the same Spartan 6 FPGA platform. As per [1], there are 188 permissible block sizes leading to 188 memory blocks required to house the address LUTs. Such

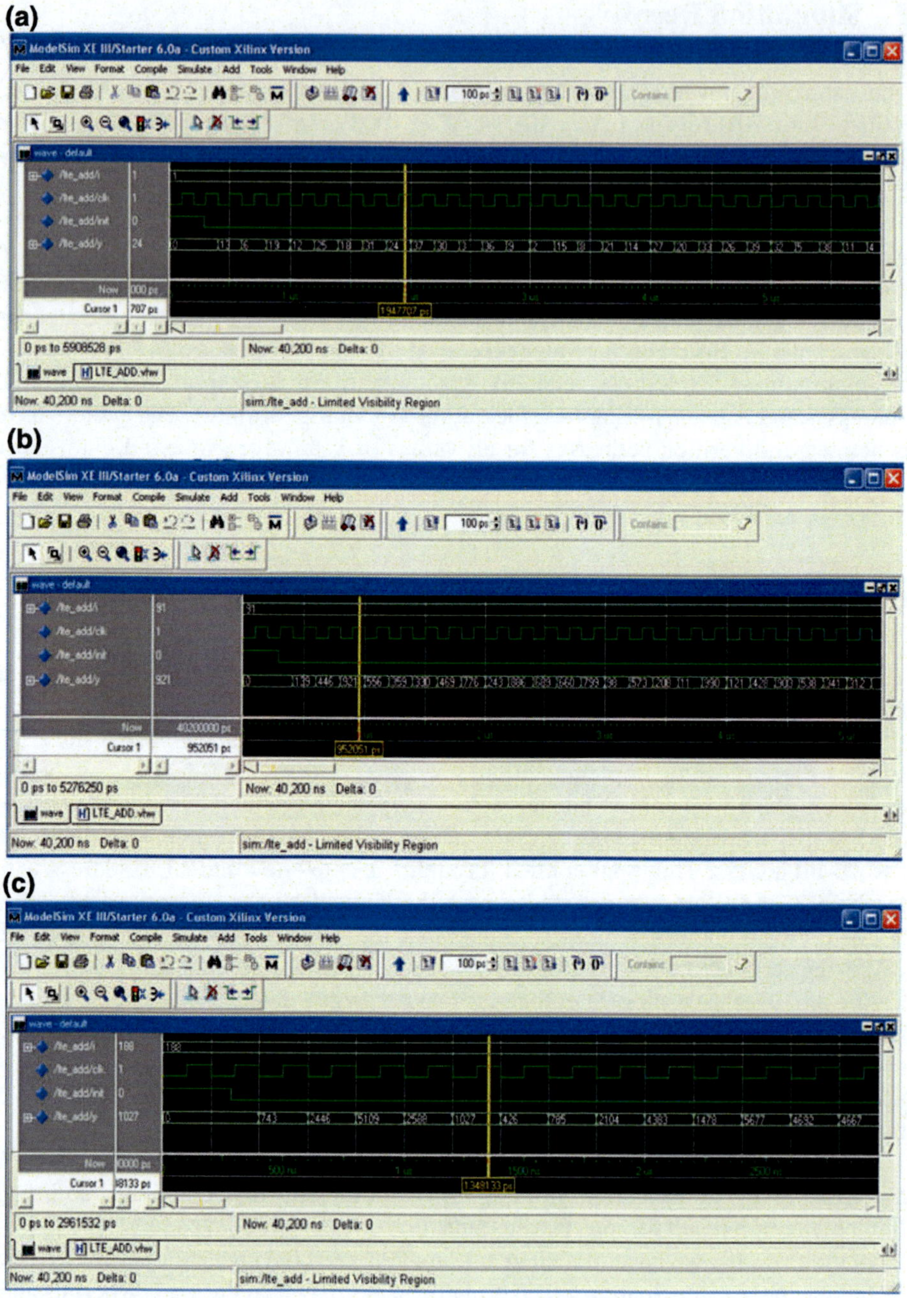

Fig. 4 (a) Timing simulation showing initial addresses for $i = 1$ ($K = 40, f_1 = 3, f_2 = 10$). (b) Timing simulation showing initial addresses for $i = 91$ ($K = 1008, f_1 = 171, f_2 = 204$). (c) Timing simulation showing initial addresses for $i = 188$ ($K = 6144, f_1 = 263, f_2 = 480$)

Table 4 HDL synthesis report

Logic circuits used	Quantity	Logic circuits used	Quantity
10-bit adder	1	1-bit register	1
13-bit subtractor	1	25-bit registers	2
25-bit adder	2	8-bit comparator greater	1
38-bit subtractor	24	38-bit comparator less equal	24
9-bit subtractor	1	25-bit 2-to-1 multiplexer	4
8-bit up-counter	1	38-bit 2-to-1 multiplexer	24

requirement cannot be catered by the target FPGA which only possess 52 BRAM blocks. To solve the problem, the authors partitioned the available memory blocks which in turn helped to reduce memory wastage as well. Implementation results of both techniques are shown in Table 5. Our proposed technique shows significant reduction in Block RAM requirement by 71.16% in comparison to improved LUT-based technique. Similarly, our proposed design shows significant improvement by 82.26% in terms of maximum operating frequency. These improvements are at the cost of minor increase in FPGA slice LUT requirement.

Table 5 Comparative device utilization summary

FPGA resources/parameters	This work		LUT-based implementation		Remarks
	utilization in number	Utilization in %	Utilization in number	Utilization in %	
Number of slice registers	59 out of 30,064	0.19	63 out of 30,064	0.21	Reduction by 0.02%
Number of slice LUTs	1128 out of 15,032	7.50	433 out of 15,032	2.88	Increase by 4.62%
Number ofbonded IOBs	23 out of 240	9.58	23 out of 240	9.58	No change
Number of block RAM	1 out of 52	1.92	38 out of 52	73.08	Reduction by 71.16%
Maximum clock speed	260.92 MHz		143.16 MHz		Speed improvement by 82.26%

7 Conclusion

This work describes a novel and efficient algorithm to model the interleaver address generator used in LTE/LTE-A. The algorithm exploits correlation between consecutive addresses of the address generator of QPP interleaver to efficiently model and implement on reconfigurable hardware. The algorithm is converted into digital hardware and implemented on Xilinx Spartan 6 FPGA using Xilinx ISE 12.1. Functionality of the address generator is tested through timing simulation using ModelSim XE-III software. The proposed work when compared with conventional LUT-based work shows significant improvement in terms of embedded memory utilization and operating speed.

References

1. 3rd Generation Partnership Project; Technical Specification Group Services and System Aspects; 3GPP System Architecture Evolution: GPRS Enhancements for LTE Access; Release 8, 3GPP, 3GPP TS 23.401
2. Ghosh A, Zhang J, Andrews JG, Muhamed R (2010) Fundamentals of LTE. Prentice-Hall
3. 3GPP TR 136.912 (2011) LTE; feasibility study for further advancements for E-UTRA (LTE-Advanced), V10.0.0
4. Agilent Introducing LTE advanced—application note. Available in http://cp.literature.agilent.com/litweb/pdf/5990-6706EN.pdf
5. Qualcomm (2013) LTE advanced: an evolution built for the long-haul, October, 2013. Available in https://www.qualcomm.com/documents/lte-advanced-evolution-built-long-haul
6. Berrou C, Glavieux A, Thitimajshima P (1993) Near Shannon limit error-correcting coding and decoding: Turbo Codes. IEEE Proceedings of the Int. Conf. on Communications, Geneva, Switzerland, pp 1064–1070
7. Sun Y, Cavallaro JR (2011) Efficient hardware implementation of a highly-parallel 3GPP LTE/LTE-advance turbo decoder, integration, the VLSI journal, Elsevier, pp 305–315
8. Kim J-H, Park I-C (2009) A unified parallel radix-4 turbo decoder for mobile WiMAX and 3GPP-LTE, in Proc. CICC, San Jose, CA, USA, pp 487–490
9. Wang G, Shen H, Sun Y, Cavallaro JR, Vosoughi A, Guo Y (2014) Parallel interleaver design for a high throughput HSPA/LTE multi-standard turbo decoder. IEEE Trans Circ Syst—I: Regular Papers. **61**(5):1376–1389
10. Al-Doori Q, Alani O (2015) A multi polynomial CRC circuit for LTE-advanced communication standard, seventh computer science and electronic engineering conference (CEEC), University of Essex, UK pp 19–23
11. Butler JT, Sasao T (2011) Fast hardware computation of x mod z,18th reconfigurable architectures workshop (RAW 2011), Anchorage, Alaska, USA

12. Takeshita OY (2006) On maximum contention-free interleavers and permutation polynomials over integer rings. IEEE Trans Inf Theory 52(3):1249–1253
13. Wong C-C, Chang H-C (2010) Reconfigurable turbo decoder with parallel architecture for 3GPP LTE system. IEEE Trans Circuit Syst-II: Express Briefs **57**(7)

Influences of Solar Activity on Food Grains Yield

D. K. Tripathi, R. P. Tripathi, and A. K. Tripathi

Abstract The increased imbalance between demand and supply of food grains in India is due to the population burst and limited land of cultivation. Every year Government used to fix a target of food grains production which is rarely achieved due to various factors. Climatic conditions are one of the factors. India has the second largest cultivable land in the world and approximately 65% of the crops cultivated are food grains in spite of that food grains per person per month are decreasing continuously. New technologies are continuously adopted to increase the food grains yield but the situation is still not good. Along with the other factors the crop yield may depend on solar activity as earth is a solar planet from whom cosmic rays are continuously coming to the earth. So, in this paper, we studied the influences of the Sun's activity on the food grains yield. For this purpose, all India food grains yield, rice yield, and yearly average number of sunspots data are used to analyze the characteristic variations and to find any possible correlation between them. A strong linkage between solar activity and all Indian food grains yield and rice yield is obtained from the analysis.

Keywords Solar activity · Food grains yield · All India rice yield · Sunspot number · Annual rainfall

1 Introduction

The research suggests that sunspots are visible when the magnetic field in the spots is greater than the critical value. There will be no dark spot formation if the field is below this critical value although the terrestrial environment will be influenced by Sun due to coronal and solar wind activity. This influence is confirmed by several Scientists

D. K. Tripathi (✉)
Department of Basic Science and Humanities (Physics), Narula Institute of Technology, Kolkata, WB, India
e-mail: dhananjay.tripathi@nit.ac.in

R. P. Tripathi · A. K. Tripathi
S. D. M. School, Dhanbad, Jharkhand, India

M. Mitra et al. (eds.), *Computational Advancement in Communication, Circuits and Systems*, Lecture Notes in Electrical Engineering 786,
https://doi.org/10.1007/978-981-16-4035-3_13

throughout the world. The cloudiness is increased due to cosmic rays as revealed by the studies of Svensmark [1]. When the solar activity is maximum (higher rate of sunspots formation), the cosmic ray radiation on Earth's orbit is less, results in less cloudiness [2–6]. Due to this the temperature increases and precipitation decreases. This led to the hot weather and drought [7–10]. Conversely, if there is a decrease in the number of sunspots, the cosmic ray radiation on Earth's orbit increases, results in high cloudiness and higher precipitation.

Solar activity may affect crop yields. So, there was a search for linkage between crop yields and solar activity since the last century. Primary level studies were conducted for 1866 through 1973 of wheat in Texas and Kansas, corn in Illinois and Nebraska, rice in Louisiana (1895−1972), and cotton in Texas [11]. The crop yields were lower when the solar activity was low and it was reported higher when high number of sunspots was formed [12–15]. Linkage is found between the state of the wheat market in medieval England and solar activity [16]. So, we expect a correlation between all India food grains yield or all India rice yield and yearly average of sunspots number.

2 Analyses and Results

In Fig. 1, Yearly mean sunspot number is plotted year wise since 1700–2019.

Figure 1 is due to the increase and decrease of solar activity which results in a lower or higher number of sunspot formations in the last three centuries.

In Fig. 2. Year wise annual Indian rainfall is plotted.

Figure 2 shows that the average annual Indian rainfall is 1149.17 mm/hr from 1813 to 2012.

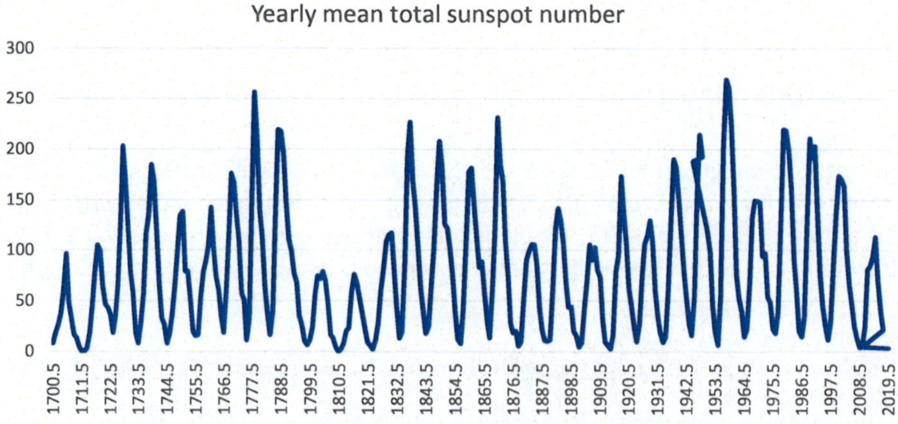

Fig. 1 Year wise yearly mean sunspot number

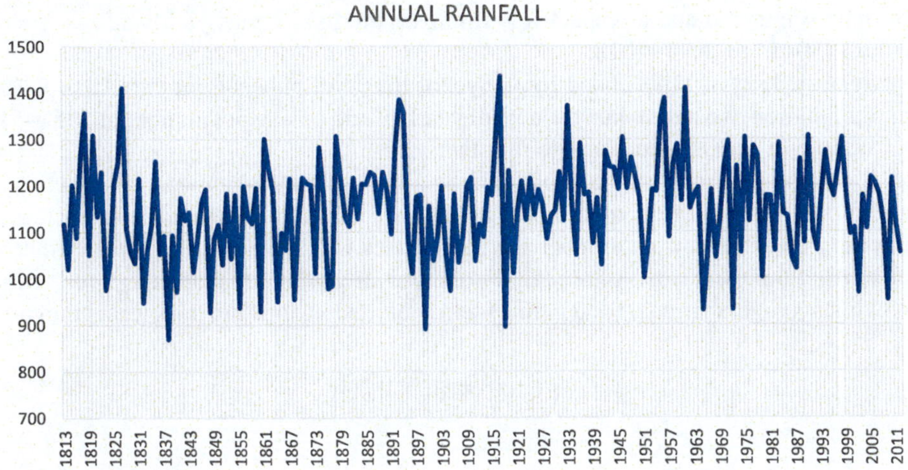

Fig. 2 Year wise annual Indian rainfall

Figures 1 and 2 suggests that the trend of variation of annual Indian rainfall is the same as the variation of the yearly mean of total sunspot number. To investigate this, yearly mean sunspot number is plotted against annual Indian rainfall from 1813 to 2012 in Fig. 3.

Figure 3, reveals that the Annual Indian rainfall is following the same trend as the yearly annual sunspot number.

Here, we want to know any linkage between food grains yield and solar activity. For this, All India food grains yield data is taken from Agricultural statistics at a glance 2018, published by Ministry of Agriculture and Farmers Welfare, Government

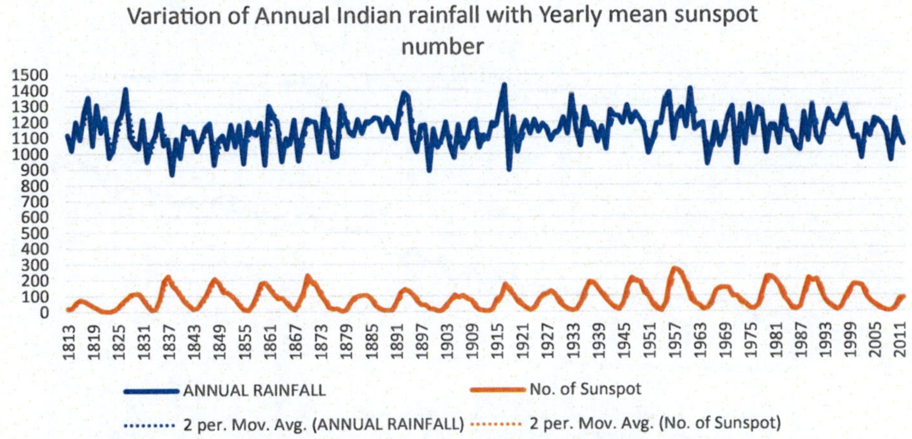

Fig. 3 Variation of Annual Indian rainfall with yearly mean sunspot number

of India. Figures 4 and 5 respectively discloses the food grains yield and rice yield from 1950–51 to 2017–2018.

Solar activity is represented by the yearly mean of sunspot number. Here is an approximation, we have taken the data of yearly mean of sunspot number of 1950–51 as the data of 1950.5 and likewise for other years.

Figure 6 and Fig. 7 suggest that there is a strong linkage between solar activity and food grains and rice yield. Food grains production increases with the increase in yearly mean of sunspot number and it decreases with the decrease of yearly mean of sunspot number with a phase lag. Further, to strengthen this idea the three years moving average of annual food grains yield is plotted against the three years moving

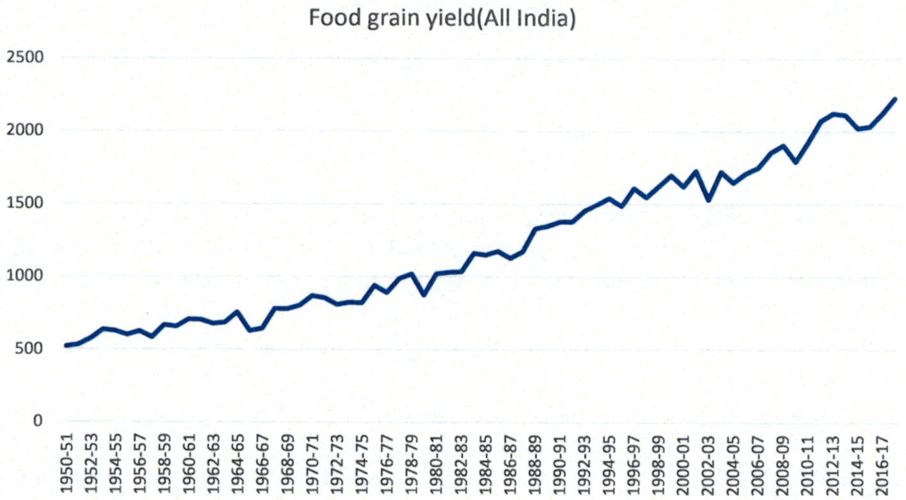

Fig. 4 Year wise food grains yield

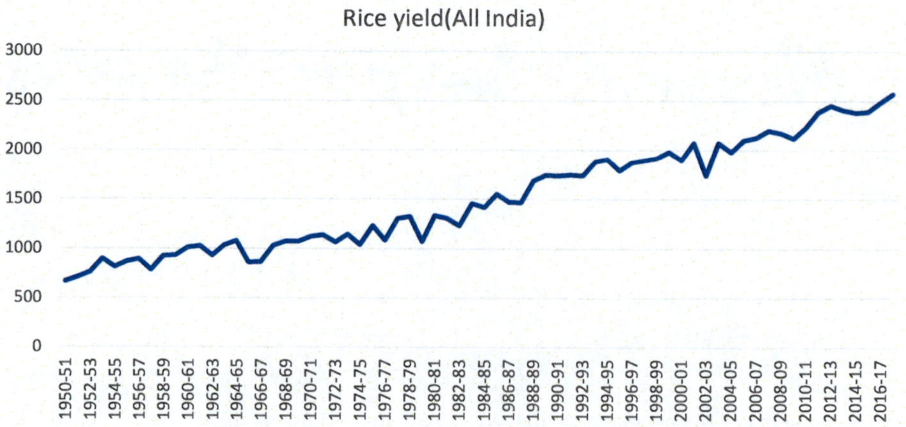

Fig. 5 Year wise rice yield

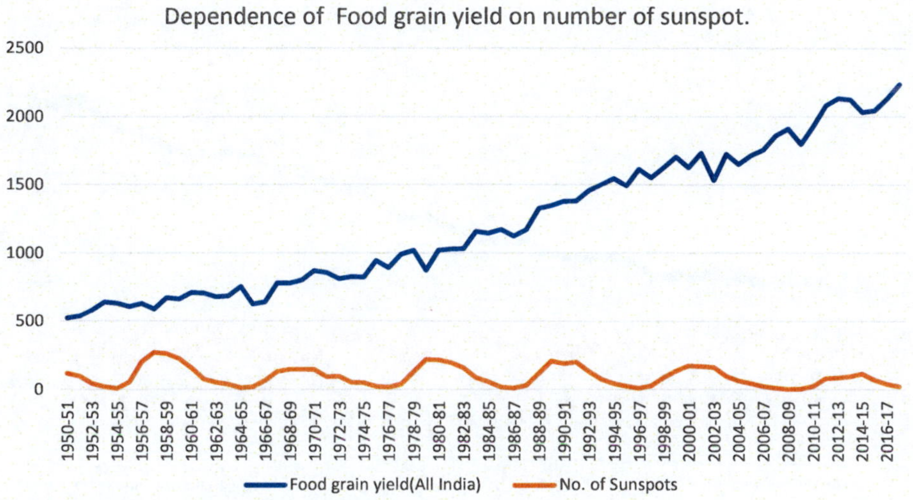

Fig. 6 Dependence of food grains yield on number of sunspots

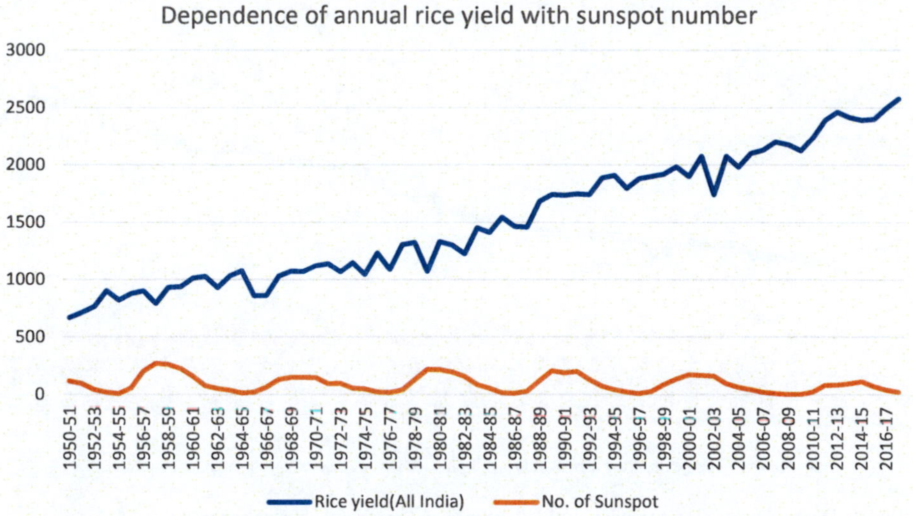

Fig. 7 Dependence of rice yield on number of sunspots

average of yearly mean sunspot number in Fig. 8 and three years moving average of annual rice yield is plotted against the three years moving average of yearly mean sunspot number in Fig. 9. Figures 8 and 9 confirms the same result as suggested by Figs. 6 and 7.

For close scrutiny, the analysis is done for solar cycles 19, 20, 21, 22, 23, and 24 separately. Following results are found.

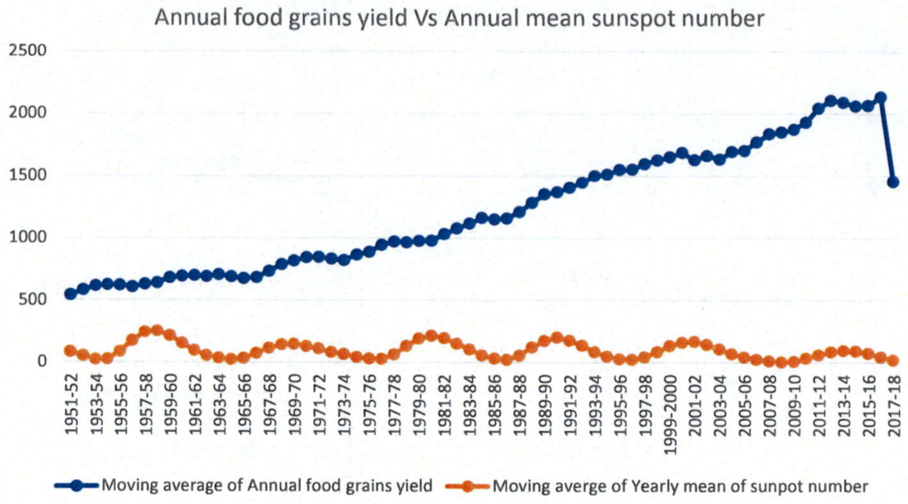

Fig. 8 Dependence of food grains yield on number of sunspots

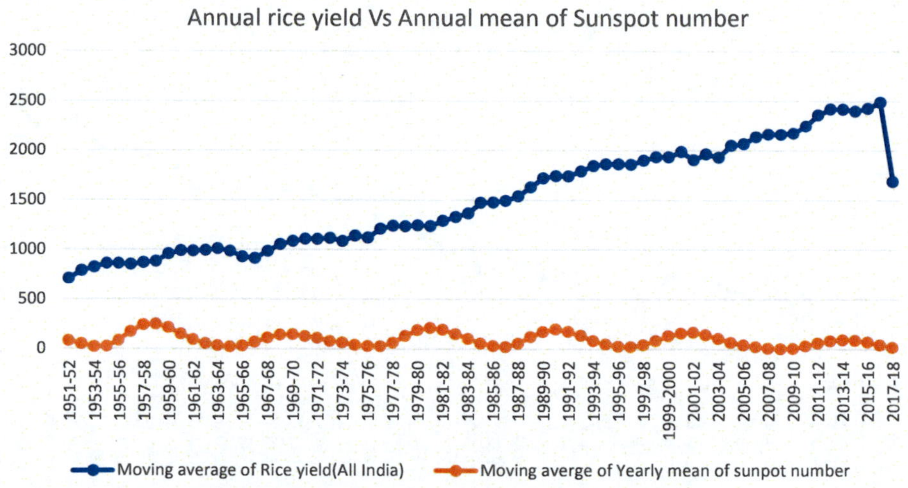

Fig. 9 Dependence of annual rice yield on annual mean of number of sunspots

(i) For Solar cycles 19, 21, and 23, the correlation coefficient is negative which suggests that these two quantities (solar activity and all India food grains yield or rice yield) are anti correlated. This means that the trend of variation of all India food grains yield and rice yield is opposite to the yearly mean sunspot number.

(ii) In contrast to the above three Solar cycles for Solar cycles 20 and 22 the correlation coefficient is positive which suggests that these two quantities (solar activity and all India food grains yield or rice yield) are highly correlated.

This means that the trend of variation of all India food grains yield and rice yield is the same as the yearly mean sunspot number.

(iii) In Solar cycle 24, the correlation coefficient is very high and positive which suggests that these two quantities (solar activity and all India food grains yield or rice yield) are highly correlated. This means that the trend of variation of all India food grains yield and rice yield is exactly the same as the yearly mean sunspot number.

3 Conclusion

The results suggest a positive correlation between solar activity and all India annual food grains yield or annual rice yield on the long-term time scale whereas for cycle to cycle (nearly 11-year time scale) it is not fixed. For some cycles, the variation of all India food grains yields or rice yields is exactly the same as the yearly mean sunspot number and in some cycles, this variation of all India food grains yield or rice yield is lagging in phase as the yearly mean sunspot number. The crop yield does not depend only on solar activity but also on precipitation, temperature, atmospheric humidity, wind velocity, atmospheric gases, soil, etc. This may be taken as a future problem.

Acknowledgments We are thankful to the Ministry of Agriculture and Farmers Welfare, the Government of India. The annual rainfall series is taken from ftp://www.tropmet.res.in/pub/data/rainseries/8-all_ind.txt website and the data for the year wise international average number of sunspot series are obtained partly by using the website: ftp://ftp.ngdc.noaa.gov/STP/SOLAR_DATA/SUNSPOT_NUMBERS/INTERNATIONAL/YEARLY.

References

1. Marsh N, Svensmark H (2000) Cosmic rays, clouds, and climate. Space Sci Rev 94:215–230
2. Tripathi DK , Das R (2019) Variations in climatic conditions due to cosmic rays. In: Nandi S, Ganguly S (eds) Proceedings of the 2nd national conference on science, technology and communication skills (NCSTCS2k19), Ideal International E-Publication Pvt. Ltd, pp 49
3. Bhattacharya AB, Tripathi DK, Roy K, Nag A (2013) Dependence of cloudiness and temperature on cosmic radiation. Int J Electron Commun Technol (IJECT) 4(SPL-1):97–99
4. Tripathi DK, Jha HS (2018) Influences of cosmic rays on Earth's climate. In: Karan S, Tripathi DK (eds) Proceedings of the 2nd national conference on science, technology and communication skills (NCSTCS2k18), Ideal International E-Publication Pvt. Ltd, pp 32
5. Tripathi DK, Bhattacharya AB (2014) Solar fingerprint in the geomagnetic storms. Int J Innov Rese Sci Eng Technol (IJIRSET), 3(SPL-6):74–77
6. Bhattacharya AB, Nag A, Roy K, Tripathi DK (2013) Total solar irradiance trend and tropospheric dynamical coupling. Int J Electron Commun Technol (IJECT) 4(SPL-1):112–113
7. Hiremath KM (2006) The influence of solar activity on the rainfall over India: cycle-to cycle variations. J Astrophys Astronomy 27:367–372

8. Hiremath KM, Mandi P (2004) Influence of the solar activity on the Indian Monsoon rainfall. New Astronomy 9:651–662. ISSN 1384–1076
9. Lockwood M (2012) Solar influence on global and regional climates. Survey Geophys 33:503–534
10. Tripathi DK, Bhattacharya AB (2014) Sunspot activity over the indian rainfall pattern. Int J Electron Commun Technol (IJECT) 5(SPL-2):43–45
11. Harrison VL (1976) Do Sunspot cycles affect crop yields. Agricultural Economic Report No. 327, Economic Research Service, U.S. Department of Agriculture
12. Huhtamaa H, Helama S, Holopainen S, Rethorn C, Rohr C (2015) Crop yield responses to temperature fluctuations in 19th century Finland: provincial variation in relation to climate and tree-rings. Boreal Environ Res 20:707–722
13. Agricultural statistics at a glance 2018, Ministry of agriculture and Farmers welfare, Government of India (2018)
14. Burakov D (2017) Do sunspots matter for cycles in agricultural lending: a VEC approach to Russian wheat market. Agris On-line Papers in Econom Info 9(1)
15. Love JJ (2013) On the insignificance of Herschel's sunspot correlation. American Geophysical Union
16. Pustilnik LA, Din GY (2003) Influence of solar activity on state of wheat market in medieval England. Solar Phys 223(1)

Different Sensors in Modern day Healthcare Service

Aritri Chakraborti, Koushik Karmakar, and Ananya Banerjee

Abstract Wireless sensor network (WSN) has a lot of applications in a wide range of fields like games, sports, environment monitoring many other fields. It is also widely used in medical field. A lot of research works are done on this topic. In the present context, we describe applications of WSN in health domain. We described the most commonly used sensor nodes in our daily life for healthcare. In this paper, we want to provide detailed information on this emerging research field.

Keywords WSN · Sensor · Healthcare

1 Introduction

A sensor network can be defined as a network consists of small autonomous devices located over a wide area that can communicate sense and report several parameters about their surrounding, such as humidity, temperature, and vibrations through a wireless link. The sensor network itself connects to the internet, WAN, LAN, or other industrial networks, to transmit data to systems and applications for analysis. Basically, Wireless sensor network consists of several nodes, which are capable of sensing several parameters (like temperature, soil qualities, humidity, etc.). Along with sensing the parameters, it can communicate with other sensor nodes, thus forming a network. It has a microcontroller, which controls the monitoring, a radio transceiver for generating radio waves, and different types of wireless communicating devices, and is also equipped with an energy source like a battery.

According to a report of World Health Organization, elderly population (65 years old and over) is increasing in the whole world [3–7]. Numbers of older people will be almost doubled by 2025 [3–7]. Also, different diseases will be increased.

In this situation Wireless Sensor Network (WSN) plays a very significant role in health monitoring [1, 2]. Specially designed sensors are used for this work. Very small sensors are either kept in wearable clothes or implanted in human body through

A. Chakraborti · K. Karmakar (✉) · A. Banerjee
Narula Institute of Technology, Kolkata, India
e-mail: koushik.karmakar@nit.ac.in

M. Mitra et al. (eds.), *Computational Advancement in Communication, Circuits and Systems*, Lecture Notes in Electrical Engineering 786,
https://doi.org/10.1007/978-981-16-4035-3_14

surgery. These small sensor nodes read the physiological conditions like heartbeat pulse, etc. from the human body and send them to a remote server. Later the physician can check them sitting in his own chamber. This is called Wireless Body area Network (WBAN). This is specially designed for health monitoring. It may be mentioned that WBAN has some differences from WSN. In the subsequent section, we will describe the differences in detail.

This paper consists of as follows. In Sect. 2 we describe the construction of a sensor node. Different applications of sensor nodes are described in Sect. 3. Section 4 describes WBAN architecture. In Sect. 5 faults and errors that may occur in WBAN, is described. A survey of the different real life sensors used in health monitoring has been discussed in Sect. 6. Section 7 concludes the paper.

2 Sensor Nodes Construction

A sensor node used in health monitoring consists of the following components [1–4]: a sensor part, memory module, microprocessor, and radio transmitter (Fig. 1). Sensor part senses the physiological information from human body. Information includes pulse rate, heart rate, blood pressure, and other relevant details. Information thus sensed by the sensor module is stored in the memory module. Microprocessor is the core part of the sensor that performs the entire task. Radio transmitter unit transmits the information to the outside world.

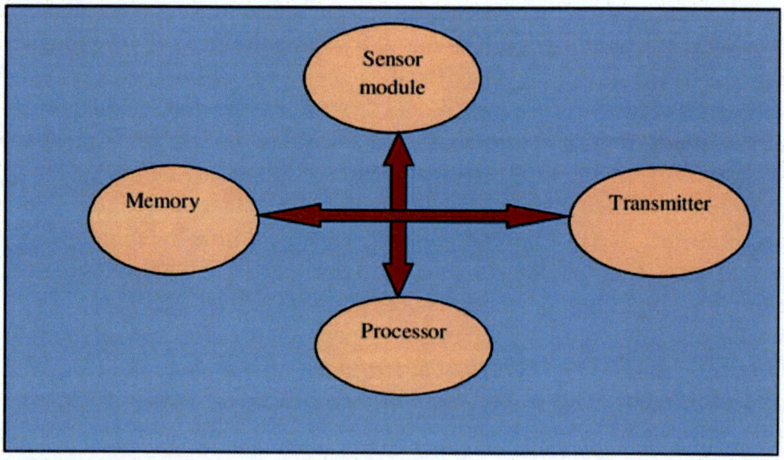

Fig. 1 Sensor module

Table 1 Difference between WSN and WBAN

WSN	WBAN
Used for environmental monitoring, battlefield and other areas	Used to measure human physiological conditions
Node placement and replacement is easy	Node placement and replacement is difficult
Data toss may not be very harmful	Data loss may be fatal as it consists of sensitive health data
Node size is generally big	Node size must be very small
Number of nodes can be hundreds in a single application	Number of nodes can be maximum of 10 to 12 in a single application

3 Application of WSN in Healthcare

Wireless Sensor Network (WSN) has huge application in healthcare monitoring systems [1–12]. However, a special sensor is used for this purpose. WBAN is a special application of WSN. So, WSN and WBAN have similarities as well as differences. Important differences are described (Table 1).

WBAN has several applications in different fields [1–5]. Though its most prominent application is in healthcare domain, it is used in other fields too. Application of WBAN can be classified into the following types:

A. Healthcare: WBAN is widely used in healthcare domain. It is used to capture the physiological conditions of the patients. Information thus captured will be transmitted to a remote server. Doctors sitting at their own chamber will read and analyze this data. Thus remote sensor based healthcare monitoring is possible.
B. Assistance for the disabled people: WBAN may be used for giving special assistance to disabled people and elderly persons. It has a vital role for ambient assistant living.
C. Non-medical application: WBAN has huge applications in other fields also like sports, military, fire fighting. So, like medical application, it has wide application in non-medical filed too.

4 WBAN Architecture

Wireless sensor network based health monitoring system or Wireless Body Area Network (WBAN) (Fig. 2) works in three layers [1–6]. It is described as follows:

Tier—1:

This communication is called Tier-1 communication. It is the primary layer and also deals with intra BAN communication. Data is transmitted from patient's body to local server also called a sink. It behaves like a gateway of information to the outside

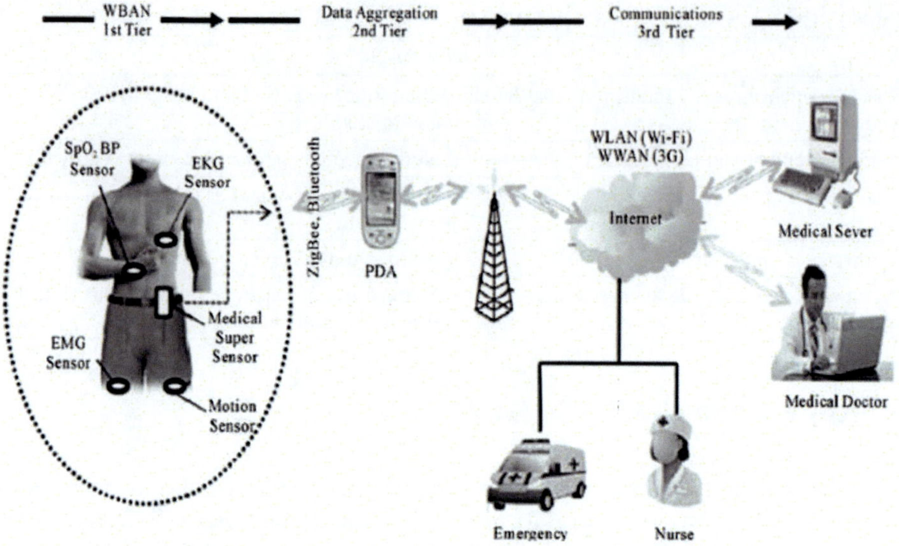

Fig. 2 WBAN architecture

world. Transmission is restricted to 2 m in and around the patient's body. From this sink, the processed physical data is sent to the Access Point (AP) in tier 2.

Tier—2:

According to this layer, information is transmitted to the next server called the access point (AP) through the network. Outside the patient's body, this AP is located. Range coverage area lies between 2 and 100 m. AP behaves like a server and also does some computations. Multiple such APs can interact with each other using the network.

Tier—3:

Basically, tier-3 is the ultimate layer of the communication process. All information passes through the AP to reach a medical server situated in a cloud. Only authorized people can access the server. Data is checked regularly in the server to get the health condition of the patient. Database is one of the important components in this tier design as it includes the necessary medical history and other relevant details. Also, doctors and the patient party can be informed immediately in an emergency.

5 Faults in WBAN

Like all electronic devices, faults and failures may also occur in WBAN. It may occur at different layers of WBAN as mentioned in the previous section. In the following section, we describe them in detail.

Tier-1[13, 14]**:**

This is the first layer which consists of different hardware components like memory, sensors, microcontrollers, and other components. Faults may occur at any one of them. Faults mean data faults; software faults and also hardware faults.

Tier-2 [13, 14]**:** In this layer data communication occurs between the sink and the access point (AP). Like tier-1, in this layer also faults may occur at sensor node, software, hardware, and also in networking. Moreover, channel through which transmission occurs may be influenced in different ways. It means data transmission problems, interference as well as other factors which include things like data integrity, confidentiality, and security.

Tier-3 [13, 14]**:** Faults in tier-3 are almost similar to that of tier-2. Additionally, it must ensure that only the authorized person can have access to this sensitive physiological data. Authentication is therefore a vital factor in this layer.

6 Different Sensors Used in Healthcare

Different sensors that are generally used for healthcare and patient monitoring in clinical settings are described below [3–5]. They are small in size, simple and easy to use and also affordable.

- Memscap—Wireless Pressure Transducer
- Millicore—DigiVent Pulmonary Air Leakage
- Novosense—CardioPatch ECG sensor
- Novelda—Medical UWB-IR radar
- VTT—Heart Monitoring Accelerometer
- SINTEF—SpO2 & Temperature sensors

Collection of long-term databases of clinical data:

Healthcare applications that leverage wireless sensor networks analyze the data gathered by sensors to infer and make decisions about the state of a patient's health and wellbeing. By improvement in monitoring consistency, continuous monitoring enhances data quality and precision for decision support leading to better titration of therapeutic interventions.

The continuously gathered data can be analyzed utilizing computational intelligence techniques to find solutions to the unsolved problems in the healthcare system.

In sensor based healthcare monitoring systems different types of sensors are widely used. In the following section (Table 2) we made a comparative study of them. A small survey is given in a tabular form.

Table 2 Different sensors used in healthcare

Device name	Sensor name	IC used	Manufacturer name	Price
Pulse oximeter	Heart rate sensor	BH1792GLC	Accusure, United States; Dr. Morepen India, etc	Approx. Rs. 1500/ =
Digital thermometer	Temperature sensor	MAX30205,	OMRON, Japan; Dr. Moreopen, India; FTA, Australia, etc	Approx. Rs. 150/ =
Blood pressure measurement	Pressure sensor	ILD233T, BM1383AGLV, etc	OMRON, Japan; Dr. Moreopen, India; etc	Approx. Rs. 2500/ =
Monitoring fetal movements in womb	Motion sensor, Heart rate sensor, Pressure sensor, etc	MAX30205, BH1792GLC, 555 timer, etc	Tigertech, India; MODOO, China, etc	Approx. Rs. 55,000/ =
Breathing test sensor	Breathing sensor	555 timer, ILD233T, etc	Manicom, India; OMRON, Japan, etc	Approx. Rs.15000/ =
BMI calculator	BMI sensor	AT89S52	Apex It, UK; Savvy, Ottawa, etc	Approx. Rs. 7000/ = (for hospital use), Rs. 2000/ = (for own use)
Non-invasive air bubble level detector	Bubble sensors		Ashoke engineering, India; Moog Inc., America, etc	Approx. Rs. 7000/ = to 11,000/ =
Finger heart rate and pulse oximeter	Heart rate sensor	BH1792GLC	OMRON, Japan; Meditive, India; etc	Approx. Rs.1000/ =
Evaluation system for integrated biopotential and bioimpedance AFE	ECG (Electrocardiogram) sensor, R-to-R peak sensor	MAX30001 EV Kit	Digi key, America; AVNET, Japan; RochesterElectronics, China, etc	Approx. Rs.8500/ =

7 Conclusion

Use of wireless sensor networks (WSN) in healthcare applications is growing at a fast pace. Numerous applications such as heart rate monitor, blood pressure monitor, and endoscopic capsule are already in use. Low-cost technologies are expected to aid in the delivery of services while simultaneously reducing costs. Wireless sensor networks have the potential to assist in meeting some of these future challenges, by simplifying the use of medical equipment, advancing at-home medical care, and displaying health and wellness information to both providers and patients. The design of better wireless medical sensor networks seems to be a good solution to part of the problem.

In India almost every sensor is available with an affordable budget, so with the help of some investors, we can implement wireless sensor networked based medical equipment in healthcare. In this paper, we found different available sensors in India and made a survey of their use in healthcare.

References

1. Ko J, Lu C, Srivastava MB, Stankovic JA, Terzis A, Welsh M (2010) Wireless sensor networks for healthcare. IEEE Proceed 98(11):1947–1960
2. Alemdar H, Ersoy C (2010) Wireless sensor networks for healthcare: a survey. Comput Netw 54(15):2688–2710
3. Movassaghi S, Lipman J, Smith D, Jamalipur AJ (2014) Wireless body area networks: a survey. IEEE Commun Survey Tutorials 16(3)
4. Chen M, Gonzalez S, Vasilakhos A, Cao H, Leung VCM (2011) Body area networks: a survey. In Springer: Mobile Netw Appl 16(2):171–193
5. Salayma M, Al-dubai A, Romdhani I (2017) Wireless body area network (WBAN): A survey on reliability, fault tolerance and technologies coexistance, ACM Comput Survey 50(1), article 3
6. Ullah S, Higgins H, Braem B, Latre B, Blondia C, Moerman I, Saleem S, Rahman Z, Kwak KS (2012) A comprehesnsive survey of wireless body area network. In Springer: J Med Syst 36(2):1065−1094. https://doi.org/10.1007/s10916-010-9571-3
7. Xing J, Zhu Y (2009) A survey on body area network. In: 5th international conference on wireless communications, networking and mobile computing (WiCom'09), Sept, pp 1–4
8. Astrin AW, Li HB, Kohno R (2009) Standardization for body area network. ICICE Trans Commun 2:366–372
9. Patel M, Wang J (2010) Applications, challenges and prospective in emerging body area networking technologies. Wireless Commun 17:80–88
10. Khan JY, Yuce MR, Bulger G, Harding B (2012) Wireless body area network design and performance evaluation. J Med Syst 36(3):1441–1457
11. Sahndhu MM, Javaid N, Jamil M, Khan ZA, Imran M, Ilahi M, Khan MA (2015) Modeling mobility and psychological stress based human postural changes in wireless body area networks. In Elsevier Comput Human Behav 51(Part B):1042–1053
12. Kumar P, Lee H-J (2011) Security issues in healthcare applications using wireless medical sensors networks: a survey. Sensors 12(1)
13. Paradis L, Han Q (2007) A survey of fault management in wireless sensor networks. Springer J Netw Syst Manage 15(2):171–190. https://doi.org/10.1007/s10922-007-9062-0
14. Salayma M, Al-dubai A, Romdhani I (2017) Wireless body area network (WBAN): A survey on reliability, fault tolerance and technologies coexistence. ACM Comput Survey 50(1), article 3

Contingency Analysis and Ranking for a 30 Bus System to Maintain Its Stability and Reliability

Parnab Saha, Suman Moitra, Bishaljit Paul, and Chandan Kumar Chanda

Abstract Contingency analysis is performed to maintain secured operation of a power system. In this analysis technique, probabilistic prediction is made for the outage of each transmission line components and to take necessary actions to regain the system security, reliability and stability. Contingencies can be broadly classified into two types, viz. simple type where outage of a single component takes place or complex type where outage of multiple components has taken place. The process of determining the severity of these contingency, contingency sorting is done, i.e. to calculate the performance indices (PI) for each case and sort them according to their performances. The main objective of this paper is to carry out contingency analysis of a 30-Bus system by calculating the performance indices for single generator with single loadline outage, double generator outage and double loadline outage with help of Newton–Raphson load flow contingency analysis on MATLAB environment and ranking of the contingency according to their respective performance indices (PI) . This type of ranking provides an effective mean to rank the different abnormal cases according to their severity and take necessary actions beforehand to prevent total system failure

Keywords Contingency analysis · Load flow · Performance index

1 Introduction

The power system being dynamic in nature will be prone to failure due to contingency even in the presence of system operators. Thus, it is of prime importance for them to perform a beforehand security assessment of the system and to take remedial actions

P. Saha (✉) · B. Paul
Narula Institute of Technology, Agarpara, Kolkata, India

S. Moitra
Elitte College of Engineering, Sodepur, Kolkata, India

C. K. Chanda
Indian Institute of Engineering Science and Technology, Shibpur, Howrah, India

M. Mitra et al. (eds.), *Computational Advancement in Communication, Circuits and Systems*, Lecture Notes in Electrical Engineering 786,
https://doi.org/10.1007/978-981-16-4035-3_15

for any possible contingency in the system [1]. This type of security analysis is vividly classified into two type's, viz. N-k, i.e. 'k' number of elements in the system has failed simultaneously ($k = 1, 2\ldots$), and N-k-k (1, 2...), i.e. k elements leads to contingency of another k elements outage ultimately to total blackout of the system. This entire security analysis is important for providing reliable power supply to the consumers at any load level [2].

Generally, all N-1 contingency analysis is done by NERC standards [3], where only one single element has gone out, but since our system has got complex with added elements, N-2 contingency analysis has got higher priority. An increase of number of transmission line component outage occurring simultaneously the total number of contingency cases is given by [4]

$$N_k = \frac{N!}{k!(N-k)!} \tag{1}$$

where N is the total number of elements (either transmission lines or generators) and k is the number of element on outage. Since with increase of 'k' the number of contingency increases, it becomes complex and time-consuming to perform all these cases. There are different types of algorithms to sort out the possible contingency as presented by Silverstein et al. [5], Youcef et al. [6] and Caglar et al. [7]. The contingency selection algorithm used in this paper was first proposed by Ejebe et al. [8] while upgradation was done later by Al-Shaalan et al. [9] and Kengkla et al. [10]. In this paper, the algorithm proposed by Ejebe et al. was used on a 30-Bus system, and the N-2 contingency analysis was computed relating to outage of generators and transmission lines. Resultant performance indexes (P.I) for each case are computed and ranking of the cases are performed. Graphical representations of P.I. for each case are also shown.

2 Theoretical Background

The basic task of power system security analysis performed by system operator is to monitor the real-time data of the power system by using SCADA system coupled with telemetry devices. The entire security assessment operation is summarized as flowchart in Fig. 1. After the basic operation is completed, the next most important task is to perform contingency analysis where the simulation is carried out for entire system for outage of equipment (N-1) at a time followed by two equipment's outage (N-2) at a time to find all sort of outages and rank them according to their credibility so as to get a clear indication of unscheduled events which might come in future and remedial actions to take at that moment.

As stated previously in Eq. 1 that for k numbers of components, outages occurs; simultaneously, N_k number of contingency cases shall rise, among which all cases will not have same impact factor on the system, thus the cases needed to be sorted in

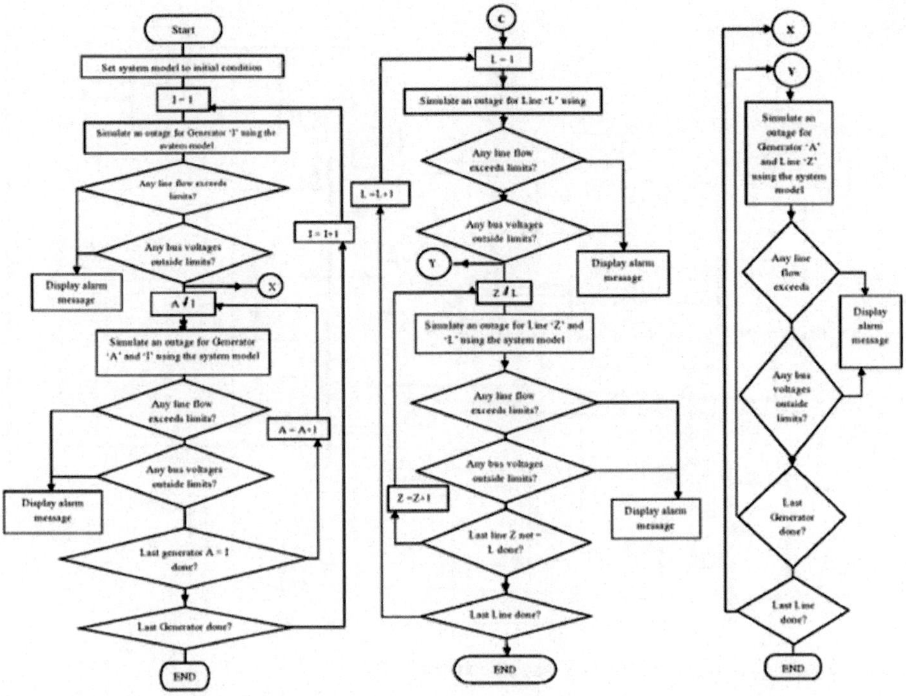

Fig. 1 Flowchart representation of total contingency analysis

descending order with the case of highest impact factor at top and reduces as we go below. This sorting covers up both the economic as well as the technical requirements for system.

3 Mathematical Formulation

In this form of sorting, each contingency case, i.e. for this paper, both N-1 and N-2 is obtained and sorted according to the impact of the occurrence of fault. A parameter known as performance index (PI) is used to calculate the impact of the outage of fault on entire system. The expression to obtain the PI is stated in Eq. 2.

$$PI_l = \sum_{i=1}^{N} \frac{w_i}{2a} \left(\frac{P_{\text{flow}}}{P_i} \right)^{2a} \tag{2}$$

where w_i is the weightage factor, N is total number of lines in the system, P_{flow} is the flow after occurrence of outage, P_i is the flow before any fault took place and a is the order of PI (exponent of function generally considered as 1).

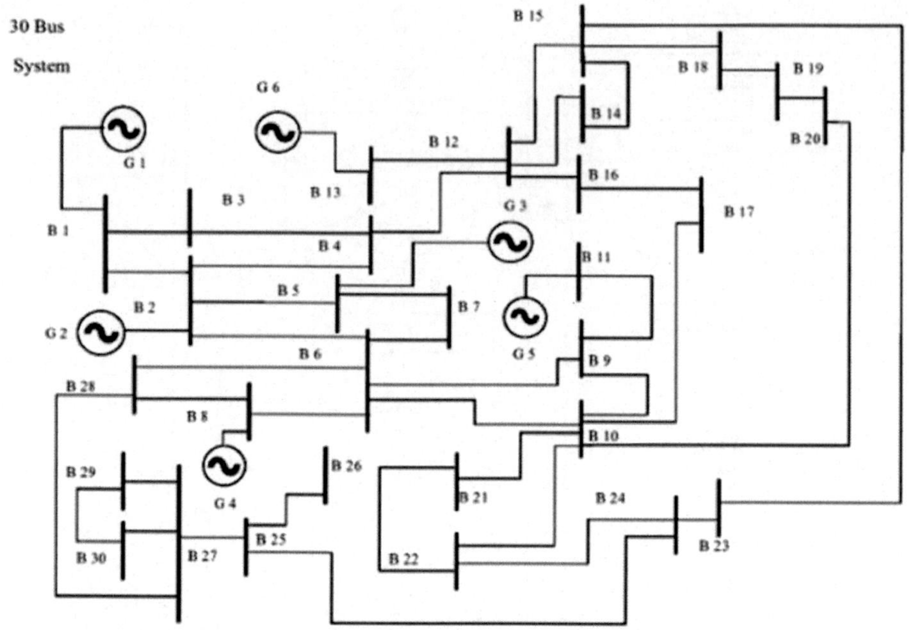

Fig. 2 Single line diagram of 30 bus test case system

4 Case Study

In this present paper, a 30-bus system is taken into consideration; N-2 contingency analysis is operated on it. Figure 2 represents the single-line diagram of entire system. In this test system, there are total of 30 buses, 41 lines and 6 generators.

5 Results

The total N-2 contingency analysis of this work is summarized into three cases, i.e. double components at outage simultaneously comprises of single generator along with single-line outage (Case A), double generator outage (Case B) and double-line outage (Case C).

Case A: In this portion, the simultaneous outage of both the loadline and the generator are considered for N-2 contingency selection. Total 6 generators and 41 loadlines are analysed for their simultaneous outages using Newton–Raphson method, which gives multiple outcomes of 244 cases. N-2 contingency ranking yields absolute cases of conditional outage, considering symmetrical fault situation with steady-state analysis of active power flow. The six most severe cases are shown in Table 1 and Fig. 3 as per their contingency ranking.

Table 1 Ranking of contingency cases according to data

Most severe N-2 contingency cases and their rankings			
Cases	Findings no:	Rank	Total P. I. at N-2 contingency selection
Case A: One generator and one loadline on outage (N-2)	N-2 L-G 15,13	1	174.308
	N-2 L-G 24,11	2	161.451
	N-2 L-G 24,8	3	161.116
	N-2 L-G 24,13	4	161.031
	N-2 L-G 24,5	5	159.472
	N-2 L-G 24,2	6	158.941
Case B: two generators on outage (N-2)	N-2 G-G11,13	1	31.146
	N-2 G-G 8,13	2	30.596
	N-2 G-G 8,11	3	30.578
	N-2 G-G 5,11	4	28.732
	N-2 G-G 5,13	5	28.725
	N-2 G-G 5,8	6	28.206
Case C: two loadline on outage (N-2)	N-2, L-L 1,24	1	3962.202
	N-2, L-L 2,24	2	2096.946
	N-2 L-L 4,24	3	2095.41
	N-2 L-L 5,24	4	2094.948
	N-2 L-L 9,24	5	2092.685
	N-2, L-L 1,35	6	1656.361

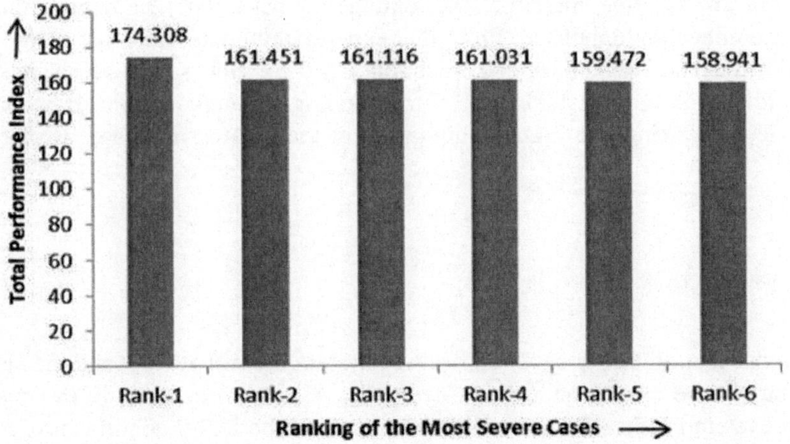

Fig. 3 The most severe cases with their corresponding total P.I. values

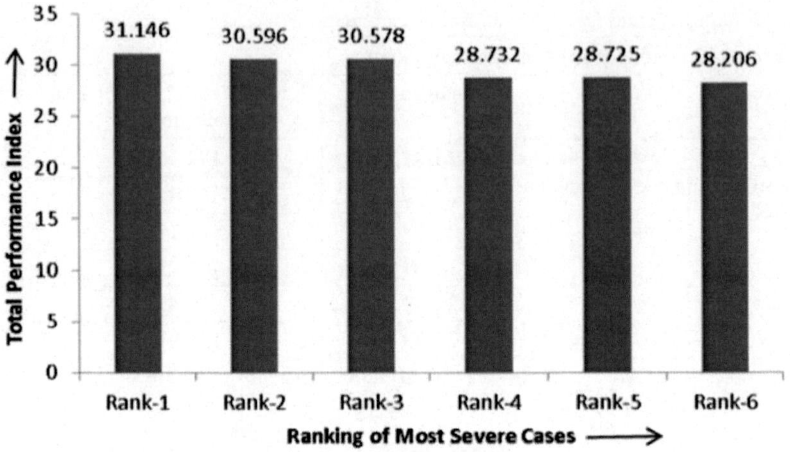

Fig. 4 The most severe cases with their corresponding total P.I. values

Case B: In this case, simultaneous outages of any two generators of the 30-bus system are considered for (N-2) contingency selection purpose. Total 6 generators are there in the test system with their different ratings. Total 15 experiments performed for every N-2 contingency conditions. The data referring to all the experimentation are recorded and tabulated and ranked. The six most severe cases are shown in Table 1 and Fig. 4 as per their contingency ranking.

Case C: In this part, two different loadlines are considered for outages. There are total 820 different cases of experiments performed to analyse each and every outage of any two loadlines for N-2 contingency selection. The data of the total P.I. is examined and tabulated. From that experimental data, the most severe cases are prioritized and ranked for N-2 selection. Only six most severe cases are shown in the following Table 1 with their corresponding findings number. Based on the contingency ranking, the 2-D column graph of the most severe cases is plotted in Fig. 5.

6 Conclusion

It is found that all faulty cases of a power system are not severe, and the severity of a faulty case can be understood by its PI. As a test scenario, in this paper, a 30-bus system is taken into consideration on which both N-2 contingency analysis is performed. Ranking of all the cases depending upon its PI is performed which helps to understand most severe faulty conditions thus proper precautions can be taken beforehand. This entire result will help the system operator to mitigate the problems at an early stage thus helps to maintain reliability and stability of system. As a conclusion, it is found that all faulty cases of a power system are not severe, and

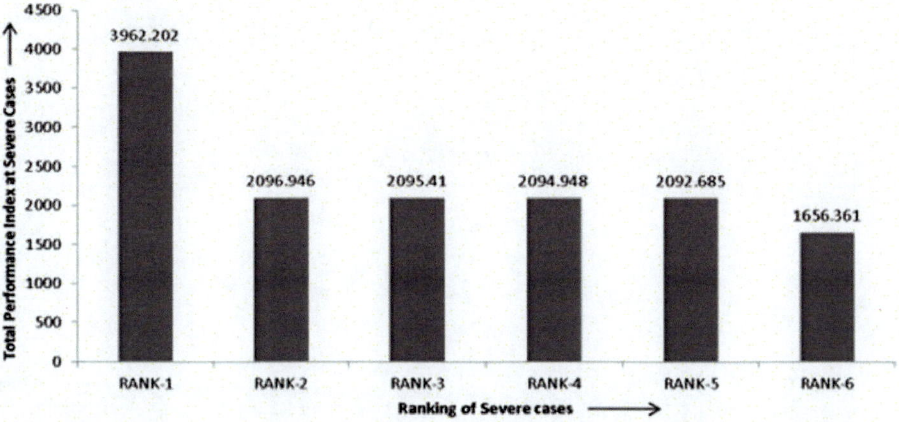

Fig. 5 The most severe cases with their corresponding total P.I. values

the severity of a faulty case can be understood by its PI. However, there are several other factors like topography, wind spend, span length which can change the severity of a line's impact; further, dynamic and transient state conditions are needed to be considered along with N-1–1 contingency cases which can be performed as a future work on similar or higher number of buses.

References

1. Zaag N, Restrepo JF, Banakar H, Galiana FD (2007) Analysis of contingencies leading to islanding and cascading outages. In: Power Tech, 2007 IEEE Lausanne 1–5, July, pp 63–67
2. Billinton R, Allan RN (1984) Reliability evaluation of power systems. In: Pitman advanced publishing programh, London
3. NERC, Transmission Planning (TPL) Standard. [Online]. http://www.nerc.com/files/TPL-003-0.pdf
4. Davis CM, Overbye TJ (2011) Multiple element contingency screening. In: IEEE Trans Power Syst 26(3)
5. Silvestein BL, Porter DM (1992) Contingency ranking for bulk system reliability criteria. IEEE Trans Power Syst 7(3):956–964
6. Bereriche Y, Ait Kadi D (2015) Reliability analysis of supply chain for contingency operations. In: International conference on industrial engineering and systems management (IESM), IEEE Conference Publications, pp 509–519
7. Caglar R, Ozdemir A (1999) Composite electric power system adequacy evaluation via transmission losses based contingency selection algorithm. In: International conference on electric power engineering, power technology, Budapest, 29 August–2 September
8. Ejebe GC, Wollenberg BF (1979) Automatic contingency selection. In: IEEE transactions on power apparatus and systems. PAS-98(1):92–104
9. Al- AM (2018) Contingency selection and ranking for composite power system reliability evaluation. J King Saud Univer—Eng Sci. https://doi.org/10.1016/j.jksues.2018.11.004
10. Kengkla N, Hooncharoen N (2018) Risk-based N-2 Contingency ranking in transmission system using operational condition. 978–1–5386–4956–5/18/$31.00 ©2018 IEEE

Swarm Intelligence-Based Reactive Power Constrained Generator and Load Scheduling in Smart Grid with Renewable Energy Sources

Sudhangshu Sarkar, Sandip Chanda, and Abhinandan De

Abstract This paper presents an effective methodology and algorithm to optimal schedule of the generators and load dispatch centres for economic integration of renewable energy sources adhering reactive power constraints of the network. The algorithm proposes optimal utilization of all the resources available in the smart grid and considers meticulously the important aspects of power system like voltage profile improvement and the power loss to develop an efficient optimal schedule for both generators and load dispatch centres. The methodology proposes the development of suitable indices which is helpful in monitoring and maintaining the system performance even in the worst possible contingency. For proper management of voltage profile in smart grid, the algorithm within the methodology identifies the weak generators and the load dispatch centres (LDCs) causing problems and alters their generation and active and reactive load schedule for desired operating conditions. It is also of great concern that how the intermittent energy sources will be scheduled with the available thermal power plants. The developed methodology maximises the conversion of renewable power into active power so that those sources are competent enough with the thermal power plants at their availability. The proposed algorithm also sheds the peak load demand in order to attain an optimal generation schedule of these sources. Simulations were carried out in the updated IEEE 30 bus system, and the results were found to be quite encouraging.

Keywords Reactive power constraints · Smart grid · Load dispatch centres · Intermittent energy sources · Load curtailment

S. Sarkar (✉) · S. Chanda
Department of Electrical Engineering, Narula Instituite of Technology, Kolkata 700129, India
e-mail: sudhangshu.sarkar@nit.ac.in

A. De
Department of Electrical Engineering, Indian Institute of Engineering, Science & Technology, Shibpur, Howrah 711103, India

M. Mitra et al. (eds.), *Computational Advancement in Communication, Circuits and Systems*, Lecture Notes in Electrical Engineering 786,
https://doi.org/10.1007/978-981-16-4035-3_16

1 Introduction

In present scenario, it can be seen that all the data of the power system is processing through smart grid. The generation, transmission, distribution companies and consumers are getting all the information about the whole power system through smart grid. Integration of potential for a solar wind-based power system has been investigated by the power system engineers through the development of an optimal distribution search algorithm, in combination with parametric physical simulation of hourly power mismatches [1, 2]. Use of energy storage may also help an improved stability of power and give a chance to form a sustainable power supply to build the power grid more consistent particularly with great proportion of grid-connected renewable sources [3]. Flexibility is the facility in power system to act in response to changes in power demand and generation. Integrating huge shares of changeable renewable energy sources, in particular wind and solar, can lead to a strong augment of flexibility that needs for the balancing system. Accommodating high shares of wind and solar power is advantageous for power system [4, 5], but maintenance of standard operating conditions is also an important issue of future smart grid. The intrinsic unpredictability of bulk-size solar generation introduces consequence challenges to the smart grid energy management. Precise forecasting of solar power is significant to secure financial process of the smart grid [6]. Solar PV and wind power are becoming increasingly cut-throat against predictable power generation due to decrease of unit speculation costs and increasing capacity for active and reactive power demand. Efforts are therefore needed to improve and optimize the grid integration of wind and photovoltaic energy in terms of LDC demands and to maintain long-term reliance on electricity supply throughout the conversion process [7]. Researchers on the power system are continually searching for a unique methodology and algorithm that will use the quasi-potential of renewable energy sources in the current power system infrastructure. References [8, 9] suggested methodologies for maximizing the operation of the power grid subscribing to a scrupulous supply of renewable energy. Reference [10] is a unique algorithm for the management of system parameters in recurring alternating sources. The simulation has optimized the scheme method by constructing a power hub, while implementing the parallel process that would require capable infrastructure which is not feasible, and the model will not work in the current power system environment. The recent strategies for load curbing were discussed in [11, 12] while some new strategies were implemented in [13]. Some authors have been planning to use approximate dynamic programming methods to increase smart grid astuteness [14–16]. Most of these strategies are optimal load curtailment (OLC) systems in an operational form which could not guarantee a reliable supply of usable model conditions for the consumer. Considering this review, it is possible to feel the need for an algorithm that will be able to ensure an effective parametric state with the aim of minimizing the price of electricity with optimal load catering without infringing reactive power generation and constraints. Methodology should be backed by an algorithm that not only combines demand response and generation functionality but also provides voltage profile reactive power responsiveness and load curtailment. The current cost forecasting models with projected active and reactive power generation constraints in [17–20] are optimistic in nature, built to suggest solutions under specific operational constraints, and these models are only fruitful for a single target area. Therefore, they do not find the competitive price balance as described in [21–25] and so cannot provide an insight into the stability of the power market. The goal of the research presented in this paper was to extend a state-space-based model of a power system equipped with smart metering service that not only predicts the optimal price but also to curtail the same without sacrificing social benefit, power market stability and successful reactive power constraint, thus ensuring prolific operating conditions. In choosing a stochastic optimization method such as particle swarm optimization (PSO), the convex existence of the solution algorithm with non-linear working plane was persuasive. An optimal power flow (OPF) paradigm [26–30] was adopted to evaluate the solution obtained. The simulations were conducted in the IEEE 30-bus system, and the results obtained were very promising. So, here a method is projected for reduction of unnecessary load shedding and voltage collapse using PSO-based algorithm. Systems with all thermal generating plants and the integrated system of wind, solar and thermal have been coupled. It has been focus that with the proposed methodology and algorithm, the integrated system offers same reliability and effectiveness as the system with thermal power plant monopoly. The planned method is implemented in MATLAB functioning platform, and the performance is veteran in IEEE 30-bus bench mark system. Also, the comparison shows that, the proposed method ensures voltage profile consistency. In view of the result obtained, it is felt that there is a scope of both enhancement and economizing the optimal solution for generation and voltage stability with renewable energy sources with the proposed methodology and algorithm.

2 Description of the Methodology and Algorithm

See Fig. 1.

2.1 Development of Voltage Profile Index

The results of conventional economic load dispatch may not be optimal as it does not consider reactive power flow constraints, demand response and intermittency of the renewable energy sources. For optimisation of the whole operation, a monitoring index is required which will calculate the quality of the solution for optimisation and also penalize the objective function so that the solution not only economically but also qualitatively viable.

Solar PV cell generates active and reactive powers. The wind power is induction generation type; it also generates active and reactive power. To enhance the system

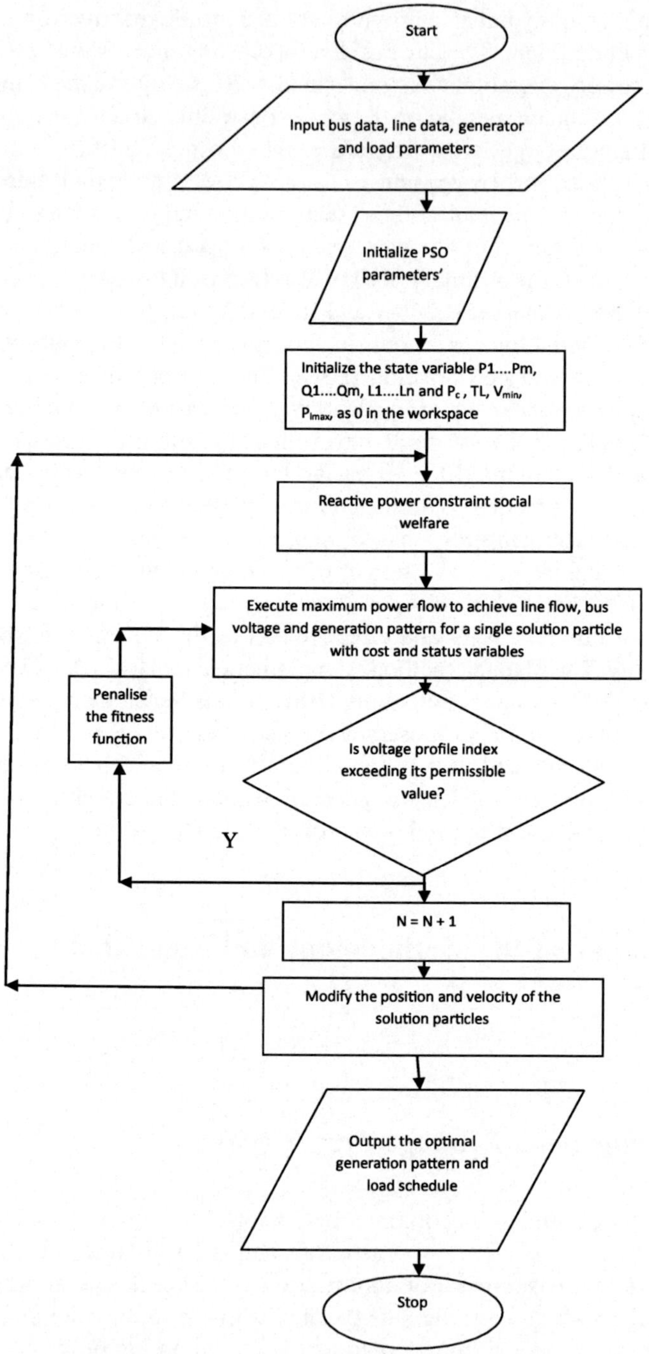

Fig. 1 Flowchart of the proposed algorithm

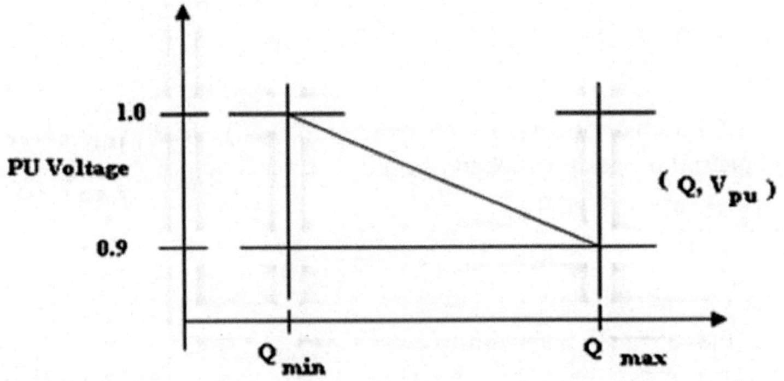

Fig. 2 Voltage profile improvement index curve

efficiency in both solar and wind power, the active power flow needs to be increased and reactive power flow needs to be controlled. For controlling the intermittency of generation reactive power, the voltage profile index has been developed.

A straight line approximation can be made for small variation of reactive power (Q) demand with respect to the voltage profile. Needless to maintain the voltage sensitivity of reactive power (Q) demand is higher than active power (P) demand of the load. A flexible band of variation can be assumed and the solution can be driven towards the same by suitable penalty to the objective function. This is of utmost importance in case of power grids harnessing renewable energy sources. The objective is to convert maximum power from the nature (solar, wind, etc.) into active power for a substantially suitable voltage profile. In the proposed voltage profile index, for maximum reactive power demand ($Q_{\max}$), the minimum allowed per unit voltage is 0.9. For minimum reactive power demand ($Q_{\min}$), however, the per unit voltage should settle at unity power factor. Thus, the straight line representation between reactive powers (Q) demands and per unit voltage has been proposed in Fig. 2. The voltage band proposed can be altered in situation by the system operator in accordance with the requirement. The voltage profile index is thus developed as under:

$$V_{\text{pu}} = mQ + C$$

$$V_{\text{pu}} = \frac{(1-0.9)Q}{Q_{\max} - Q_{\min}} + C$$

$$\text{when, } Q = 0,$$

$$\text{then, } V_{\text{pu}} = 0,$$

$$V_{\text{pu}} = \frac{0.1Q}{Q_{\max} - Q_{\min}} + C$$

$$\text{then, } C = \frac{Q_{\max} - Q_{\min}}{0.1Q}$$

Let, the maximum total capacity for reactive power generation = $Q_{\max}$
And the overall minimum capacity to produce reactive power = $Q_{\min}$

$$\text{Then, } Q_C = \frac{V_{\text{pu}}\left(Q_{\max} - Q_{\min}\right)}{0.1Q} \tag{1}$$

where Q_C is the reactive power limit.

For any value of reactive power demand (Q), reactive power limit (Q_C) is calculated in the proposed algorithm and the same appears in the objective function with a penalty. This drags the solution towards the voltage profile with the proposed band of variation.

$$P = P_{\max} - \frac{P_{\max}(S - S_{\max})}{S_{\max} - S_{\min}} \tag{2}$$

where $S_{\max}$ = Sum of maximum generation possible—Sum of minimum demand cap.

$P_{\max}$ = Sum of maximum demand limit-Sum of minimum demand limit.

Where S represents surplus generation and P represents the corresponding curtailment upper limit or ISO load curtailment limit. The method proposed is using Eq. (2) determining the maximum permissible load curtailment at any given price equilibrium or ISO curtailment limit.

2.2 *Modelling of the Proposed Algorithm and Its Mathematical Formulation*

The objective function has been model as social welfare optimization as depicted, but along with active power constraints in this work, reactive power constraint has also been taken into account. Both this constraints by using curtailment limit and the proposed voltage profile index are of utmost importance for smart grid environment, where more penetration of renewable energy sources is expected and the same has to be supported by higher reliability of power and sustainability of the grid. The objective function and the constraints of the proposed algorithm are as under.

$$C(d) = \sum_{j=1}^{n_d} r_j P_{dj}^2 + s_j P_{dj} = \text{ demand cost benefit function} \tag{3}$$

Herer_j and s_j is the cost co-efficient of the demand function.

$$C(\text{glv}) = \sum_{i=1}^{n_g} \left(a_i P_{g_i}^2 + b_i P_{gi} + c_i\right) + P_c.C_1 + T_L.C_2 + V_{\min}.C_3 + P_{l\max}.C_4 + Q_c.C_5 \quad (4)$$

P_{dj} = load catered to jth consumer, 'C(glv)' specifies the cost of generation with line limit violation.

$a_i, b_i, c_i, \alpha_i, \beta_i, \gamma_i$ = Cost coefficient of the ith generator, P_{gi} = real power generation of the ith generator, Q_{gi} = reactive power generation of the ith generator, P_c = real power load curtailed, T_L = Total line loss, $V_{\min}$ = minimum bus voltage, $P_{l\max}$ = maximum line flow, Q_c = reactive power limit, and C_1,C_2, C_3, C_4, C_5 = ISO limit violation penalties for P_c, T_L, $V_{\min}$, $P_{l\max}$, Q_c, respectively.

The objective function of the proposed optimization algorithm is therefore appropriate

$$\text{Maximize}\left(\sum_{j=1}^{n_d} r_j P_{dj}^2 + s_j P_{dj}\right) - \left(\sum_{i=1}^{n_g} a_i P_{g_i}^2 + b_i P_{gi} + c_i\right) + P_{c.}C_1 + T_L.C_2 + V_{\min}.C_3 + P_{l\max}.C_4 + Q_c.C_5 \quad (5)$$

Inequality or generator output constraints

$$P_{gi}^{\min} \le P_{gi} \le P_{gi}^{\max} \quad (6)$$

$$Q_{gi}^{\min} \le Q_{gi} \le Q_{gi}^{\max} \quad (7)$$

where P_{gi}, Q_{gi} = Active and reactive power of generator i respectively, $P_{gi}^{\min}$, $Q_{gi}^{\min}$ = Lower limit of active and reactive power of the generators, $P_{gi}^{\max}$, $Q_{gi}^{\max}$ = Upper limit of active and reactive power of the generators

$$\text{Voltage constraint}: V_i^{\min} \le V_i \le V_i^{\max} \quad (8)$$

$V_i^{\min}$, $V_i^{\max}$ are upper and lower limit of voltage V_i

$$\text{Transmission constraint}: P_{ij\max} \le P_{ij} \le P_{ij\min} \quad (9)$$

$P_{ij\max}$, $P_{ij\min}$ are the maximum and minimum line flow limits of limits of P_{ij}

$$\text{Curtailment constraints}: 0 < P_c < P_{ci} \quad (10)$$

P_{ci} is the maximum permissible curtailment limit set by ISO set by Eq. (2) Take a look at it.

$$\text{Constraint of Transmission Loss}: 0 < T_L < T_{LM} \quad (11)$$

T_{LM} is the maximum power loss value in MW.

For depicting the relevance of the proposed methodology in context of the system operation under smart grid, it has been compared with the traditional cost optimization technique. The objective function of the cost optimization technique is as below:

$$C = \sum_{i=1}^{n_g} a_i P_{g_i}^2 + b_i P_{gi} + c_i \quad (12)$$

The constraints to the objective function for this traditional optimization, however, remains same as (1–12).

2.3 Application of the Proposed IEEE 30 Bus System Methodology

To show the efficiency of the proposed approach and algorithm, the same has been checked with respect to conventional optimization techniques in the worst possible condition of reactive power loading and contingency. For reactive power loading, bus no 30 has been chosen for the same is the weakest bus of the system [31, 32]. The contingency criterion has been adopted from [3].

2.4 Comparison Between Traditional Optimization Technique and the Proposed Methodology in All Thermal System

The results have been depicted in Table 1 where it has been found that for the same generation cost, the maximum line flow has been reduced considerably in case of the proposed algorithm. This reduction not only will enhance the stability margins of the line but also will enable the system operator to plan for additional power transmission even in case of the most in the adverted condition of the system. Reduction of transmission loss is another benefit of the proposed methodology and algorithm as depicted in Table 1 also demonstrated the intelligent load management capability of the proposed algorithm through a significant reduction in the load curtailment.

This result was quite encouraging in testing the proposed algorithm in the integrated renewable energy environment of the grid. In the grid model proposed in [3] along with the generation cost, the active power demand response has been considered. For this consideration, a solar power plant and a wind power plant has been incorporated in bus no 8 and 11, respectively. This work has been extended towards reactive power control as the same is necessary to convert more of the energy of the nature from solar or wind sources to active power (Table 2).

Table 1 Comparison between traditional optimization technique and the proposed methodology in all thermal system

MVAR demand on bus 30	Traditional optimization with all thermal				Propose optimization with all thermal			
	Gen cost ($/MWH)	Line loss in MW	Load curtailment	Max line flow in MW	Gen cost ($/MWH)	Line loss in MW	Load curtailment	Max line flow in MW
1.9	1095.20	5.9904	61.75	67.5641	1136.50	6.8737	51.4239	71.684
2.9	1107.60	5.9988	60.31	65.9714	1140.30	6.6952	50.30	71.8846
3.9	1090.30	5.9987	63.21	67.4059	1090.10	5.9976	62.94	69.7764
4.9	1136.00	7.0901	51.91	71.5896	1110.30	5.9932	59.85	65.6561
5.9	1099.50	5.9977	62.65	65.1546	1137.00	6.7822	51.43	71.5861
6.9	1138.50	7.0232	51.23	71.5603	1085.00	5.9595	65.71	64.754
7.9	1138.50	7.0526	50.99	71.9971	1124.60	6.9209	53.90	71.8654
8.9	1133.10	6.974	52.31	71.5603	1134.20	6.9846	52.24	71.9633
9.9	1139.40	7.3159	51.08	71.7759	1139.90	7.3657	50.92	71.9971
10.9	1136.90	7.4548	51.56	71.8903	1135.70	7.2995	52.15	71.6454
11.9	1134.70	7.3816	52.13	71.8461	1137.80	7.3767	51.85	71.6615
12.9	1137.80	7.4047	51.79	71.8529	1110.90	7.3747	56.76	71.8202
13.9	1133.50	7.6111	52.82	71.9913	1136.90	7.6857	51.90	71.7919
14.9	1141.00	7.8432	51.06	71.8725	1138.80	7.7536	51.68	71.7927
15.9	1138.80	7.8775	51.93	71.6039	1136.70	7.8478	51.96	71.8701
16.9	1129.70	8.2257	53.54	71.9469	1139.80	8.0914	51.79	71.7952
17.9	1139.40	8.3728	51.69	71.9313	1134.90	8.109	53.07	71.7008
18.9	1139.10	8.3472	52.35	71.9363	1094.40	8.226	59.63	71.9125
19.9	1140.30	8.5227	52.11	71.9226	1136.30	8.3542	52.94	71.8468

Table 2 Comparison between traditional optimization technique and the proposed methodology in all thermal system under the contingency

Traditional optimization with all thermal					Proposed optimization with all thermal		
Sl. no	Contingency	Max line flow	Load curtailment	Total power loss	Max line flow	Load curtailment	Total power loss
1	1–2,	71.979	95.05	5.7399	171.9412	10.05	21.2359
2	1–2, 2–4	71.3366	95.62	5.6942	180.3259	5.01	23.6135
3	1–2, 2–4, 2–5	71.504	98.44	8.5147	184.625	7.03	30.9389

2.5 Comparison Between Traditional and Proposed Optimization Algorithm on Integrated Renewable Energy Systems

This table shows the ability of the proposed algorithm also in managing transmission line flow congestion, transmission line loss and load curtailment in the integrated thermal-wind-solar system. This result will help the system operator to add benefit to the wind and solar power plants for the promotion of their use (Tables 3 and 4).

Table 3 Comparison between traditional and proposed optimization algorithm on integrated renewable energy systems

MVAR demand on bus 30	Traditional optimization with all thermal			Propose optimization with renewable energy sources		
	Line loss in MW	Load curtailment	Max line flow in MW	Line loss in MW	Load curtailment	Max line flow in MW
1.9	5.9904	61.75	67.5641	13.3003	7.11	128.2467
2.9	5.9988	60.31	65.9714	12.8205	9.34	128.6246
3.9	5.9987	63.21	67.4059	13.3507	11.95	126.6722
4.9	7.0901	51.91	71.5896	15.0048	9.26	140.8051
5.9	5.9977	62.65	65.1546	14.1639	7.80	139.9852
6.9	7.0232	51.23	71.5603	15.1848	10.18	142.9006
7.9	7.0526	50.99	71.9971	13.1252	7.83	124.2216
8.9	6.974	52.31	71.5603	13.2339	11.16	129.9922
9.9	7.3159	51.08	71.7759	15.1031	7.21	145.8502
10.9	7.4548	51.56	71.8903	17.0028	4.33	155.7942
11.9	7.3816	52.13	71.8461	14.1659	6.14	129.1432
12.9	7.4047	51.79	71.8529	13.6121	12.69	126.6343
13.9	7.6111	52.82	71.9913	13.9511	7.61	126.0074
14.9	7.8432	51.06	71.8725	14.1543	12.03	128.6409
15.9	7.8775	51.93	71.6039	14.6038	8.35	129.2767
16.9	8.2257	53.54	71.9469	13.7839	14.08	130.3281
17.9	8.3728	51.69	71.9313	14.8915	7.69	128.5636
18.9	8.3472	52.35	71.9363	17.8672	5.24	152.7965
19.9	8.5227	52.11	71.9226	15.0654	9.24	128.4822

Table 4 Comparison between traditional optimization technique and the proposed methodology in all thermal system under the contingency

Traditional optimization with all thermal					Propose optimization with non-renewable energy sources		
Sl. no	Contingency	Max line flow	Load curtailment	Total power loss	Max line flow	Load curtailment	Total power loss
1	1–2,	71.979	95.05	5.7399	197.5253	10.26	27.0287
2	1–2, 2–4	71.3366	95.62	5.6942	193.9713	13.92	26.4956
3	1–2, 2–4, 2–5	71.504	98.44	8.5147	181.8862	25.97	29.7702

2.6 Comparison Between All Thermal System and Integrated Renewable Energy System with the Proposed Optimization Technique

The efficacy of the proposed algorithm has been substantiated in Table 5 where both the thermal power system and integrated power system have been simulated with the proposed methodology, and algorithm in the entire dimension shows improvement in result. Thus, the proposed algorithm will assist the system operator in benefiting the generator companies, the distribution companies and the transmission companies with additional profit in terms of more reliability and less congestion (Table 6).

The proposed algorithm pivots around the minimization of the reactive power flow (in the previous work in [3], the active power flow was effectively controlled by curtailment index) and thus to improve the quality of solution of the optimization for all the three conditions considered in Tables 1, 3 and 5 and comparison in voltage profile especially the minimum bus voltage has been shown in Fig. 3 The criteria of the voltage profile index developed in Eq. 1 was to ratify a band of voltage variation even in worst possible loading and contingencies. Figure 4 depicts that the same has been negotiated with a minimum peak shaving of reactive power demand. This result also ensures a higher degree of reliability of the proposed algorithm as depicted in Fig. 1.

3 Conclusion

One of the major concerns in the present smart grid is the effective and efficient incorporation of renewable energy sources. For the reduction of costs involved in power transmission, the potential of local renewable energy use needs to be maximized. This also helps in producing clean energy. In this pursue the work presented in this paper has focused on utilizing the smart grid resources that can be converted to active power, controlling the flow of reactive power so that their use can be effective in terms

Table 5 Comparison between all thermal system and integrated renewable energy system with the proposed optimization technique

MVAR demand on bus 30	Proposed optimization with all thermal			Proposed optimization with integrated renewable sources		
	Line loss in MW	Load curtailment	Max line flow in MW	Line loss in MW	Load curtailment	Max line flow in MW
1.9	6.8737	51.4239	71.684	13.3003	7.11	128.2467
2.9	6.6952	50.30	71.8846	12.8205	9.34	128.6246
3.9	5.9976	62.94	69.7764	13.3507	11.95	126.6722
4.9	5.9932	59.85	65.6561	15.0048	9.26	140.8051
5.9	6.7822	51.43	71.5861	14.1639	7.80	139.9852
6.9	5.9595	65.71	64.754	15.1848	10.18	142.9006
7.9	6.9209	53.90	71.8654	13.1252	7.83	124.2216
8.9	6.9846	52.24	71.9633	13.2339	11.16	129.9922
9.9	7.3657	50.92	71.9971	15.1031	7.21	145.8502
10.9	7.2995	52.15	71.6454	17.0028	4.33	155.7942
11.9	7.3767	51.85	71.6615	14.1659	6.14	129.1432
12.9	7.3747	56.76	71.8202	13.6121	12.69	126.6343
13.9	7.6857	51.90	71.7919	13.9511	7.61	126.0074
14.9	7.7536	51.68	71.7927	14.1543	12.03	128.6409
15.9	7.8478	51.96	71.8701	14.6038	8.35	129.2767
16.9	8.0914	51.79	71.7952	13.7839	14.08	130.3281
17.9	8.109	53.07	71.7008	14.8915	7.69	128.5636
18.9	8.226	59.63	71.9125	17.8672	5.24	152.7965
19.9	8.3542	52.94	71.8468	15.0654	9.24	128.4822

Table 6 Comparison between all thermal system and integrated non-renewable energy system with proposed optimization technique under the contingency

Proposed optimization with all thermal					Propose optimization with integrated non-renewable energy sources		
Sl. No	Contingency	Max line flow	Load curtailment	Total power loss	Max line flow	Load curtailment	Total power loss
1	1–2	171.9412	10.05	21.2359	197.5253	10.26	27.0287
2	1–2, 2–4	180.3259	5.01	23.6135	193.9713	13.92	26.4956
3	1–2, 2–4, 2–5	184.625	7.03	30.9389	181.8862	25.97	29.7702

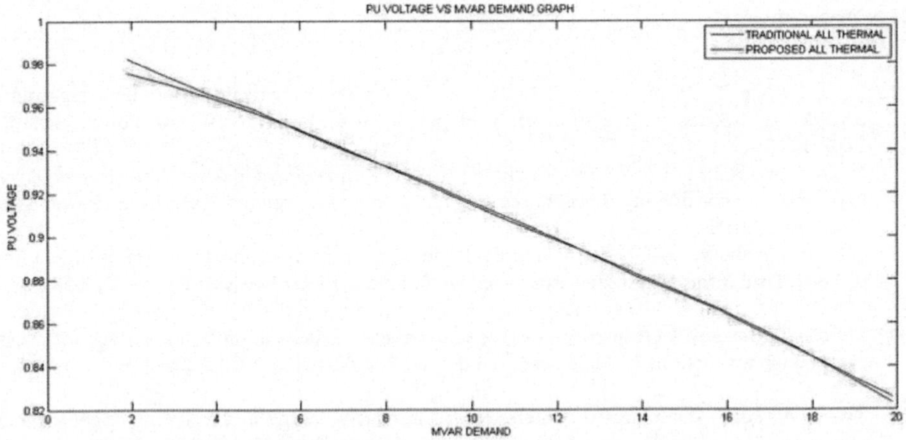

Fig. 3 Comparison of voltage profile index between traditional all thermal system and proposed all thermal

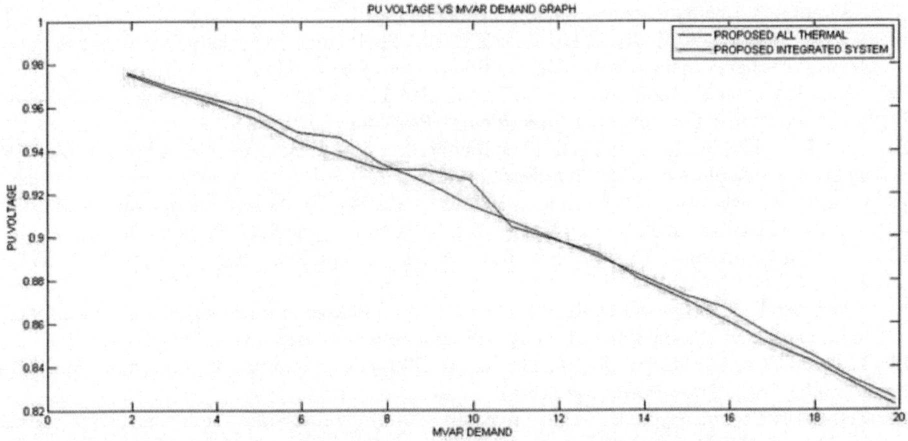

Fig. 4 Comparison of voltage profile index between proposed all thermal system and proposed integrated system

of cost. It has been demonstrated by computer simulation that with the assistance of the proposed methodology and algorithm even in worse possible reactive power loading and transmission line contingency the system can maintain a prospective improvement in presence of renewable energy sources with a minimum increment of cost of generation. The improvement of operating conditions in integrated renewable energy systems may compensate for the escalation of the costing of generation. The methodology and algorithm developed can be useful for both generation and load management in smart grids.

References

1. Wang R, Xie Y, Zhang H, Li C, Li W, Terzija V (2017) Dynamic power flow algorithm considering frequency regulation of wind power generators. IET Renew Power Generat 11(8):1218–1225
2. Xu X, Zhang H, Li C, Liu Y, Li W, Terzija V (2017) Optimization of the event-driven emergency load-shedding considering transient security and stability constraints. IEEE Trans Power Syst 32(4):2581–2592
3. Chanda S, Abhinandan DE (2014) A multi-objective solution algorithm for optimum utilization of Smart Grid infrastructure towards social welfare. Int J Electr Power & Energy Syst 58:307–318
4. Mareda T, Gaudard L, Romerio F (2017) A parametric genetic algorithm approach to assess complementary options of large scale wind-solar coupling. IEEE/CAA J Automatica Sinica 4(2)
5. Zahedi A (2013) International conference on alternative energy in developing countries and emerging economics, Elsevier, ScienceDirect, Energy Procedia 52:642–650
6. Huber M, Dimkova D, Hamacher T (2014) Integration of wind and solar power in Europe: assessment of flexibility requirements. Energy 69:236–246
7. Ramos A, Chatzopoulou MA, Guarracino I, Freeman J, Markides CN Hybrid photovoltaic-thermal solar systems for combined heating, cooling and power provision in the urban environment. Energy Convers Manage J Homepage
8. Wan C, Zhao J, Song Y, Xu Z, Lin J, Hu Z (2015) Photovoltic and solar power forecasting for smart grid energy management. CSEE J Power Energy Syst 1(4)
9. Weida S, Kumar S, Madlener R (2016) Financial viability of grid-connected solar PV and wind power systems in Germany. Sci Direct Energy Procedia 106:35–45
10. Yann R, Seddik B, Stephene P (2011) Optimal power flow management for grid-connected PV systems with batteries. IEEE Trans Sustain Energy 2(3)
11. Chandel M, Agrawal GD, Mathur S, Mathur A (2014) Techno-economic analysis of solar photovoltaic power plant for garment zone of Jaipur city. Case Stud Thermal Eng 2:1–7
12. Geidl M, Anderson G (2007) Optimal power flow of multiple energy carriers. IEEE Trans Power Syst
13. Sortomme E, El-Sharkawi Optimal power flow for a system of micro-grids with controllable loads and battery storage. In: Power system conference and exposition. PSCE '09. P. 1–5
14. Majid O, Gerd B, Hasan S, Mohammad S (2004) Market-based transmission expansion planning. IEEE Trans Power Syst 19(4)
15. Fernandes Thelma SP, Lenzi JR, Mikilita MA (2008) Load shedding strategies using optimal load flow with relaxation of restrictions. IEEE Trans Power Syst 23(2)
16. Momoh JA (2006) Optimal methods for power system operation and management. In: Proceeding IEEE PES power system conference exposition, pp 179–187
17. Werbos PJ (2009) Putting more brain-like intelligence into the electric power grid: what we need and how to do it. Proc Int Joint Conf Neural Netw 3356–3359
18. Venayagamoorthy G (2011) Dynamic, stochastic, computational, and scalable technologies for smart grids. IEEE Comput Intell Mag 6(3):22–35
19. Kåresen Kjetil F, Egil H (2002) A joint state-space model for electricity spot and futures prices. Norwegian Computing Center
20. Ofuji K, Kanemoto S (2007) Price forecasting of japan electric power exchange using time-varying AR model. In: Proceedings of the 14th international conference on intelligent system applications to power systems, ISAP 2007, November 4–8
21. Ramteen S, Walter S (2009) Evaluating the impacts of real-time pricing on the usage of wind generation. IEEE Trans Power Syst 24(2)
22. Venayagamoorty GK, Sharma RK, Gautam PK, Ahmadi A (2016) Dynamic Energy Management System for a Smart Microgrid. IEEE Trans Neural Netw Learning Syst 27(8)

23. Naga Lathish TLV, Naresh Babu AV, Sivanagaraju S (2013) A solution to optimal power flow problem using artificial bee colony algorithm incorporating FACTS device, IOSR J Electr Electron Eng (IOSR-JEEE) 7(6):26–31. e-ISSN: 2278-1676, p-ISSN: 2320-3331
24. Cuffe P, Keane A Voltage responsive distribution networks: comparing autonomous and centralized solutions. IEEE Trans Power Syst 30(5):2234–2242
25. Wang L, Thi MSN (2014) Stability enhancement of large-scale integration of wind, solar, and marine-current power generation fed to an sg-based power system through an LCC-HVDC Link. IEEE Trans Sustain Energy 5(1):160–170
26. Shi Q, Cui H, Li F, Liu Y, Ju W, Sun Y (2017) A hybrid dynamic demand control strategy for power system frequency regulation. CSEE J Power Energy Syst 3(2):176–185
27. Lu SY, Wang L, Lo TM, Prokhorov AV (2015) Integration of wind power and wave power generation systems using a DC Microgrid. IEEE Trans Ind Appl 51(4):2753–2761
28. Orfanos GA, Georgilakis PS, Hatziargyriou ND (2013) Transmission expansion planning of systems with increasing wind power integration. IEEE Trans Power Syst 28(2):1355–1362
29. Faias S, de Sousa J, Reis FS, Castro R (2012) Assessment and optimization of wind energy integration into the power systems: application to the portuguese system. IEEE Trans Sustain Energy 3(4):627–635
30. Vallee F, Lobry J, Deblecker O (2008) System reliability assessment method for wind power integration. IEEE Trans Power Syst 23(3):1288–1297
31. Sandip C, Abhinandan D (2011) Improvement of the economic aspect of power network congestion management by swarm intelligence based multi-objective algorithm. Int J Eng Sci Technol 3(5)
32. Chanda S, Bose D, Chakraborty S, De A (2012) A self healing methodology for economic integration of wind farms to modern power networks. Int J Electr Electron Eng (IJEEE) 1:Iss-3. ISSN (PRINT): 2231–5284

Impact of Atmospheric Features for COVID-19 Prediction

Debpuja Dhar, Tamasree Biswas, and Mousumi Saha

Abstract In the context of contagious diseases, recent advances in experimental techniques have not only generated a dramatic increase in the amount and diversity of data but also an ever increasing and complexifying molecular biology with context to meteorological parameters. To combat this probable inefficiency, decision tree-based methods have emerged to be one of the finest data ensembles showcasing excellent accuracy in combining interpretability. For past infectious diseases like influenza and severe acute respiratory syndrome (SARS), etc., direct correlations were spotted with respect to meteorological parameters including temperature, humidity and air pollution among others. The present study targets to explore the association between COVID-19 mortality rates and weather parameters for which the daily death numbers of corona virus disease 2019, meteorological parameters and air pollution data from March 28, 2020 to April 22, 2020 of different states of India were collected. To explore the effect of the minimum temperature, maximum temperature, minimum humidity and maximum humidity on the infection count of COVID-19, the gradient boosting model (GBM) has been implemented thereby achieving optimal performance by tuning its parameters. For prediction of active cases in Maharashtra, the GBM results stand at its best accuracy of R^2 as 0.95. For the prediction of recovered cases of COVID-19 in Rajasthan and Kerala, R^2 equals 0.98. The present study explores the correlation between atmospheric parameters and transmission rate of COVID-19 in different states of India thereby predicting the active and recovered cases of COVID-19 and establishing an efficient tree-based machine learning approach to explore the effect of temperature and humidity on the transmission rate of the said disease.

Keywords Gradient boosting model · SARS · Meteorological factors

D. Dhar · T. Biswas (✉) · M. Saha
Narula Institute of Technology, Agarpara, India
e-mail: tamasree.biswas@nit.ac.in

M. Mitra et al. (eds.), *Computational Advancement in Communication, Circuits and Systems*, Lecture Notes in Electrical Engineering 786,
https://doi.org/10.1007/978-981-16-4035-3_17

1 Introduction

A novel coronavirus disease (COVID-19) epidemic was reported in Wuhan, China, in December 2019. Caused by severe acute respiratory syndrome coronavirus-2 (SARS-CoV-2), the COVID-19 has been affirmed to have human-to-human transmission which raised high attention worldwide thereby resulting in a still consistent lockdown as of October 11, 2020. With 50.9 million positive cases and 1.26 million deaths worldwide, the COVID-19 has not only blemished the socio-economic structure but also disturbed the mortality and fatality rate to an obscure number. In some studies, the local seafood market of Wuhan, China has been identified to be a source of COVID-19 consequently indicating its transmission from bat to human, though the exact transmission route is still under research. In most of the infected cases, it spread due to humans to human transmission through the respiratory tract by human contact in social gatherings, meetings, link between patients and healthcare workers, etc. The transmission of coronavirus was found to have multiple routes besides surface because of its presence in blood and fecal swabs and in the air around the hospital area.

Upon genetic analysis of COVID-19, the virus was found to be similar to severe acute respiratory syndrome coronavirus (SARS-CoV). But the transmission of COVID-19 occurs during the prodromal period when those infected are mildly ill and therefore are capable to carry on usual activities. This not only contributes to the spread of the infection but also establishes its domain as lethal from that of SARS- CoV. In 2003, the outbreak of SARS in Guangdong gradually disappeared with the increase in temperature and was practically over within a few months. It has been documented that the temperature and its variations might have affected the SARS outbreak as the risk of influenza incidence was reviewed to have significantly increased with low daily temperature and low relative humidity, resulting to a positive significant association for diurnal temperature range (DTR). Though it is hard to find any study based on the impact of the atmospheric factors including temperature and humidity on the transmission rate of the COVID-19 in India, some recent studies have established the effect of climatic conditions on the same.

In order to reduce the workload of doctors and health workers, machine learning-based approaches have been widely implemented in the health care system for disease diagnosis, monitoring and prediction which has recently been proven to be extremely useful in the identification of the COVID-19. The prediction of COVID-19 in India utilizing the atmospheric factors using machine learning is yet to be explored. Therefore, to signify and exhibit the relationship among the atmospheric factors (humidity and temperature) and the daily proliferation rate of COVID-19 in different states of India, we contrivance a gradient boosting machine (GBM) learning approach.

The present study aims to dwell into the following and results into fruition:

1. explore the correlation between atmospheric parameters and transmission rate of COVID-19 in different states of India.
2. showcase predictability between the active and recovered cases of COVID-19.

3. to investigate the effect of temperature and humidity on the transmission rate of COVID-19, we plan to establish an efficient tree-based machine learning approach.

2 Collection and Compilation of COVID-19 and Atmospheric Conditions' Data

The meteorological parameters including minimum temperature, maximum temperature, minimum humidity and maximum humidity of all states of India were collected from the Indian Central Pollution Control Board (CPCB) and Indian Meteorological Department (IMD) on a daily basis from March 28, 2020 to April 22, 2020.

In addition to an open source, the COVID-19 data were collected from the Ministry of Health and Family Welfare, Government of India which includes daily new infection cases, active cases (accumulated total cases up to previous days-recovered cases-deceased cases), recovered cases and mortality till the end date. Finally, the meteorological parameters and COVID-19 information were combined for further analysis thereby imputing the missing values of the said parameters by replacing with the median values. The variations of imputed values of minimum and maximum temperature, and minimum and maximum humidity are shown in Figs. 1 and 2, respectively.

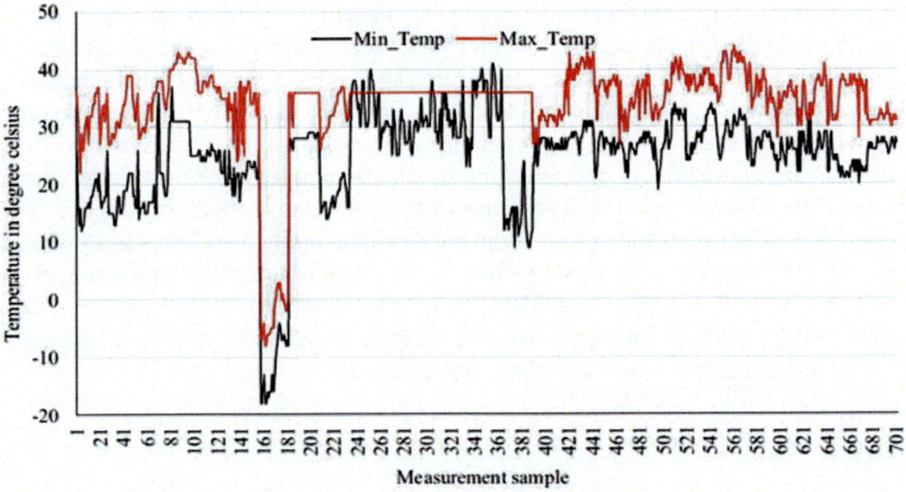

Fig. 1 The variation in the minimum and the maximum temperature

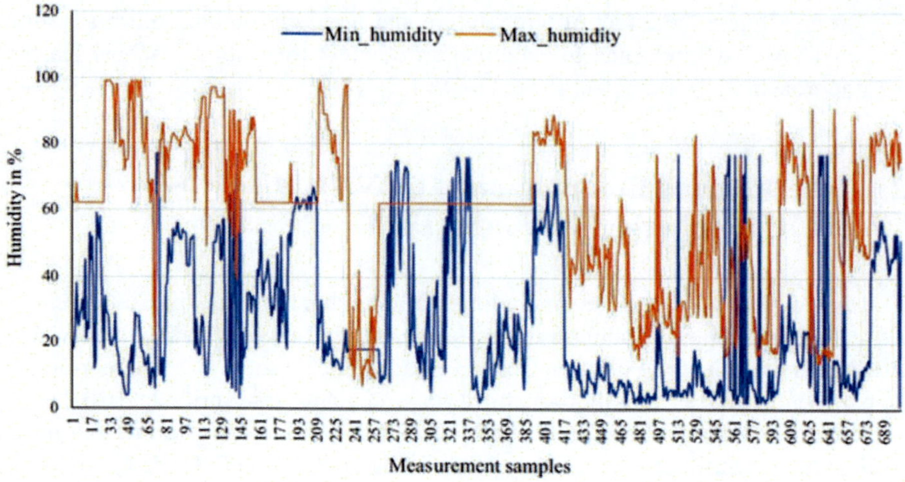

Fig. 2 The variation in the minimum and the maximum humidity

3 Gradient Boosting Machine (GBM) Learning Approach

Gradient boosting is used for regression and classification problems. It is a machine learning technique which produces a prediction model accumulating weaker prediction models like decision trees. Though regression gradient boosting algorithms were subsequently developed by Jerome H. Friedman, gradient boosting generally discards all weaker predictions and picks the stronger one thereby making it immensely useful for decision trees of a fixed size as base learners. It comparatively analyzes every successor using the structure score, gain calculation and increasingly refined approximations to build a set of the optimally satisfying structure of the tree. By invoking an additional classifier, the prediction performance of a GBM can be boosted, thereby optimizing the accuracy of the tree without affecting its speed. This version of GBM is capable to handle the big data with optimal accuracy and provides an easily distributable and parallelizable feature with an effortless environment for model tuning and selection. The GBM model was optimized for the number of trees $(K) = 1, 2, 3, \ldots 50$, the maximum number of trees $K = 50$ being selected arbitrarily.

Algorithm Table for Gradient Boosting Machine:

1. Initialization $f_{ko} = 0$, k = 1, 2, , K
2. Repeating m = 1 to M
   a. $P_k(x) = \frac{e^{f_k(x)}}{\sum_{l=1}^{K} e^{f_l(x)}}$ for all k = 1, 2, ……., K
   b. Repeating k=1 to K
      i. Calculation $r_{ikm} = y_{ik} - p_k(x_i)$, i = 1, 2,… …, N
      ii. Fitting regression tree to the targets r_{ikm}, i = 1, 2, … …, N
      iii. Calculation $r_{ikm} = \frac{K-1}{K}\left(\frac{\sum_{x_i \in R_{jkm}}(r_{ikm})}{\sum_{x_i \in R_{jkm}} |r_{ikm}|(1-|r_{ikm}|)}\right)$, where j = 1, 2.
      iv. Upgrading $f_{km}(x) = f_{k,m-1} + \sum_{j=1}^{J_m} r_{ikm}\, I(x \in R_{jkm})$
3. Result $f_k(x) = f_{km}(x)$, where k = 1,2, … …, K

In the GBM model, the average of the maximum and minimum temperature and the average of the maximum and minimum humidity among the four atmospheric parameters: minimum temperature, maximum temperature, minimum humidity and maximum humidity, were used as input to predict the number of recovered and active cases in all states and also for some individual states of India thereby reducing the computational complexity.

4 Analysis of Outcome

The GBM model was tuned with the number of trees, learning rate, number of folds, and distribution functions (Gaussian, Tweedie, Huber, Laplace, Poisson, Quan-tile and Gamma) with the performance of the GBM being evaluated on the basis of mean square error (MSE), root mean square error (RMSE), mean average error (MAE), mean residual deviance (MRD) and coefficient of determination (R^2).

To demonstrate the predicted results of GBM for the active and recovered cases of COVID-19, we take the data of Delhi and probe into the matter (Fig. 3; Table 1).

5 Discussions

Tree-based algorithms are considered to be one of the best and mostly used supervised learning methods that are capable of empowering predictive models with high accuracy, stability and ease of interpretation. We selected the GBM approach for predicting the transmission rate of COVID-19 as it proved exceptionally efficient when the different atmospheric factors were used for the prediction of the pandemic. On the basis of the performance measures, the various distribution functions in the

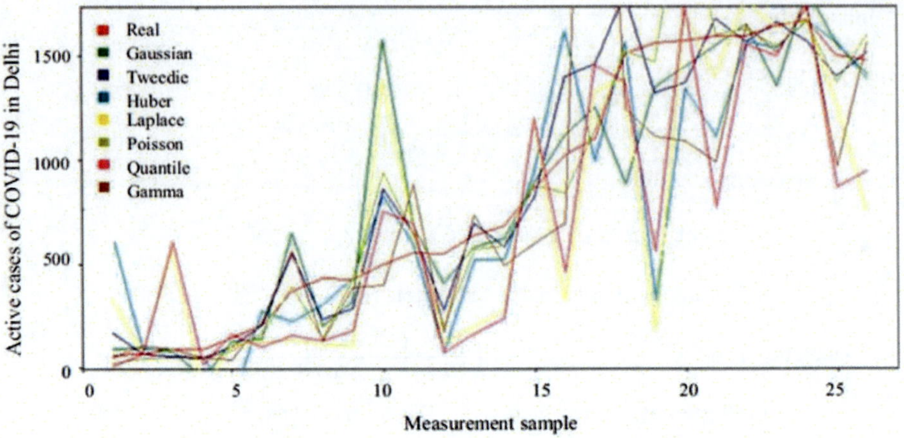

Fig. 3 Performance of GBM in the prediction of active cases of COVID-19 of Delhi, India

Table 1 Performance of GBM in the prediction of recovered and active cases in Delhi, India

Performance parameter	Distribution						
	Gamma	Quintile	Poisson	Laplace	Huber	Tweedie	Gauss-ian
Recovered cases							
MSE	9500.73	25,376.32	533.36	23,019.17	16,594.05	1129.63	2476.93
RMSE	97.47	159.30	23.09	151.72	128.82	33.61	49.77
MAE	48.19	66.21	13.45	58.38	49.86	18.69	32.87
MRD	9.14	33.11	−821.58	58.38	3717.56	1.39	2476.93
R^2	0.70	0.21	0.98	0.28	0.48	0.96	0.92
Active cases							
MSE	132,171.40	124,920.40	55,572.56	139,083.10	58,240.21	57,000.78	73,851.49
RMSE	363.55	353.44	235.74	372.94	241.33	238.75	271.76
MAE	243.64	230.35	117.85	201.10	151.60	139.85	160.40
MRD	15.39	115.18	−10,037.81	201.10	31,323.97	2.65	73,851.49
R^2	0.62	0.64	0.84	0.60	0.83	0.84	0.79

Note R^2 (coefficient of determination), MSE (mean square error), RMSE (root mean square error), MAE (mean average error), and MRD (mean residual deviance)

training of the GBM has shown the ensuing trends: Poisson > Gaussian > Tweedie > Gamma > Huber > Quantile. The COVID- 19 dataset used in the present analysis has high dimensionality and relationality in nature which might be the reason for the deprived performance of a deep neural network method. Though tree-based learning has better performance in the dense numerical datasets, in the analysis of categorical feature dataset deep learning has better performance.

6 Conclusion

The present study investigates the effects of temperature, DTR and humidity on the daily mortality of COVID-19 and establishes an association between the number of COVID-19 cases and meteorological parameters in different states of India. From the exampled study of the capital city Delhi, the maximum achieved values of the R^2 and minimum values of the errors of the GBM suggest a certain association between the transmission rates of COVID-19 and the climatic factors. That a correlation between the cases of COVID-19 and atmospheric parameters can be captured by this GBM model was proven by the outcome of the experiment. In case of further research, the additional meteorological parameters would be included, and the handling of deep neural network would have to be improved so that the subservience of the transmission rate of COVID-19 on atmospheric conditions can be better understood.

Bibliography

1. Xu X, Chen P, Wang J, Feng J, Zhou H, Li X, Zhong W, Hao P (2020) Evolution of the novel coronavirus from the ongoing Wuhan outbreak and modeling of its spike protein for risk of human transmission. Sci China Life Sci 63(3):457–460. https://doi.org/10.1007/s11427-020-1637-5
2. Lin K, Fong DYT, Zhu B, Karlberg J (2006) Environmental factors on the SARS epidemic: air temperature, passage of time and multiplicative effect of hospital infection. Epidemiology & Infection 134(2):223–230. https://doi.org/10.1017/S0950268805005054
3. Ministry of Health and Family Welfare Government of India (2020) COVI-19 India. https://www.mohfw.gov.in. Accessed 2 May 2020
4. Ma Y, Zhao Y, Liu J, He X, Wang B, Fu S, Yan J, Niu J, Zhou J, Luo B (2020) Effect of temperature variation and humidity on the death of COVID-19 in Wuhan. China Sci Total Environ 724:138226.https://doi.org/10.1016/j.scitotenv.2020.138226
5. Srivastav, LK, Jha SK (2020) A gradient boosting machine learning approach in modeling the impact of temperature and humidity on the transmission rate of COVID-19 in India. https://doi.org/10.1007/s10489-020-01997-6
6. Tosepu R, Gunawan J, Effendy DS, Lestari H, Bahar H, Asfian P (2020) Correlation between weather and Covid-19 pandemic in Jakarta. Indonesia Sci Total Environ 725:138436. https://doi.org/10.1016/j.scitotenv.2020.138436
7. Caliskan A, Yuksel ME, Badem H, Basturk A (2018) Performance improvement of deep neural network classifiers by a simple training strategy. Eng Appl Artif Intell 67:14–23. https://doi.org/10.1016/j.engappai.2017.09.002
8. World Health Organization (2020) Coronavirus disease (COVID-19) Pandemic. https://www.who.int/emergencies/diseases/novel-coronavirus-2019

New Sorting Algorithm—RevWay Sort

Swarna Saha, Soumyadip Sarkar, Rituparna Patra, and Subhasree Bhattacharjee

Abstract Sorting provides a method of rearrangement of elements in ascending or descending order. In this paper, we are introducing a new sorting algorithm called RevWay sort in which the two consecutive numbers are compared from left and then from right. This process is repeated $((n/2) + 1)$ times, where n is the total number of elements. We have compared running time of the proposed algorithm with other sorting algorithms. We run the algorithm starting from 10,000 to 50,000 elements. We found that the newly proposed RevWay sort yields lesser running time compared to bubble and selection sort. For 10,000 elements, RevWay sort takes 203.636 ms, whereas bubble sort takes 364.8243ms and selection sort consumes 337.5543 ms.

Keywords Complexity · RevWay · Algorithm

1 Introduction

Data is raw fact, and sorting is the processing that converts the data into information. The data structure is vital for the organization of data, and it includes every single detail [1]. Right from the definition to processing all come under data structure, and sorting comprises a very important part of data structure. Searching is a vital job for any data processing. One of the most efficient searching algorithms works on the principle of divide and conquers which in turn can only be applied if the data is sorted [2]. Sorting usually uses the concept of compare and sort or radix sort [3].

Compare sort is to compare and check which input value is greater or smaller and then depending upon that we swap the values so that the value can reach their appropriate position in a particular data structure. Radix sort checks and compares depending upon the position of the digits individually in a number or it is more appropriate to say that it uses counting sort as the base. Repeated calls to the counting subroutine are done to get the required processed data [4].

S. Saha · S. Sarkar · R. Patra · S. Bhattacharjee (✉)
Narula Institute of Technology, Kolkata, India

M. Mitra et al. (eds.), *Computational Advancement in Communication, Circuits and Systems*, Lecture Notes in Electrical Engineering 786,
https://doi.org/10.1007/978-981-16-4035-3_18

If we talk about a real-life instance, fingerprint scanning is one application that uses sorting in its core coding part. Many other examples are there that uses sorting [5]. Library management to school management all require sorting. When we talk about any database management system (DBMS), the definition says data insertion–deletion and retrieval are the major processes, among which all the processes need a prior sorted data set. Mathematical calculations and computations also need sorting.

Complexity defines the efficiency of a code, and the unit for that measurement is big O in terms of n where n is the number of basic operations performed in the code. Few examples of values are $O(n \log n)$, $O(n)$ and $O(n^2)$ [6]. Among these values, $O(n)$ is the lowest and best [7, 8].

2 Working Procedure and Algorithm of RevWay Sort

2.1 *Procedure*

The process of this sort is given as follows:

1. In pass 1:
   i. $A[0]$ and $A[1]$ are compared and swap if required and also at the same time $A[n-1]$ and $A[n-2]$ are compared and swap if required.
   ii. $A[1]$ and $A[2]$ are compared and swap if required and also at the same time $A[n-2]$ and $A[n-3]$ are compared and swap if required.
   iii. $A[3]$ and $A[4]$ are compared and swap if required and also at the same time $A[n-3]$ and $A[n-4]$ are compared and swap if required.
2. In pass 2, we have to continue the same method as well as the method which is described in 1.

Let us explain the procedure with an example (Fig. 1).

2.2 *Algorithm*

STEP 1: INITIALIZE
STEP 2: Repeat STEPS 3,4 and 5 for k=0 till k<(n/2)+1
STEP 3: Repeat STEPS 4 and 5 for i=0 and j=n-1 till i<n/2
STEP 4: if(arr[i]>arr[i+1]) swap arr[i] and arr[i+1]
STEP 5: if(arr[j]<arr[j-1]) swap arr[j] and arr[j-1]
STEP 6: END

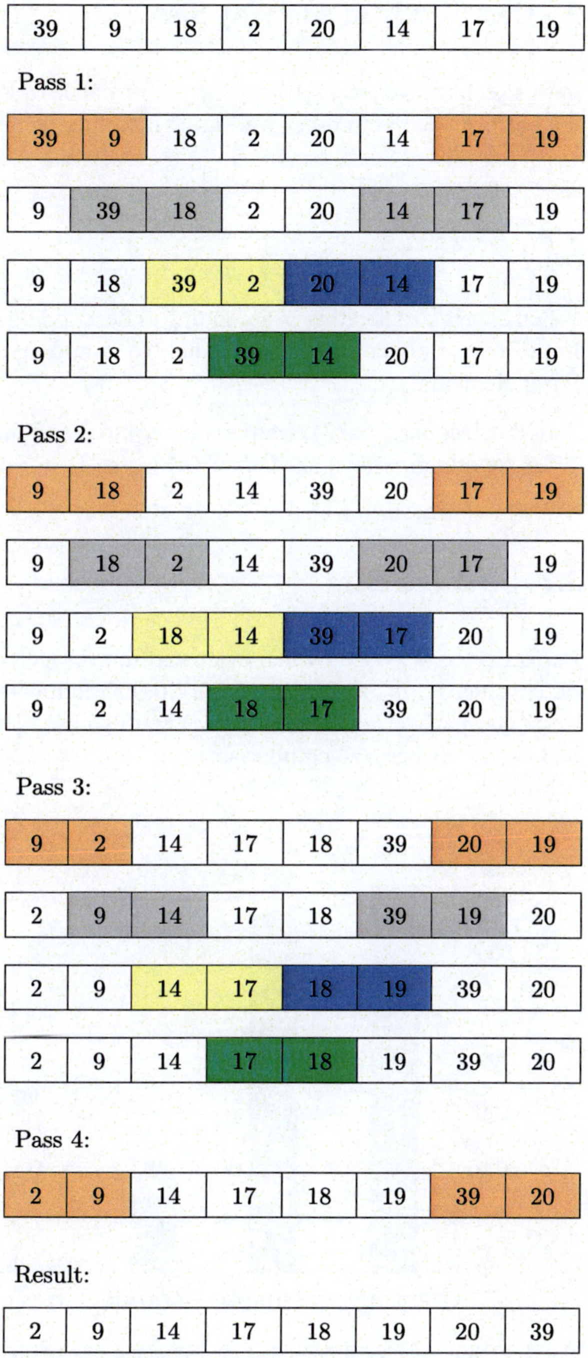

Fig. 1 Example of RevWay sort

3 Case Study of RevWay Sort

In this section, we are analyzing the performance of RevWay Sorting and comparing the performance with other well-known sorting techniques.

3.1 Best Case

When data is sorted either in ascending or descending order, then best case will occur. In this case, only comparison is done. No swapping is required.
For n elements:

In the first pass, $(n/2)$ comparison required and the array is sorted.
So, total comparison needed $=T(n) = (n/2) = O(n)$.

3.2 Average Case

Performance of sorting when evaluated in average case, then random data needs to be considered. In random case, for 10,000 elements time required is 203.636 ms. Total comparison required is $O(n^2)$ in this case. There are total $(n/2) + 1$ passes and each pass has $n/2$ comparisons.

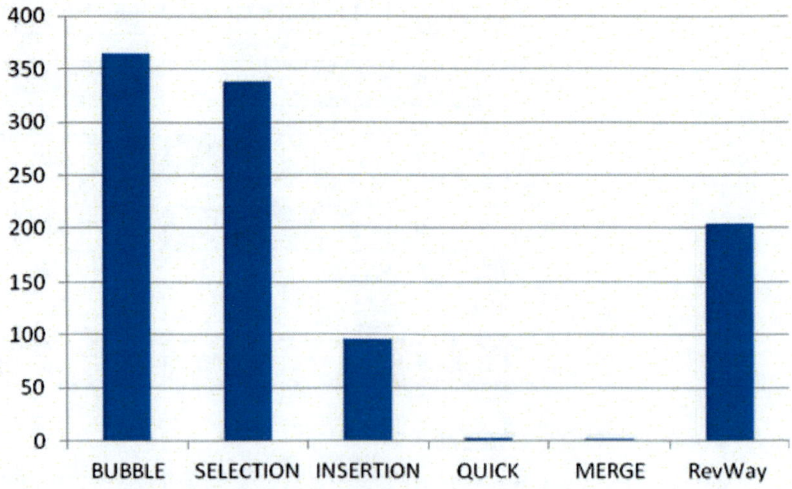

Fig. 2 Comparison of different sorting techniques for 10,000 elements

3.3 Comparison with Well-Known Sorting Algorithm

See Figs. 2, 3, 4, 5, and 6; Tables 1, 2, 3, 4, and 5.

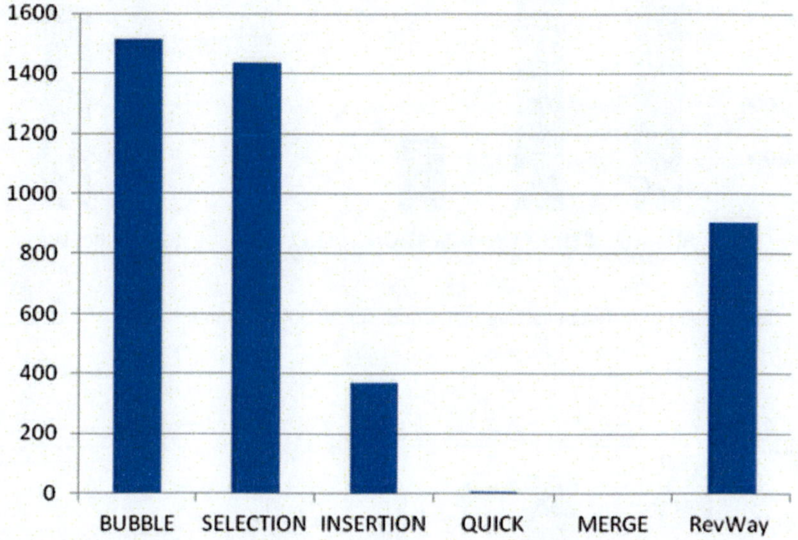

Fig. 3 Comparison of different sorting techniques for 20,000 elements

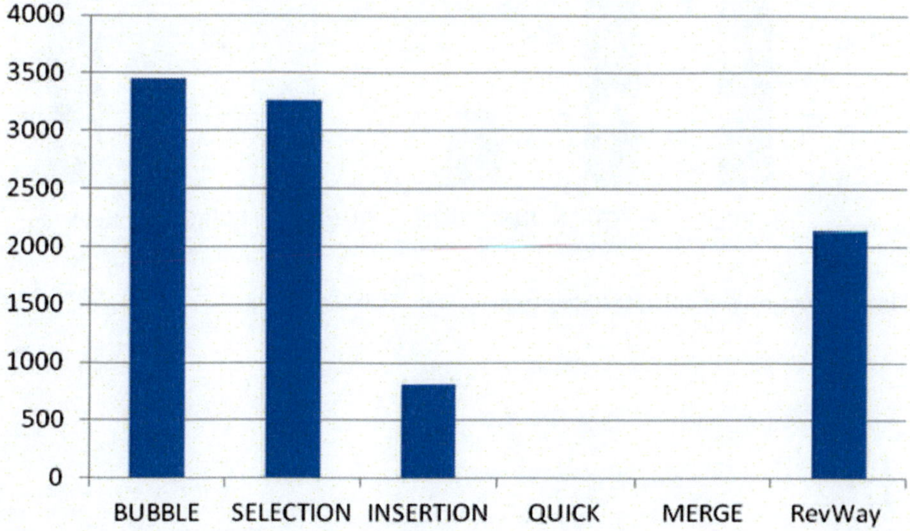

Fig. 4 Comparison of different sorting techniques for 30,000 elements

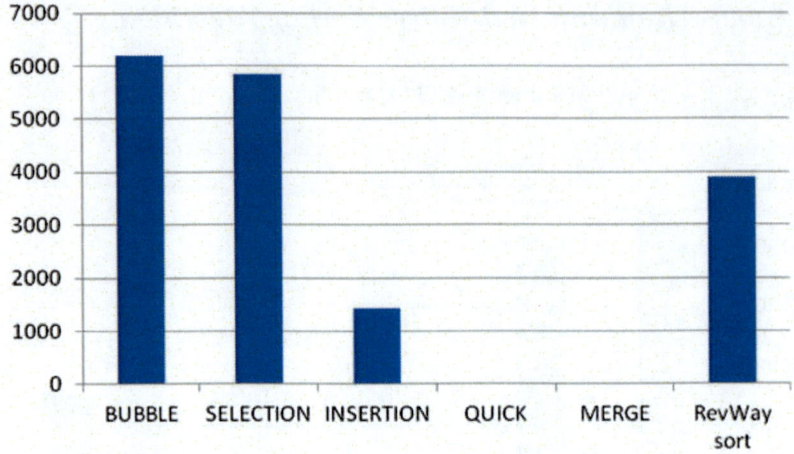

Fig. 5 Comparison of different sorting techniques for 40,000 elements

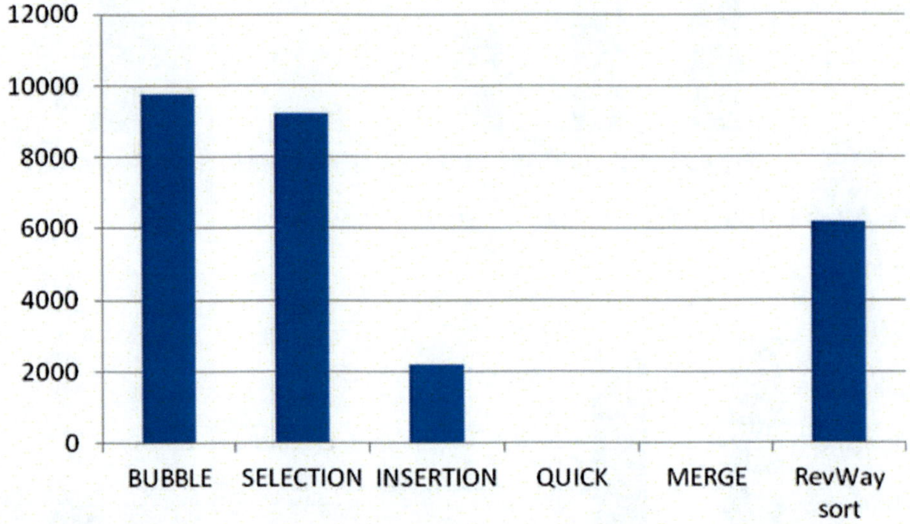

Fig. 6 Comparison of different sorting techniques for 50,000 elements

Table 1 For 10,000 elements

Name of sorting	Time taken by sorting techniques (in ms)						
	1	2	3	4	5	6	Average
Bubble	376.753	358.828	358.252	364.703	365.358	365.052	364.8243
Selection	334.543	327.684	328.554	331.487	374.777	328.281	337.5543
Insertion	98.069	93.161	97.662	94.244	93.373	96.826	95.5558
Quick	3.117	1.753	2.161	1.648	4.083	2.917	2.6131
Merge	2.875	1.608	1.850	1.503	1.511	1.782	1.8548
Revway	201.811	205.408	204.221	204.428	203.205	202.743	203.636

Table 2 For 20,000 elements

Name of sorting	Time taken by sorting techniques (in ms)						
	1	2	3	4	5	6	Average
Bubble	1533.363	1504.652	1517.370	1507.285	1514.762	1511.423	1514.809
Selection	1431.013	1444.578	1435.315	1412.978	1463.910	1428.764	1436.093
Insertion	377.105	361.889	370.183	366.849	370.190	368.208	369.0706
Quick	5.0	6.240	6.403	6.510	3.476	5.430	5.5098
Merge	3.334	4.017	3.990	3.547	3.546	4.384	3.8046
Revway	904.605	901.073	904.045	921.911	904.099	902.018	906.2918

Table 3 For 30,000 elements

Name of Sorting	Time taken by sorting techniques (in ms)						
	1	2	3	4	5	6	Average
Bubble	3451.182	3478.834	3439.239	3440.158	3469.195	3458.047	3456.1091
Selection	3290.609	3248.376	3265.231	3235.718	3251.780	3271.521	3260.5391
Insertion	786.645	799.991	798.396	818.218	802.178	828.525	805.6588
Quick	9.564	9.526	8.590	8.886	10.593	10.446	9.60
Merge	6.201	5.921	7.763	7.651	6.158	5.7	6.5656
Revway	2146.786	2415.218	2139.874	2134.667	2138.735	2143.985	2141.545

Table 4 For 40,000 elements

Name of sorting	Time taken by sorting techniques (in ms)						
	1	2	3	4	5	6	Average
Bubble	6179.753	6208.213	6208.184	6170.811	6189.394	6188.102	6190.7428
Selection	5846.303	5867.388	5830.152	5824.327	5855.041	5842.735	5844.3243
Insertion	1434.848	1425.219	1411.604	1448.907	1464.596	1418.316	1433.915
Quick	7.808	11.825	10.245	12.506	12.335	11.901	11.103
Merge	6.061	8.256	9.072	7.025	9.066	10.092	8.262
Revway	3895.826	3899.884	3911.356	3900.310	3901.050	3892.453	3900.147

Table 5 For 50,000 elements

Name of sorting	Time taken by sorting techniques (in ms)						
	1	2	3	4	5	6	Average
Bubble	9709.333	9772.463	9739.506	9712.120	9743.102	9788.132	9744.1093
Selection	9234.060	9258.496	9223.994	9210.438	9246.588	9217.992	9231.928
Insertion	2257.245	2165.444	2225.018	2179.746	2193.386	2195.712	2202.7585
Quick	13.923	13.935	13.595	15.066	14.628	12.771	13.9863
Merge	9.292	10.464	10.085	9.101	10.808	10.257	9.6326
Revway	6170.053	6168.222	6178.021	6162.432	6169.823	6170.146	6169.78283

4 Conclusion

In this paper, comparison is done between our newly proposed algorithm RevWay sort with existing algorithms. In this sort, the two consecutive numbers are compared from left and then from right. This process will be repeated $((n/2) + 1)$ times, where n is the total number of elements. In this sorting, we have illustrated time taken by different sorting methods with different number of elements. It is found that running time of RevWay sort is 203.636 ms which is lesser than bubble sort and selection sort.

References

1. (2013) Int J Eng Adv Technol (IJEAT) 3(1) ISSN: 2249-8958
2. Int J Innovat Technol Explor Eng (IJITEE) 8(12), October 2 ISSN: 2278-3075
3. (1996) Commun ACM SIGPALN, ACM 31(3):28–36
4. Kunth DE (1975) The art of computer programming. Sort. Search 3, 2nd printing. Addison-Wesley, Reading, MA
5. Sedgewick R (1988) Algorithms, 2nd edn. Addison-Wesley Series in Computer Science, pp 111–113
6. Jadoon S, Solehria S, Rehman S, Jan H (2011) Design and analysis of optimized selection sort algorithm 11(1):16–21, 2011. Available: http://www.ijens.org/IJECS
7. Williams JWJ (1964) Algorithm 232: Heapsort. Commun ACM 7:347–348
8. Hoare CAR (1961) Algorithm 64: quicksort. Commun ACM 4:321

Price Sensitivity in a 30 Bus Congested Power System

Parnab Saha, Sujit Pani, Bishaljit Paul, and Chandan Kumar Chanda

Abstract For the well being of a system, different parameters are needed to be tested by system operators, among which congestion management is of one prime importance. Different forms of economic parameters are required to signals the congestion management, and the most sensitive signal being the locational marginal price (LMP). These LMPs are the change of price of energy at each bus in the congested power market. LMPs are solved using of shift factor (SF) techniques on DC-OPF (DC-Optimal Power Flow). The LMPs are primarily comprises of three parts viz. marginal energy price (MEP), marginal congestion price (MCP) and marginal loss price (MLP). In this paper, nodal prices that are actually the LMPs are calculated in a thirty (30) bus test case system, and it shows that the LMPs vary from bus to bus when the system is congested.

Keywords Congestion · DC-optimal power flow · Locational marginal price · Shift factors

1 Introduction

In our modern deregulated power system for the transmission of power the locational marginal pricing market approach is an efficient economic market signal. Since the generation capacity and transmission line limits are fixed upto a limit thus for the transfer of power congestions occurs. The cost of energy, the cost of congestion and the cost of losses accountable to the Locational Marginal Price [1] at any node. An accurately and clear picture of power market is provided by the locational marginal price for each and every busses.

P. Saha (✉) · B. Paul
Narula Institute of Technology, Agarpara, Kolkata, India

S. Pani
CESC, Kolkata, India

C. K. Chanda
Indian Institute of Engineering Science and Technology, Shibpur, Howrah, India

M. Mitra et al. (eds.), *Computational Advancement in Communication, Circuits and Systems*, Lecture Notes in Electrical Engineering 786,
https://doi.org/10.1007/978-981-16-4035-3_19

For a complex power system where there are multiple sellers and multiple buyers, the congestion management [2–6] provides a market based solution with economic efficiency. For a deregulated structure, generation, transmission and distribution are within a control of a central agency (ISO). Generation is dispatched at least cost accordingly. Security constrained economic dispatch (SCED) [7–10] provides optimal solution and eliminates congestion. So the power flow limits are not exceeded in transmission lines, and the generators are dispatched accordingly. Congestion management [11–14] states that the transactions are to be prioritized and committed to work in such a schedule by the system operator (ISO) that does not overload the network.

Though the work of system operators look simple, they are actually not in a deregulated complex environment. It is because all buyers always wish to buy power at the cheapest price from the cheapest generator irrespective of relative geographical location. The transmission corridors would get overloaded and congestion will takes place if all such transactions are approved. The overloading of the transmission lines cannot be approved by the system operators. Therefore, congestion management strategies [15–17] and takes the remedial action are needed to be taken beforehand. Only that set of transactions taken together are approved quickly in real time so that the transmission lines may not get overloaded due to unplanned flow. A set of definite rules are to be ensured and enforced so that the security and reliability of the system can be ensured.

2 Terminology

2.1 Shift Factor (SF)

A flow pattern matrix whose each element specify the change of flow in any link, owing to the insertion of 1 MW power at one node and retreating it at the reference node. For instance, a flow on the line 1 to 6 resulting from injecting 1 MW at node 3 and withdrawing it at node 5 is given by

$$SF_{(1-6),\,3} - SF_{(1-6),\,5}$$

2.2 Shadow Price (SP)

It is the maximum dispatch cost saved due to an increment increase in flow capacity in the line with violating transmission constraints. In an integrated power system, shadow prices which are Lagrange multipliers [2, 3] for line inequality constraints, occur due to line congestion.

2.3 Locational Marginal Price (LMP)

LMP at any bus is defined as the marginal cost of the marginal generators of supplying the next increment of electric energy at that bus, while maintaining the physical aspects of transmission system. Congestion in the transmission line creates different LMPs at the buses, otherwise LMP at every node will be the same. LMP at the reference bus is the Lagrange multiplier of generation-demand equality constraint.

3 Mathematical Formulation

To have a secured constrained economic dispatch (SCED) operation of power system, an optimal power flow (OPF) is commonly applied and LMP at any bus or node can be calculated (Fig. 1).

Let us consider that the line connecting the nodes '1' and '2' for flow F1 is constrained, and the power flow is at its limit. So this has violated the security of the system. These cause two marginal generators to operate. Let the marginal generators be *A* and *B*. To determine the LMP at any node '5', it is required to calculate incremental inputs ΔA and ΔB of the marginal units at nodes '1' and '2' so as to deliver 1 MW to node '5', without increasing the flow between the nodes '1' and '2' of the congested line.

From the power transfer distribution matrix (PTDF), if 1 MW of power is injected at node '3' where there is a marginal generator and withdrawn at node '4' where LMP is to be found, there may be an increase or decrease in flow in line connecting between nodes '1' and '2'.

$$\Delta \text{Flow}_{(1-2)3} = \text{PTDF}_{(1-2)3} - \text{PTDF}_{(1-2)5} \quad (1)$$

Fig. 1 Diagram of a 5 bus system

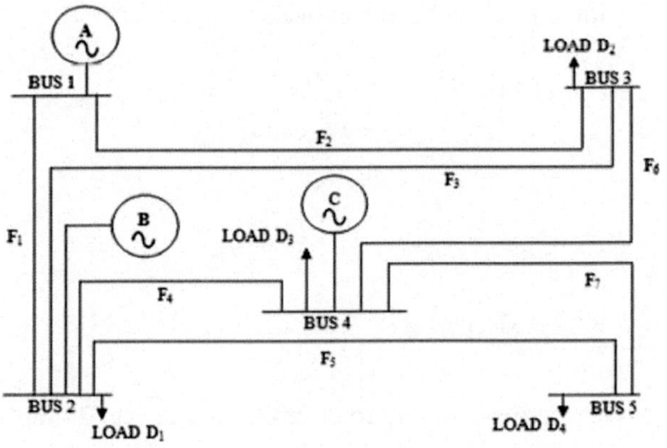

Similarly, from the power transfer distribution matrix (PTDF), if 1 MW of power is injected at node '*m*' where there is a marginal generator and withdrawn at node '*k*' where LMP is to be found, there may be an increase or decrease in flow in line connecting between nodes '*i*' and '*j*'.

$$\Delta \text{Flow}_{(1-2)4} = \text{PTDF}_{(1-2)4} - \text{PTDF}_{(1-2)5} \tag{2}$$

In general, considering that the line connecting the nodes '*i*' and '*j*' is constrained and the power flow is at its limit. So this has violated the security of the system. These cause two marginal generators to operate. Let the marginal generators be *A* and *B*. To determine the LMP at any node '*k*', it is required to calculate incremental inputs ΔP_A and ΔP_B of the marginal units at nodes 'l' and '*m*' so as to deliver 1 MW to node '*k*', without increasing the flow between the nodes '*i*' and '*j*' of the congested line.

The quantities must satisfy the system of equations

$$\begin{bmatrix} \Delta \text{Flow}_{(i-j)l} & \Delta \text{Flow}_{(i-j)m} \\ 1 & 1 \end{bmatrix} * \begin{bmatrix} \Delta A \\ \Delta B \end{bmatrix} = \begin{bmatrix} 0 \\ 1 \end{bmatrix} \tag{3}$$

A similar calculation is analysed to determine the shadow price, which are direction specific, in the congested line connecting between nodes '*i*' and '*j*'. The aim is to perturb the outputs of the marginal generators at nodes 'l' and '*m*' by incremental amounts ΔA and ΔB so as to increase the flow in the congested line between nodes '*i*' and '*j*' by 1 MW, while maintaining the energy balance.

The quantities must satisfy the system of equations

$$\begin{bmatrix} \Delta \text{Flow}_{(i-j)l} & \Delta \text{Flow}_{(i-j)m} \\ 1 & 1 \end{bmatrix} * \begin{bmatrix} \Delta A \\ \Delta B \end{bmatrix} = \begin{bmatrix} 1 \\ 0 \end{bmatrix} \tag{4}$$

For any pair of nodes, '*i*' and '*j*' a relation between locational marginal price and shadow price can be formulated as

$$\text{LMP}_j - \text{LMP}_i = \sum_{\text{congested lines }(h-k)} \text{SP}_{\text{congested lines }(h-k)} * \left(\text{PTDF}_{(h-k)j} - \text{PTDF}_{(h-k)i}\right) \tag{5}$$

4 Case Studies

a. Generating unit and load information (Fig. 2; Tables 1 and 2).

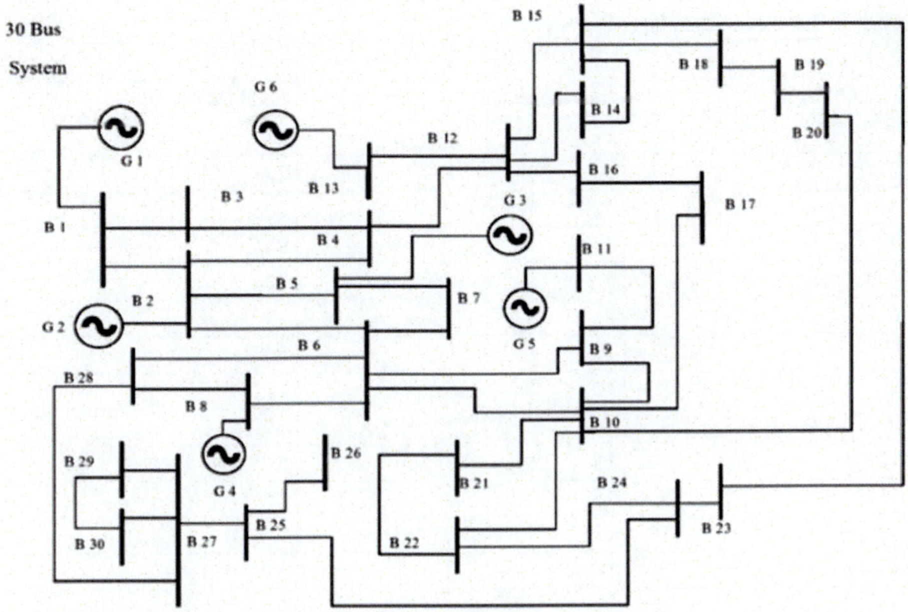

Fig. 2 30 bus test system

Table 1 Generator bus data of 30 bus test case system

Generator data			
Gen No.	Generator cost	Bus connected	Power generated (MW)
1	10	1	200
2	20	2	80
3	15	5	50
4	15	8	35
5	15	11	30
6	15	13	40

5 Results

The simulation is performed on MATLAB environment, and the simulated outputs, i.e., LMPs for each buses are given in Table 3.

Table 2 Load data of 30 bus test system

Load data		
Load No.	Bus connected	Load (MW)
1	2	21.7
2	3	2.4
3	4	7.6
4	5	94.2
5	7	22.8
6	8	30
7	10	5.8
8	12	11.2
9	14	6.2
10	15	8.2
11	16	3.5
12	17	9.0
13	18	3.2
14	19	9.5
15	20	2.2
16	21	17.5
17	23	3.2
18	24	8.7
19	26	3.5
20	29	2.4
21	30	10.6

Table 3 Calculated LMPs of 30 bus test system

Bus No.	1	2	3	4	5	6	7	8	9	10
LMPs	10	15.266	11.385	14.565	15	14.734	14.844	14.734	14.703	14.687
Bus no	11	12	13	14	15	16	17	18	19	20
LMPs	14.7	14.637	14.637	14.646	14.654	14.658	14.678	14.666	14.672	14.676
Bus no	21	22	23	24	25	26	27	28	29	30
LMPs	14.7	14.688	14.679	14.689	14.704	14.704	14.714	14.732	14.714	14.714

6 Conclusion

In a complex power market at different phases of time, the LMPs at different system buses reflect the price sensitivities, which are analyzed by congestions of the transmission networks. The DC model is superior to AC because of fast accurate nodal price calculation but the losses are neglected. The ISO (Independent System Operator) can prioritize the transaction so as to maintain the security in the system. Depending upon the type of generator used different generator bids are obtained which further generates different forms of LMPs. The value of the obtained LMP can be small, large or even negative depending upon the congestive nature of the system. In the absence of congestion the LMP will be flat in nature. The primary reason for different LMPs is because of transmission congestion in the system.

References

1. Litvinov E (2010) Design and operation of the locational marginal prices-based electricity markets. IET Gener Transm Distrib 4(2):315–323
2. Singh H, Hao S, Papalexopoulos A (1998) Transmission congestion management in competitive electricity markets. IEEE Trans Power Syst 13(2):672–680
3. Lo KL, Yuen YS (2000) Congestion management in deregulated electricity markets. In: The international conference on electric utility deregulation and restructuring and power technologies 2000. City University, London, 4–7 April 2000
4. Ott AL (2003) Experience with PJM market operation, system design and implementation. IEEE Trans Power Syst 18(2):528–534
5. Christie RD, Wollenberg BF, Wangensteen I (2000) Transmission management in the deregulated environment. Proc IEEE 88:170–195
6. Momoh JA (2000) Electric power system applications of optimization. Marcel Dekker, New York
7. Shahidehpour M, Yamin H, Li ZY (2002) Market operations in electric power system. John Wiley and Sons Inc., New York
8. Kirschen DS, Strbac G (2004) Fundamentals of power system economics. John Wiley & Sons Ltd.
9. Alaywan Z Facilitating the congestion management market in California. California Independent System
10. Gomez T, Marnay C (1999) Transmission pricing and congestion management in California
11. Chao H, Peek S (1996) A market mechanism for electric power transmission. J Regul Eco 10(1):25–59
12. Hsu M (1997) An introduction to the pricing of electric power transmission. Utilities Policy 6(3):257–270
13. Ma X, Sun DI, Ott A (2002) Implementation of PJM financial transmission rights auction market system. IEEE Power Eng Soc Summer Meet 3:221–225
14. Ma X, Sun DI, Cheung KW (2003) evolution toward standardized market design. IEEE Trans Power Syst 18(2):460–469
15. Paul B, Pathak M, Chanda CK, Pal J (2017) On transmission congestion and financial rights. IJESRT ISSN: 2277-9655
16. Paul B, Pathak M, Chanda CK, Pal J (2018) On transmission congestion management strategies and forecasting locational marginal prices in a deregulated competitive power market. In: Universities power engineering conference (AUPEC), Melbourne, 2017. IEEE
17. Paul B, Pathak M, Chanda CK, Pal J (2018) A comparison of locational marginal prices and locational load shedding marginal prices in a competitive deregulated power market. In: Calcutta conference (CALCON), Calcutta, 2017. IEEE

Security of Load Flow Analysis with Photovoltaic Energy Sources

Dipu Mistry, Bishaljit Paul, and Chandan Kumar Chanda

Abstract Photovoltaic energy sources are the most reliable renewable energy sources. The major limitations of solar energy are its Weather-Dependency and availability at daytime only. As photovoltaic energy storage is very expensive hence it is smarter to use solar energy during the day and take energy from the grid during the night. So, it is a good solution to connect the solar power systems with the grid which is linked with an infinite bus. In this paper, a microgrid system, based on two photovoltaic generating stations and an infinite bus system that is capable to export or import power to the grid is simulated. The microgrid system investigated in this paper represents a study based on an eight-bus system. The output voltage of photovoltaic generation system and infinite bus with constant voltage is calculated in our study. Computation of bus voltage and power flow of the microgrid for green power importation or exportation to the local power grid is carried out using Newtown—Raphson algorithm.

Keywords microgrid · Photovoltaic · Renewable energy · Newtown—Raphson power flow algorithm

1 Introduction

The coal and oil production in the world will reach its peak in 2019. The world's Uranium production is expected to reach its peak in 2035. Solar power is the key to a green energy future. Every day, the sun gives off far more energy than we need to power everything on earth [1]. That's why it needs to concentrate on solar power and its distribution system. An electrical grid, electric grid, or power grid, is an interconnected network for delivering electricity from producers to consumers. It consists of generating stations that produce electric power' electrical substations for stepping

D. Mistry (✉) · B. Paul
Department of Electrical Engineering, Narula Institute of Technology, Kolkata, India

C. K. Chanda
Department of Electrical Engineering, IIEST Shibpur, Kolkata, India

M. Mitra et al. (eds.), *Computational Advancement in Communication, Circuits and Systems*, Lecture Notes in Electrical Engineering 786,
https://doi.org/10.1007/978-981-16-4035-3_20

electrical voltage up for transmission, or down for distribution, high voltage transmission lines that carry power from distant sources to demand-centers and distribution lines that connect individual customers [2]. A microgrid is a group of interconnected loads and distributed energy resources within clearly defined electrical boundaries that acts as a single controllable entity with respect to the grid. A microgrid can connect and disconnect from the grid to enable it to operate in both grid-connected or island modes [3]. A grid is a centralized technology whereas microgrid is a localized system.

Day by day cost of decentralized power generation technologies like solar power, wind power, etc. is falling. These local means of producing electricity is distributed through microgrid network system [4]. A microgrid is typically composed of fuel generators, Renewable energy sources, Energy storage solutions (batteries generally Lithium or Lead Acid), Power Grid (if available), and Load. A load can be commercial, industrial, or residential. It is designed to operate independently or in synchronization with the national network, within a defined area. The present study is an example of microgrid that synchronizes between local PV generations to a national network that is presented here as infinite bus system. Microgrid energy resources allow electricity to be generated closer to where it is used, protecting businesses and institutions from unexpected power outages, decreases transmission losses, and protect environment by utilizing Renewable energy sources.

A review paper [5] has summarized energy management approaches for microgrids from a new perspective and classified it into four categories, namely non-renewable, Energy Storage System, Demand-side Management, and hybrid-based Energy Management System. A survey report concludes that "Microgrid" is winding up increasingly normal in the present vitality designing engineering. It is as yet an idea in its early stages, yet it has incredible potential for particular offices and geographic impressions, for example, petrochemical plants, college grounds, mechanical parks that require solid vitality quality, and army installations that require secure vitality supplies. Micro-microgrids will give enhanced electrical administration dependability and power quality for end clients, and can likewise profit nearby organizations by giving a delivery load to be utilized among crest periods and helping or deferring updates of the framework dispersion [6].

Sivaramadurai et al. [7] developed a power flow algorithm to control among three sources in the DC Microgrid. The algorithm was tested for various load conditions and for fluctuations in solar and wind power in MATLAB/SIMULINK environment. The literature [8] designed the operating state of the energy storage module by detecting and comparing the load demand power and photovoltaic power generation, i.e., using the battery and the super capacitor as energy relaxation. Here a simulation model based on MATLAB is used to verify the effectiveness of the control strategy.

Chaphekar et al. [9] proposed a new algorithm for connecting the Microgrid to distribution network and determining the optimal location of Microgrid in the system. In order to locate the optimal placement of Microgrid, the power flow was carried out by considering different penetration ratios of Microgrid. A case study consists of a small hydro generation unit and a wind farm that contains nine variable-speed, double-fed induction generator-based wind turbines was carried out [10]. The study modeled and simulated to identify the technical issues involved in the operation of a microgrid system based on renewable power generation units.

In the literature [11] a generic three phase load flow algorithm has been formulated for Microgrid. This algorithm is based on a convergent Newton-Trust region method and has incorporated all operating modes of Microgrid. It concluded that generators in a power system are normally run with the help of a P-Q control through a centrally run control center. Nowadays due to various reasons such as the availability of renewable sources, technical and economic factors, Distributed generators are increasingly being installed. Such systems are generally small and sufficient only to meet the local demand.

1.1 The Power Flow Analysis

The terms power flow studies and load flow studies are used interchangeably and it refers to the flow of power from the generating units to the loads. The power flow analysis means determination of the voltage of the load buses for a specified system model with a schedule of generation and loads.

The fundamental problem in a power grid is the power flow analysis. It needs to ensure the designed power grid can deliver adequate electric energy to the power grid loads at acceptable voltage and frequency. It is done by solution of the power flow problem [12].

To obtain stable operation it needs to maintain system bus load voltages at 1 p.u with a deviation of no more than 10% in emergency operating conditions.

For scheduled transmission system, scheduled generation system, and scheduled bus loads it needs to determine bus voltages. There are three most important types of bus (Fig. 1)

- load bus,
- a generator bus, and
- a swing bus.

Load buses. At each non generator bus, called a load bus.

Voltage-controlled buses. Any bus of the system at which the voltage magnitude is kept constant is said to be voltage controlled usually called a voltage-controlled or PV bus.

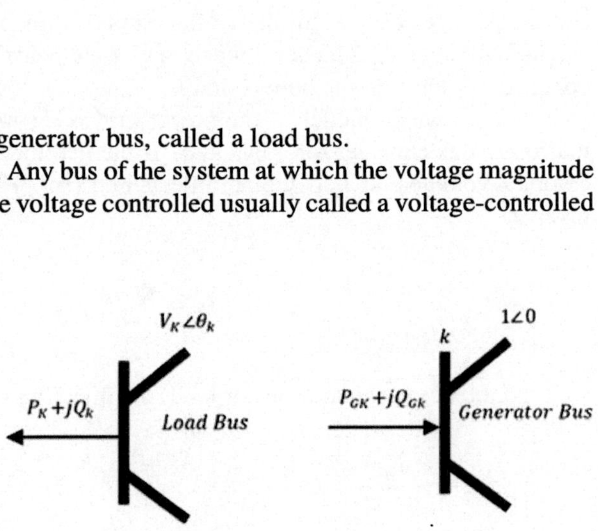

Fig. 1 Load bus and generator bus

Slack bus/swing bus. The voltage angle of the slack bus serves as a reference for the angles of all other bus voltages. slack bus is always the generation's bus.

For each bus, four variables are the active power, the reactive power, the magnitude of bus voltage, and the phase angle.

2 Mathematical Foundation

The power flow problem can mathematically be stated as given by a bus admittance matrix.

$$[I_{\text{Bus}}] = [Y_{\text{Bus}}][V_{\text{Bus}}] \tag{1}$$

where, I_{Bus} is bus injected current vector, Y_{Bus} is bus admittance matrix as in Eq. 2 and V_{Bus} is bus voltage profile vector [13]. All bus may be interconnected and hence Y_{Bus} is $n \times n$ squire matrix where n is the number of bus in the microgrid.

$$Y_{\text{bus}} = \begin{bmatrix} Y_{11} & Y_{12} & \dots & Y_{1j} & \dots & Y_{1n} \\ Y_{21} & Y_{22} & \dots & \dots & \dots & Y_{2n} \\ Y_{31} & Y_{32} & \dots & \dots & \dots & Y_{3n} \\ \dots & Y_{i2} & \dots & Y_{ij} & \dots & \dots \\ \dots & \dots & \dots & \dots & \dots & \dots \\ Y_{n1} & Y_{n2} & \dots & \dots & \dots & Y_{nn} \end{bmatrix} \tag{2}$$

Denoting the elements in row j and column k as Y_{nn} which is not the admittance but rather the elements of the admittance matrix. The diagonal admittance Y_{kk} is the self-admittance at bus k. Similarly, the mutual admittance between the buses i and j can be written as Y_{ij}.The admittance matrix elements are zero if there are no direct connections between the buses i and j.

The Y_{Bus} matrix model of the power grid is a symmetric, complex, and sparse matrix. Let us consider for a bus k, S_k is the net injected complex power, V_k is the complex voltage and I_k is the net injected current. The current injected at bus-k is given as

$$I_k = \sum_{j=1}^{n} Y_{kj} V_j \tag{3}$$

Then the bus (generator bus or load) net injection power is described as

$$S_k = V_k . I_k \tag{4}$$

Substituting bus net injected current from Eq. 3 in Eq. 4, obtain the residue form of the equation for each bus k as

$$S_k = V_k\left(\sum_{j=1}^{n} Y_{kj} V_j\right)$$

$$S_k - V_k\left(\sum_{j=1}^{n} Y_{kj} V_j\right) = 0 \quad (5)$$

For bus number 1, $k = 1$,

$$S_1 - V_1\left(\sum_{j=1}^{n} Y_{1j} V_j\right) = 0 \quad \text{or} \quad f_1(X) = 0 \quad (6)$$

For bus number 2, $k = 2$,

$$S_2 - V_2\left(\sum_{j=1}^{n} Y_{2j} V_j\right) = 0 \quad \text{or} \quad f_2(X) = 0 \quad (7)$$

For bus number n, $k = n$,

$$S_n - V_n\left(\sum_{j=1}^{n} Y_{nj} V_j\right) = 0 \quad \text{or} \quad f_n(X) = 0 \quad (8)$$

Where.
Here X is the bus voltage vector and $x_1, x_2, \ldots,$ is the $V_1, V_2, \ldots, V_n$.

$$X = \begin{bmatrix} x_1 \\ x_2 \\ \vdots \\ x_n \end{bmatrix} \quad (9)$$

So for bus 1, Eq. 6 can be expressed as

$$f_1(X) = f_1(x_1, x_2, \ldots, x_n) = 0$$

Substituting initial velocity vector input X_0 values are, $V_{10}, V_{20}, \ldots, V_{n0}$ or x_{10}, $x_{20}, \ldots, x_{n0}$

$$X_0 = \begin{bmatrix} x_{10} \\ x_{20} \\ \vdots \\ x_{n0} \end{bmatrix} \quad (10)$$

Imagine writing a Taylor series for the solution $g(x)$, say initially about $x = x_0$. Then

$$x = x_0 + \Delta x$$

$$g(x) = g(x_0 + \Delta x) = g(x_0) + \Delta x \frac{\partial g(x_0)}{\partial x} + \text{higher order terms} \quad (11)$$

Comparing Eq. 6 with Eq. 11
$g(x) = f_1(X)$.
$g(x_0) = f_1(X_0)$.
$X = X_0 + \Delta X$

$$f_1(X) = f_1(X_0 + \Delta X) = f_1(X_0) + \Delta X \frac{\partial f_1(X)}{\partial X}_{|X_0} = f_1(X_0) + \Delta X \frac{\partial f_1}{\partial X}_{|X_0} \quad (12)$$

Expanding Eq. 12, about a guess solution X_0., we have

$$f_1(X) = f_1(x_{10}, x_{20}, \ldots, x_{n0}) + \Delta x_1 \frac{\partial f_1}{\partial x_1}_{|X_0} + \Delta x_2 \frac{\partial f_1}{\partial x_2}_{|X_0} + \ldots + \Delta x_n \frac{\partial f_1}{\partial x_n}_{|X_0} \quad (13)$$

where,
$\Delta x_1 = x_1 - x_{10}$; $\Delta x_2 = x_2 - x_{20}$ and $\Delta x_n = x_n - x_{n0}$.
Equation 13 can be expressed in compact form as

$$f_1(X) = f_1(X_0) + \sum_{j=1}^{n} \frac{\partial f_1}{\partial x_j}_{|X_0} \Delta x_j \quad (14)$$

$$f_2(X) = f_2(X_0) + \sum_{j=1}^{n} \frac{\partial f_2}{\partial x_j}_{|X_0} \Delta x_j \quad (15)$$

$$f_n(X) = f_n(X_0) + \sum_{j=1}^{n} \frac{\partial f_n}{\partial x_j}_{|X_0} \Delta x_j \quad (16)$$

Equation 14 to 16 can be written as single matrix equation

$$F(X) = F(X_0) + \begin{bmatrix} \frac{\partial f_1}{\partial x_1} & \frac{\partial f_1}{\partial x_2} & \cdots & \frac{\partial f_1}{\partial x_n} \\ \frac{\partial f_2}{\partial x_1} & \frac{\partial f_2}{\partial x_2} & \cdots & \frac{\partial f_2}{\partial x_n} \\ \cdots & \cdots & \cdots & \cdots \\ \frac{\partial f_n}{\partial x_1} & \frac{\partial f_n}{\partial x_2} & \cdots & \frac{\partial f_n}{\partial x_n} \end{bmatrix}_{|X_0} \begin{bmatrix} \Delta x_1 \\ \Delta x_2 \\ \vdots \\ \Delta x_n \end{bmatrix} = 0 \quad (17)$$

The matrix of Eq. 17 is known as the Jacobian matrix $[J]$. The above equation can be rewritten as

$$F(X_0) + [J]_{|X_0}[\Delta X] = 0 \quad (18)$$

For the bus k the Eq. 18 is

$$f_k(X_0) = S_k - V_k \sum_{j=1}^{n} (Y_{kj} V_j) = 0 \quad (19)$$

Equation 18 is a power equation, therefore, the term $F(X_0)$ represents the power mismatch at each bus. When the term $f_k(X_0)$ is very small the power flow solution has been obtained.

$$S_{k(\text{calculated})} = V_k \sum_{j=1}^{n} (Y_{kj} V_j) \quad (20)$$

$S_{k(\text{calculated})}$ is the calculated power flowing away from bus k to all the other buses j. From Eq. 18,

$$[\Delta X] = -[J]^{-1}_{|X_0} F(X_0) \quad (21)$$

$$f_k(X_0) = \Delta S_k = \Delta P_k + j\Delta Q_k = S_{k(\text{Scheduled})} - S_{k(\text{Calculated})} \quad (22)$$

$$\Delta P_k = P_{k(\text{Scheduled})} - P_{k(\text{Calculated})} \quad (23)$$

$$\Delta Q_k = Q_{k(\text{Scheduled})} - Q_{k(\text{Calculated})} \quad (24)$$

The flow chart of Newton Raphson method is given below (Fig. 2).

3 Mathematical Formulation

To perform load flow analysis using Newton Raphson method, the algorithm developed is as follows:

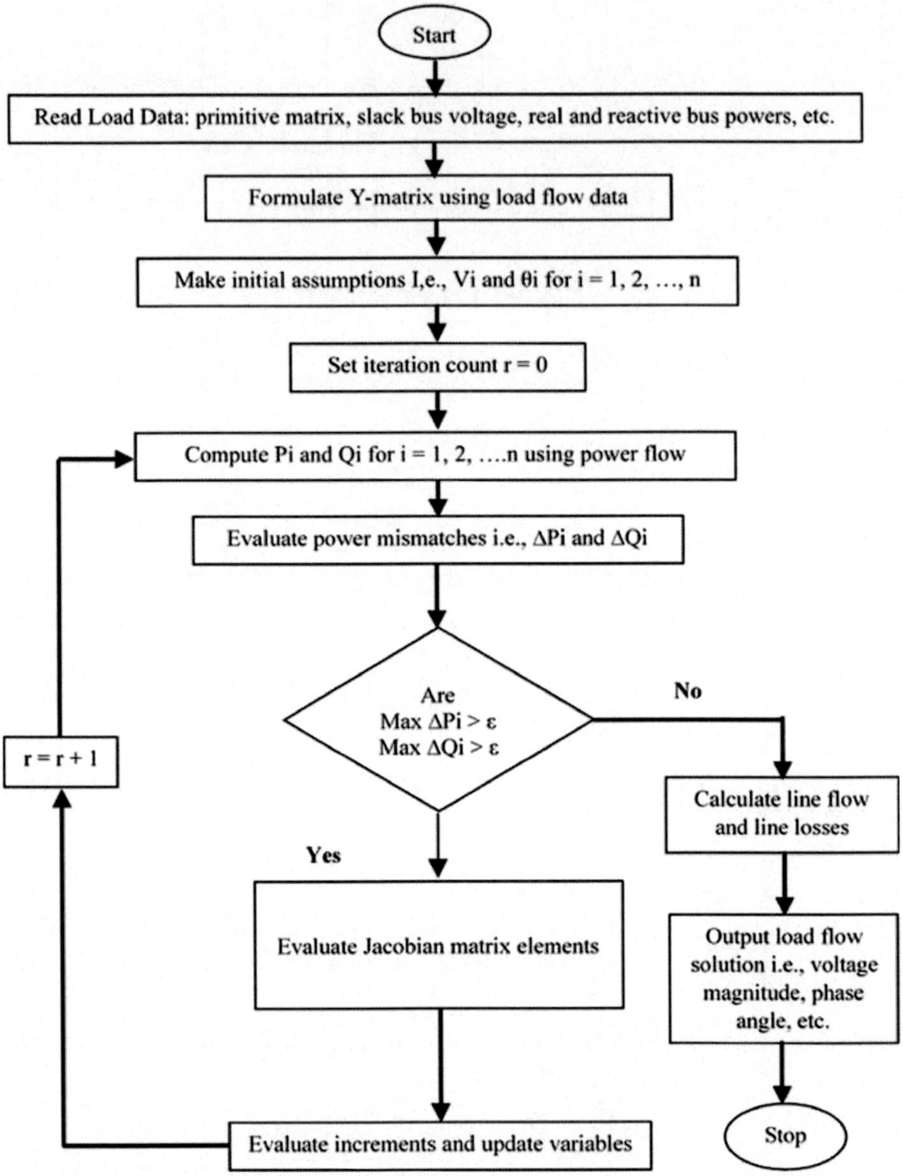

Fig. 2 Detailed flow chart of Newton Raphson method

- Write load flow equations in residual form: $F\ (X) = 0$
- Guess a solution vector, i.e., X_0 and evaluate $F\ (X_0)$
- Calculate J at $X\ (0)$

$$[J]_{|X_0} = \begin{bmatrix} \frac{\partial f_1}{\partial x_1} & \frac{\partial f_1}{\partial x_2} & \cdots & \frac{\partial f_1}{\partial x_n} \\ \frac{\partial f_2}{\partial x_1} & \frac{\partial f_2}{\partial x_2} & \cdots & \frac{\partial f_2}{\partial x_n} \\ \cdots & \cdots & \cdots & \cdots \\ \frac{\partial f_n}{\partial x_1} & \frac{\partial f_n}{\partial x_2} & \cdots & \frac{\partial f_n}{\partial x_n} \end{bmatrix}_{|X_0} \tag{25}$$

Consider an initial guess for $X = X_0 = V_{10}, V_{20}, \ldots, V_{n0}$. Now the ΔX has been computed from the following steps

Step 1. $[\Delta X] = -[J]^{-1}_{\Delta X_0} F(X_0)$.
Step 2. Update $X^{(i+1)} = X^i + \Delta X$.
Step 3. Calculate $F\ (X)$ at $X^{(i+1)}$.
Step 4. Check if $F\ (X^{(i+1)}) < 10^{-6}$, then the solution has converged. Store the solution vector $X^{(i+1)}$.
Step 5. Update i to $i + 1$.
If not, go to Step 1.

In context to various steps involved in carrying out load flow studies with Newton Raphson method, the following detailed flow chart has been designed.

4 Case Study

Here loadability or the extent of load which can flow through the line without exceeding the limitations of transmission lines of a microgrid has been studied. The single line diagram of a system of microgrids under this study is given in Fig. 3. It is an eight-bus system consists of two PV (Power-Voltage) bus connected with Photo Voltaic generating station (Bus 2 and Bus 3), one slack bus (Bus 1), and five load bus or P-Q bus (4, 5, 6, 7 and 8).

The following assumption has been made in this paper.

- Capacity of all transmission lines is 2.5 MW.
- The base value of the volt—amp $S_b = 10$ MVA and the voltage base on the transmission line side is specified as $V_b = 13\ 2.$ kV then compute the p.u model.
- PV generating station 1 is rated at 0.75 MW and PV generating station 2 is rated at 3 MW. Assume PV generating stations are operating at unity power factors.
- Power factor in all P-Q bus is 0.9 lagging/leading.

In the present research work, three cases have been studied and analyzed using MATLAB software [14]. To compute the results NR load flow methods have been used. The input data required for carrying out power flow analysis are shown in tables.

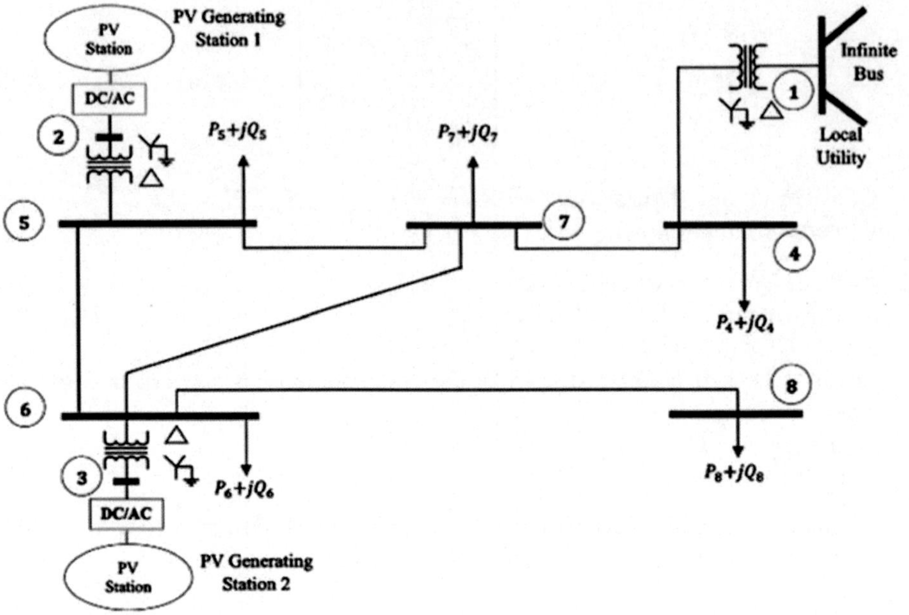

Fig. 3 The photovoltaic microgrid

4.1 *Input Parameter*

Line Parameter. Where R_{pu} is resistance per unit, X_{pu} is reactance per unit and B is susceptance per unit (Table 1).

Generator Data

For Case Studies 1 (Table 2).

For Case Studies 2 (Table 3).

Table 1 Line parameter information

L	From	To	R_{pu}	X_{pu}	$B/2$
L_1	1	4	0	0.1	0
L2	2	5	0	0.1	0
L3	3	6	0	0.1	0
L4	4	7	0.020	0.115	0.00479
L5	5	6	0.012	0.069	0.00287
L6	5	7	0.008	0.046	0.00191
L7	6	7	0.008	0.046	0.00191
L8	6	8	0.016	0.092	0.00383

Table 2 The active power (MW) and reactive power (MVAr) of each bus

Bus	2	3	4	5	6	7	8
P	0.075	0.300	0.145	0.140	0.120	0.050	0.080
Q	–	–	−0.070	−0.068	0.058	0.024	−0.039

Table 3 The active power (MW) and reactive power (MVAr) of each bus

Bus	2	3	4	5	6	7	8
P	0.075	0.300	0.200	0.150	0.180	0.100	0.120
Q	–	–	−0.097	−0.072	0.087	0.048	−0.058

Table 4 The active power (MW) and reactive power (MVAr) of each bus

Bus	2	3	4	5	6	7	8
P	0.075	0.300	0.180	0.200	0.200	0.150	0.100
Q	–	–	−0.087	−0.097	0.097	−0.072	0.048

For Case Studies 3 (Table 4).

5 Results and Discussion

For Case 1 (Table 5).

For Case 2 (Table 6).

For Case 3 (Table 7).

For Case 1 (Table 8).

Lines flow, 1–4, 2–5, and 3–6 are from generation. This paper studies line flow power to bus (4–7, 5–6, 5–7, 6–7, and 6–8) where capacity is assumed to be 2.5 MW. For lines 4–7 the active power obtained from Table 4 is 0.015 MW/unit. The base

Table 5 Calculated voltage—angle

Bus No.	1	2	3	4	5	6	7	8
Voltage mag	1	1	1	1.001	1.001	1.001	1.001	1.002
Angle deg	0	−0.008	0.005	−0.009	−0.013	−0.012	−0.014	−0.020

Table 6 Calculated voltage—angle

Bus No	1	2	3	4	5	6	7	8
Voltage mag	1	1	1	1.001	1.001	1.001	1.001	1.002
Angle deg	0	−0.032	−0.020	−0.021	−0.036	−0.037	−0.037	−0.047

Table 7 Calculated voltage—angle

Bus No	1	2	3	4	5	6	7	8
Voltage mag	1	1	1	1.001	1.001	1.001	1.001	1.001
Angle deg	0	−0.044	−0.031	−0.026	−0.049	−0.048	−0.048	−0.057

Table 8 Line flow

L	From	To	MW	MVAr
L1	1	4	0.160	−0.879
L2	2	5	0.075	−1.082
L3	3	6	0.300	−1.199
L4	4	7	0.015	−0.810
L5	5	6	−0.047	−0.449
L6	5	7	−0.018	−0.567
L7	6	7	0.053	−0.325
L8	6	8	0.080	−0.807

active power is 10 MW; hence the obtained power is $0.015 \times 10 = 0.15$ MW which is less than the assumed capacity of 2.5 MW. So, the transmission line is secured. Similarly, rest of the lines are also secured for case 1.

For Case 2 (Table 9).

The same observation is true for case 2 also.

For Case 3 (Table 10).

Here in case, three lines 5–6, 5–7, 6–7, and 6–8 are secured but that is not true for lines 4–7. For lines 4–7 the active power obtained from Table 6 is 0.275 MW/unit hence the obtained power is $0.275 \times 10 = 2.75$ MW which is greater than the assumed capacity of 2.5 MW.

Table 9 Line flow

L	Fr	To	MW	Mvar
L1	1	4	0.375	−0.902
L2	2	5	0.075	−1.072
L3	3	6	0.300	−1.183
L4	4	7	0.175	−0.806
L5	5	6	−0.021	−0.446
L6	5	7	−0.054	−0.555
L7	6	7	−0.021	−0.317
L8	6	8	0.120	−0.826

Table 10 Line flow

L	Fr	To	MW	Mvar
L1	1	4	0.455	−0.918
L2	2	5	0.075	−1.084
L3	3	6	0.300	−1.174
L4	4	7	0.275	−0.832
L5	5	6	−0.036	−0.412
L6	5	7	−0.089	−0.576
L7	6	7	−0.036	−0.389
L8	6	8	0.100	−0.720

6 Conclusion

In this work, power flow must be solved during the system planning for a large-interconnected power grid using the forecasted bus load to ensure plan generation and transmission system at an acceptable bus load voltage so that the system transmission lines are not overload. The load flow problem has been dealt with. To obtain power flow solution, the most popular method of including slack bus into the Jacobians for Newton–Raphson method has been used. The security checking is carried out for three different cases. The power flow results obtained were analyzed and discussed. The system security study shows that case 3 is not secure whereas case 1 and case 2 are secured based on line capacity.

References

1. Sarkar S, Chakrabarti U, Bhattacharyya S, Chakrabarti A (2020) A comprehensive assessment of the need and availability of smart grid technologies in an electricity distribution grid network. J Inst Eng India Ser B
2. https://en.wikipedia.org/wiki/Electrical_grid
3. The Pew Charitable Trusts (2016) Microgrids in the evolving power system. LG CNS
4. https://elum-energy.com/en/2020/09/17/what-is-a-solar-hybrid-microgrid-2/
5. Shayeghi H, Shahryari E, Moradzadeh M, Siano P (2019) A survey on microgrid energy management considering flexible energy sources. MDPI, Energies
6. Kumar A, Khan MZU, Akbar Hussain M (2018) Microgrids technology: a review paper. Gyancity J Electr Comput Sci 3(1):11–20
7. Gowtham K, Hariprasath P, Sivaramadurai CV, Indurani B (2018) A management of power flow for DC microgrid with solar and wind energy sources. In: IEEE international conference on computer communication and informatics
8. Zhang Z, Lv J, He M (2018) Design of DC micro-grid based on bus voltage grading control. IEEE, pp 3288–3293
9. Chaphekarl SN, Karad PR, Dharme AA (2016) Optimal power flow for power management in microgrid. In: IEEE international conference on power electronics. Intelligent Control and Energy Systems, pp 1–5

10. Ravichandrudu K, Manasa M, Yohan Babu P, Anjaneyulu GVP (2013) Design of micro-grid system based on renewable power generation units. Int J Sci Res Publ 3(8)
11. Chaphekar SN, Ambekar GS, Dharme AA (2015) Load flow analysis for power management of microgrid. IEEE INDICON, pp 1–5
12. Keyhani A (2011) Design of smart power grid renewable energy systems. John Wiley & Sons, Inc., Hoboken, New Jersey Published simultaneously in Canada
13. Ikule FT, Ame-Oko A, Idoko E (2019) Load flow analysis using newton Raphson Method—a case study of south-west Nigeria 330 kV network. IJARIIE 5(3):1960–1969
14. Mander DK, Virdi GS (2017) Result Analysis on Load Flow by Using Newton Raphson Method. Int J Adv Res Electr, Electron Instrum Eng 6(7):5835–5844

Evaluation of Azimuth Angle Profile for Solar Photovoltaic System in Humid Subtropical Climate of Varanasi City

Suman Moitra, Parnab Saha, Bishaljit Paul, and Chandan Kumar Chanda

Abstract This research work focuses on the prognosis of energy exploration opportunity due to geographical coordinates and celestial positioning of Sun at Varanasi city. As per the Köppen classification, the city has its humid subtropical climate with pretty higher temperature and scattered precipitation all over the year. The prime objective of this research work is to predict the solar energy security of a future smart city like Varanasi with its geographical circumstances. Computer program with MATLAB coding is used for mathematical computation of solar azimuth angle profile assessment of Varanasi city. The mathematical computation of azimuth angle profile relates to the mathematical expression of altitude angle, longitude angle and angular measurement of Sun position. The summary of experimental results shows significant variation of azimuth angle profile corresponding to different seasons of the year of the city.

Keywords Azimuth angle · Renewable energy · Solar PV energy · Solar altitude angle

1 Introduction

Developing countries are facing energy crisis in twentieth century. Future economic expansion of developing nations will depend on fruitful utilization of renewable energy sources. North Indian cities like Varanasi has enormous potential of solar energy production. Varanasi has its distinctive humid subtropical climate. The city has moist maritime tropical air with warm ambience. The warm summer months give plenty of solar energy to mankind. Being a highly ancient city of international

S. Moitra (✉)
Elitte College of Engineering, Sodepur, Kolkata, India

P. Saha · B. Paul
Narula Institute of Technology, Agarpara, Kolkata, India

C. K. Chanda
Indian Institute of Engineering Science and Technology, Shibpur, Howrah, India

M. Mitra et al. (eds.), *Computational Advancement in Communication, Circuits and Systems*, Lecture Notes in Electrical Engineering 786,
https://doi.org/10.1007/978-981-16-4035-3_21

tourist destination, Varanasi needs to fulfill its future energy crisis. Varanasi city can be upgraded as a model smart city with significant share of solar power generation. That would enable other cities of developing nations to fulfill ambitious dream of renewable energy development. Varanasi is blessed with plenty of solar energy due to its humid subtropical climate.

Previous research work suggested that solar apparatus may produce maximum power in south-west Nigeria, when aligned between 5 and 7° orientation excluding the south directional alignment which may tilted between 15 and 20° tilt angle of the solar photovoltaic installation. The investigation carried out to measure power output from solar module. Data were recorded corresponding to different orientation of azimuth angles and or tilt angles [1]. Previous research work also investigated the maximum power output condition from multiple sub-array inclined surface photovoltaic installation in research work of Osaka University. It was found that output energy was more in parallel combination of PV modules than series combination with usage of blocking diode [2]. Significant research also performed regarding statistical assessment of azimuth angle and elevation angles in the research of Military University of Technology, Poland [3]. Research work also suggested calculation regarding angle of incidence, altitude angle and azimuth angle from any specific point of observer at any time of calendar year. This work gives idea about maximum power generation corresponding to optimum positioning of PV module [4]. Research work also carried out on SAR imaging of metallic bridge with impact of azimuth angle. Simulation performed with the help of Kirchhoff approximation algorithm and Pierson-Moskowitz spectra for modeling [5]. Research work related to azimuth angle probability distribution showed that it has greater impact on inclination angle with certain orientation [6]. Research work also carried out to evaluate azimuth angle profile for man-made structures using spatial arrangements with various polarimetric indices. Thus, it concluded that the polarization orientation angle has correlation with azimuth angle [7]. Research work regarding pairing combination of azimuth angle and elevation direction of plane wave was done with an algorithm using computer simulation [8]. Investigation also carried out regarding relation establishment between azimuth angle and its corresponding characteristic value using maximum intensity imaging of a microstructure optical fiber [9]. Mathematical calculation of azimuth angle and elevation pointing vector was estimated using an angle of arrival of an object with its microphone array in similar research work of US army research laboratory [10]. Research exploration was performed based on three different commercially available photovoltaic technologies for their water-body related performance analysis. They found that the CdTe module on water body performed better than the ground based CdTe module [11]. A feature like azimuth profile was introduced in road segment digital map to decide instant driving direction of a vehicle with match-making algorithm [12]. Research work also carried out related to mathematical modeling regarding azimuth angle definition was introduced [13].

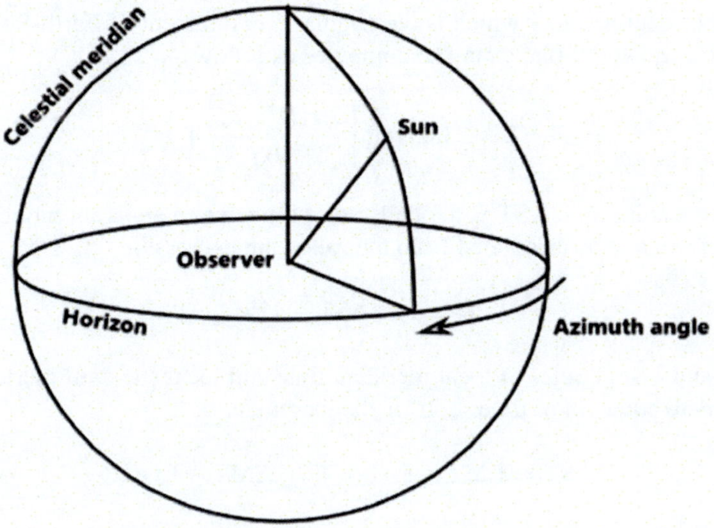

Fig. 1 The position of Sun and azimuth angle

2 Mathematical Foundation of Azimuth Angle

In spherical coordinate system, azimuth is refer to an angle which represents the angular measurement from particular object to a perpendicularly projected reference plane. It signifies a certain angle between projected vector and reference vector [14]. The angle measured in degree. For celestial positioning of Sun, the sunbeam projected at a particular angle to the particular point of observer is considered as the azimuth angle of Sun [15] (Fig. 1).

3 Mathematical Formulation

The computer program coding used in MATLAB software gives calculation of the long mathematical formulation. The prime novelty of this research work is to predict the solar energy security of a future smart city like Varanasi with its geographical circumstances. Programming code computes the angle of declination and Equation of time. Altitude angle α, the latitude of the location L and angle of declination δ may be known from geographical data which is very useful in photovoltaic energy analysis. Altitude angle which is the angular positioning of the Sun from horizontal line, may be expressed from the following relation,

$$\sin\alpha = \sin L.\sin\delta + \cos L.\cos\delta.\cos\omega \tag{1}$$

Angle of declination δ which is the angular measurement of Earth-Sun rotating vector with equatorial plane may be computed as follows,

$$\delta = 23.45^\circ \sin\left[\frac{2\pi(N-81)}{365}\right] \tag{2}$$

Apparent solar time AST is the daily solar time due to apparent motion of Sun. Hour angle ω may be calculated from the following expression,

$$\omega = 15^\circ(\mathrm{AST} - 12\mathrm{h}) \tag{3}$$

LMT and LSMT refer to local meridian time and local standard meridian time. EoT refers to equation of time. LOD is the longitude.

$$\mathrm{AST} = \mathrm{LMT} + \mathrm{EoT} \pm 4^\circ/(\mathrm{LSMT} - \mathrm{LOD}) \tag{4}$$

LSMT is the regional time zone which is very similar like prime meridian zone of Greenwich Meridian time GMT. The relation between them is expressed as follows,

$$\mathrm{LSMT} = 15^\circ T_{\mathrm{GMT}} \tag{5}$$

EoT refers to the comparison of apparent and mean solar time at a certain longitude and expressed as follows,

$$\mathrm{EoT} = 9.87\sin(2B) - 7.53\cos B - 1.5\sin B \tag{6}$$

The coefficient B may be computed as,

$$B = \frac{2\pi}{365}(N-81) \tag{7}$$

Azimuth angle θ may be considered from the following expression, where ω remain as the hour angle representing the angular displacement of Sun from point of observation.

$$\sin\theta = \frac{\cos\delta\sin\omega}{\cos\alpha} \tag{8}$$

From sunrise to sunset, a whole solar day is considered. The equal value of sunrise and sunset hour angles $\omega_{\mathrm{sr,ss}}$ may be computed as follows,

$$\omega_{\mathrm{sr,ss}} = \cos^{-1}(-\tan L.\tan\delta) \tag{9}$$

The solar time corresponding to every hour angle may be computed from the following expression, where apparent solar time with respect to sunrise and sunset refers to $\mathrm{AST}_{\mathrm{sr,ss}}$

Table 1 Geographical coordinates of Varanasi

Latitude angle (Degree)	Longitude angle (Degree)	Time difference with reference to GMT	Seasons of Varanasi (humid subtropical climate)	Experimental date recorded as per Gregorian calendar year of 2020	Days (value of N)
25.3176° N	82.9739° E	(+) 5.5 h	Winter	January 1, 2020	1
			Spring	February 1, 2020	31
			Spring	March 1, 2020	61
			Summer	April 1, 2020	92
			Summer	May 1, 2020	122
			Summer	June 2020	153
			Autumn	August 1, 2020	214
			Autumn	November 1, 2020	306

$$\frac{\omega_{\text{sr,ss}}}{15^\circ} \pm 12\text{h} = \text{AST}_{\text{sr,ss}} \tag{10}$$

The sunrise sunset hour angle and apparent solar time is computed through MATLAB coding (Table 1).

4 Results and Discussion

Computer programming through MATLAB software gives sunrise and sunset time adjustment of a complete solar day. The data related to altitude and latitude are examined from geographical information. The azimuth angle is being computed from sunrise and sunset time correction at a particular day of Gregorian calendar. Then, the calculation repeated further over different date of the same year. The investigation examined over multiple times all through the year in different seasons. All the data were recorded. The significant variation of solar azimuth angle profile for Varanasi city is only reflected in this research paper. The pivotal difference found among the dataset of recorded experimental results. Variation of results among the seasons of winter, spring, summer and autumn are classified and categorized in this research work. Figures 2 and 3 graph reflect the steady rise of azimuth angle profile of the city during winter on January 1, 2020 and onset of spring on February 1, 2020 accordingly. Figure 4 shows moderate variation of azimuth angle profile of the city during spring. Figures 5, 6 and 7 reflect the graph of azimuth angle profile of the city on April 1, May 1 and June 1, 2020, respectively. These three graphs reflect abrupt

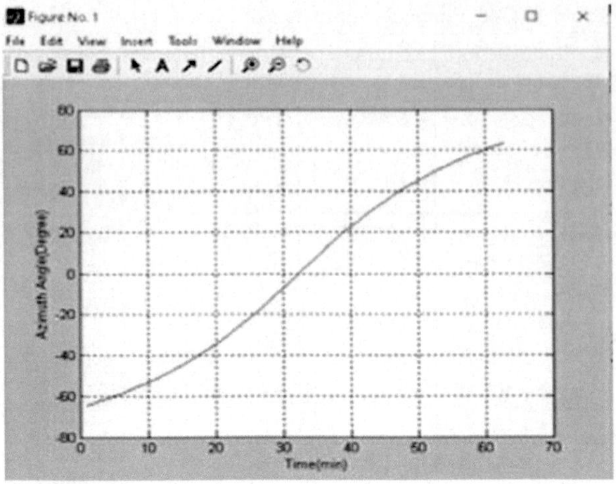

Fig. 2 Azimuth angle profile of Varanasi city on January 1, 2020 during winter ($N = 1$)

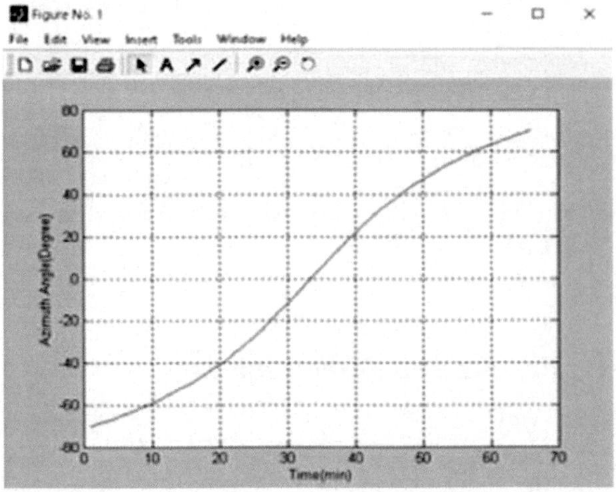

Fig. 3 Azimuth angle profile of Varanasi city on February 1, 2020 during spring ($N = 31$)

change of azimuth angle profile during onset, middle and end of summer. Figure 8 shows that prevailing of abrupt changing nature in the azimuth angle profile graph sustains even at the onset of autumn on August 1. Figure 9 shows that the graph again becomes steady at the end of autumn on November 1.

5 Conclusion

From the above experimental results, we can find the characteristics of azimuth angle profile of Varanasi city. The graphs show that the azimuth angle profile changes

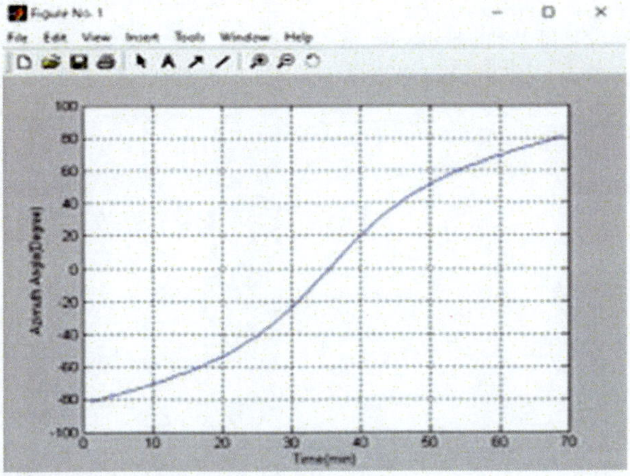

Fig. 4 Azimuth angle profile of Varanasi city on March 1, 2020 during spring ($N = 61$)

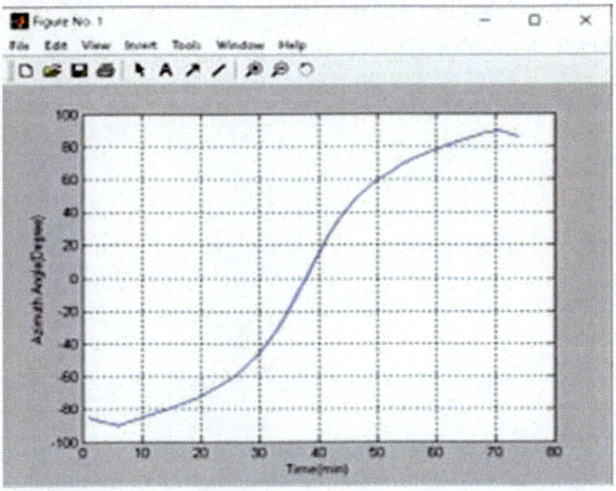

Fig. 5 Azimuth angle profile of Varanasi city on April 1, 2020 during summer ($N = 92$)

abruptly during summer. The summer season should be considered more economically crucial for PV installation technicality. This helps the engineers to envisage the solar energy scenario of the city before its up gradation to energy independent smart city. The prime novelty of this research work is to identify and explore the solar energy security of a future smart city of Varanasi to meet the craving energy demand for the future.

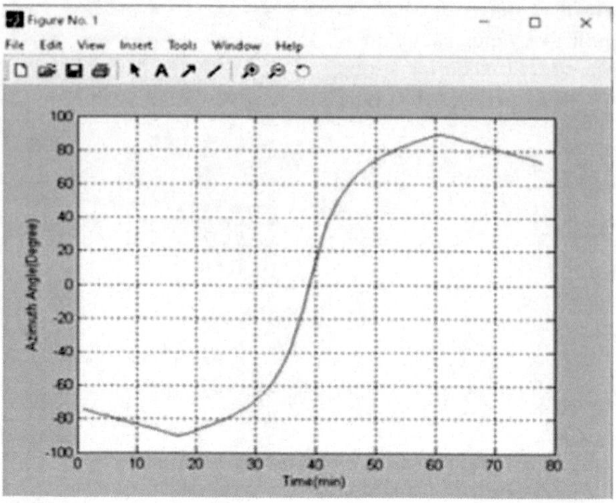

Fig. 6 Azimuth angle profile of Varanasi city on May 1, 2020 during summer ($N = 122$)

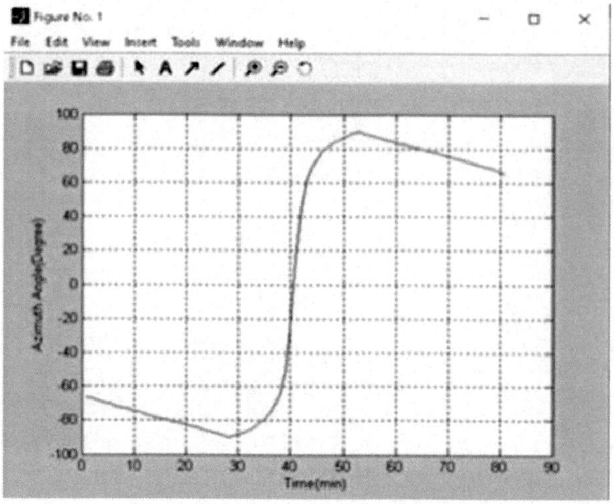

Fig. 7 Azimuth angle profile of Varanasi city on June 1, 2020 during summer ($N = 153$)

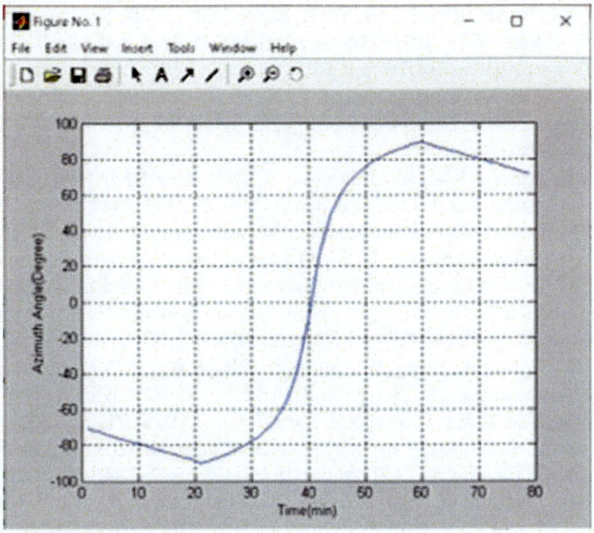

Fig. 8 Azimuth angle profile of Varanasi city on August 1, 2020 during autumn ($N = 214$)

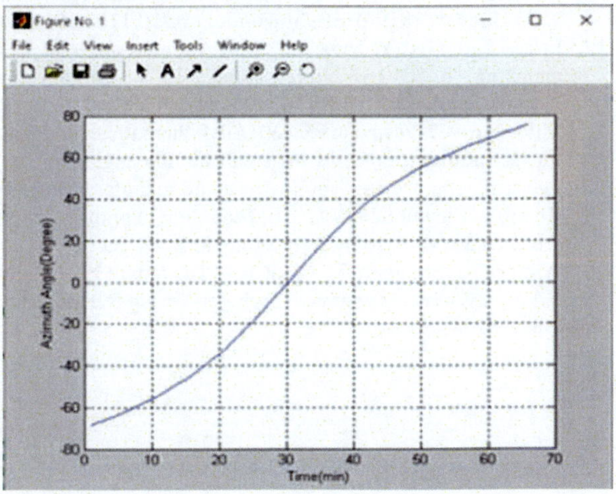

Fig. 9 Azimuth angle profile of Varanasi city on November 1, 2020 during autumn ($N = 306$)

References

1. Etukudor C, Orovwode H, Wara S, Agbetuyi F, Adoghe A, Obieje BO, Oparaocha CN (2018) Optimum Tilt and Azimuth Angles for Solar Photovoltaic Systems in South-West Nigeria, 2018 IEEE PES/IAS PowerAfrica, 978-1-5386-4163-7/18/$31.00 ©2018 IEEE
2. Takakura H, Koujitani K, Akhmad K, Belley F, Kitamura A, Hamakawa Y (1994) Simulation calculation of solar array output connecting two sub-arrays with different azimuth angles. CH3365-4/94/0000-0840 $4.00 0 1994 IEEE, First WCPEC, Hawaii, 5–9 Dec 1994
3. Ziółkowski C, Kelner JM (2018) Statistical evaluation of the azimuth and elevation angles seen at the output of the receiving antenna. In: IEEE Transactions on antennas and propagation. IEEE, pp 0018–926X (c). https://doi.org/10.1109/TAP.2018.2796719

4. Jha SK, Roy S, Singh VK, Mishra DP (2020) Sun's Position Tracking by Solar Angles Using Matlab. In: ICREISG-2020. IEEE, Bhubaneswar India, 14–15 Feb 2020. 978-1-7281-7034-3/20/$31.00 ©2020 IEEE
5. Gan X, Wang Y, Yang T, Li H (2017) Influence on azimuth angle and water surface roughness on SAR imagery of a bridge. In: IGARSS 2017, IEEE. 978-1-5090-4951-6/17/$31.00 © 2017 IEEE
6. Ahmed KM, Chairuangwit S, Rajatheva RMAP Effect of Azimuth Station and Terrestrial Radio Relay Link. In: ICPWC'99, 0-7803-4912-1/99/$10.0001999 IEEE. Angle Probability Distribution on Interference between Mobile Satellite Gateway
7. Iwasa S, Susaki J (2011) Classification of building area using azimuth angle and density indices derived from polarimetric SAR. In: Stilla U, Gamba P, Juergens C, Maktav D (Eds) JURSE 2011—Joint Urban Remote Sensing Event—Munich, Germany, 11–13 April 2011. 978-1-4244-8657-1/11/$26.00 ©2011 IEEE
8. Wang H, Wang JK, Song X (2004) Azimuth/elevation angle estimation via TST-ESPRIT technique. 0-7803-8560-8/04/$20.00 © 2004 IEEE
9. Luo Y, Xie J, Huang H, Tang J, Lu H, Yu J, Zhang J, Chen Z, Wang X, Peng J, Dai N (2015) Azimuth angle orientation of a microstructured optical fiber using its side image characteristics. IEEE Photonics Technol Lett 27(17):1837–1840
10. Damarla T Azimuth & elevation estimation using acoustic array. US Army Research Laboratory. https://doi.org/10.1109/ICIF.2010.5711874
11. Kumar M, Kumar A (2019) Performance assessment of different photovoltaic technologies for canal-top and reservoir applications in subtropical humid climate. IEEE J Photovoltaics 2156–3381 © 2019 IEEE. https://doi.org/10.1109/JPHOTOV.2019.2892520
12. Li L, Liu B, Lai Y, Tang X (2012) Azimuth angle of GPS point and its application in map-matching. In: 2012 15th international IEEE conference on intelligent transportation systems. Anchorage, Alaska, USA, 16–19 Sept 2012
13. Milovzorov D, Yasoveyev V (2016) Mathematical modeling of the determining azimuth process for inclinometric systems for small incline angles. In: 2016 2nd international conference on industrial engineering, applications and manufacturing (ICIEAM)
14. Khatib T, Elmenreich W (2016) Modeling of photovoltaic systems using MATLAB: simplified green codes. © 2016 by John Wiley & Sons, Inc.
15. Winter C-J, Sizmann RL, Vant-Hull LL (eds) (1991) Solar power plants fundamentals, technology, systems, economics. Springer-Verlag Berlin, Heidelberg. https://doi.org/10.1007/978-3-642-61245-9

Spectrum Based Prediction for Seismic Activity

Pranab Hazra, Soumashis Das, Soumendu Biswas, Pratiti Debsharma, and Krishnendu Ghosh

Abstract Based on the retrospective study of seismic activity we have analyzed prospective seismic activity based on results from Continuous Wavelet Transform (CWT) and from studies on The logged data of Very Low Frequency (VLF) transmitted sub-ionospheric signals at 16.4 kHz from Novik, Norway (Lat: 66.97° S; Long: 13.9° E), 19.8 kHz from North West Cape, Australia (Lat: 21.82° S; Long: 114.16° E) and 25 kHz from Petropavlovsk-Kamchatsky, Russia (Lat: 53.15° N; Long:158.92° E,) at Kolkata (Lat: 22.56° N, Long: 88.5° E) are studied throughout the period of April 3, 2013–April 24, 2013, when there happened 18 large earthquakes with $M \geq 5$. Here the introduction of other signals which are generated due to the seismic activity (considered as noise) is captured from the spectrum analysis using the method of CWT. In this method, we can watch a yellow region in the spectrum of blue color. Blue color spectrum is for VLF signals without any noise in fair-weather conditions and in this yellow color indicates the introduced noise which may also start to observe few hours (12 h) prior to the event of the earthquake. The identified event may have been the result of a combination of changes in seismicity patterns and the yellow color gives the forecasting of the main event as a signature of the prediction.

Keywords Earthquake · CWT · Morse wavelet · Atmosphere ionosphere coupling · Scalogram

1 Introduction

Different techniques are studied to find the probability for the prediction of earthquakes. Seismic activities radiate electromagnetic waves, and this can be used for the prediction of an earthquake. Due to tectonic plate movements, these waves are generated. This is also created under the earth's crust due to the oxidation of different reactive metals present. In our proposed method we are able to detect these waves

P. Hazra (✉) · S. Das · S. Biswas · P. Debsharma · K. Ghosh
ECE, Narula Institute of Technology, Kolkata, India
e-mail: pranabhazra2017@nit.ac.in

M. Mitra et al. (eds.), *Computational Advancement in Communication, Circuits and Systems*, Lecture Notes in Electrical Engineering 786,
https://doi.org/10.1007/978-981-16-4035-3_22

much before the earthquake occurs by its seismic radiations and take measures to be safe. We are also able to observe the frequencies of extremely low frequency (ELF) bands. Before an earthquake within the ionosphere, electromagnetic anomalies are observed and are found as evidence that processes of the earthquake. Due to the actions of various factors, i.e., electromagnetic radiation, electric fields as a result ionospheric effects can be seen. These factors can be formed by aerosols of the atmosphere. The anomalies create a circular path or waveguide through which the signals are coming. This particular path of the waveguide is known as Earth Ionosphere Waveguide. If some distortion is observed in this waveguide this will hamper the signal which will be recorded. This phenomenon is considered for the existence of "Atmosphere—Ionosphere Coupling" [1]. Due to different atmospheric discharges in the tropical regions, the seismic radiations are merged with background noises. This is because of the different ionosphere altitudes day and night. The more is the distance of the ionosphere with the earth's surface, the less will be the noise propagation [2, 3]. In our proposed idea we are to use a very simple tool named continuous wavelet transformation (CWT) to visualize the pattern of the waves. From the data sets of the seismic radiation we could predict when the earthquake happened and with how much intensity from its frequency [4].

2 Materials and Methods

Here we have proposed a simple method to process the data received for the detection of the seismic waves. ELF band EM radiations are usually represented by east-west, north-south, and vertical magnetic field components. Firstly, the EM waves of the ELF band are collected through a tri-axes loops antenna. The collected data are averaged over 6 s' intervals, i.e., 14,400 points per day. In the signals that are received, the vertical axis represents a density of a magnetic flux and the horizontal axis represents time [5, 6].

2.1 Normalization

The signal needs to be normalized to bring all the signals to a normalized range. We can do it through Matlab. For example, if we have a signal $s = \text{randn}(1,1\text{e}6)$; that is normalized in the following way:

$$s = s/\text{sqrt}(\text{sum}(\text{abs}(s.2))/1\text{e}6) \quad (1)$$

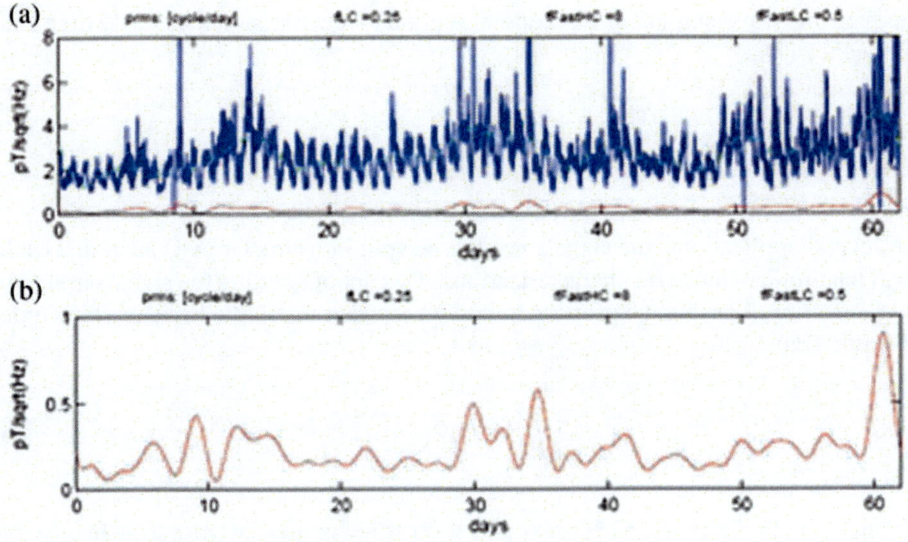

Fig. 1 **a** Received signal, **b** filtered signal by band-pass filter

2.2 *Filtering*

The observed signals contain global noise which has stronger power than local signals. Therefore, global noise distorts the results of earthquake prediction. In order to overcome this distortion, it is necessary to eliminate global noises from the observed signals. For extraction of the seismic radiation, the low frequency (LF) component with frequency smaller than 8 cycles/day is filtered out from the signal [7]. For example, if we consider a random signal in Fig. 1a, it consists of global noise. We need to filter the noise to reduce the distortion of signal for earthquake prediction. The filtering of the signal is done by a conventional band-pass filter.

2.3 *CWT*

Then a transformation of the signals and visualization is to be done by a tool called CWT (Continuous Wavelet Transform). The CWT is an integral transform which can quantity the time-frequency variation. Mathematically, CWT function $f(t)$ can be written as:

$$\omega(a, b) = \int f(t)\varphi * \left(\frac{t-b}{a}\right)\mathrm{d}t \tag{2}$$

where, $\varphi*$ signifies the complex conjugate, and ‘a’ and ‘b’ are dilation and translation parameters, respectively. The $\omega(a, b)$ function represents the wavelet coefficients. We applied this wavelet transform because it is good to localize simultaneously time and frequency [7].

2.4 Segmentation

In several applications, the signals need to be split into smaller parts such that each part has similar statistical characterizations. This act of splitting the signals is signal segmentation. For this we propose to apply spectrogram on the basis of short time Fourier Transform:

$$S(t, f) = \int_{-\infty}^{\infty} X(\tau)W(\tau - t)\mathrm{e}^{-j2\pi f\tau}\mathrm{d}\tau \tag{3}$$

where $X(\tau)$ is the analyzed process and $W(t)$ is an appropriate window. In case of real signal 3 $X_1, X_2, \ldots, X_N$ analysis the discrete short time Fourier transform takes the form:

$$S(t, f) = X_k W_{k-t}\mathrm{e}^{-j2\pi fk} \tag{4}$$

On the basis of $S(t, f)$ function, the signal can be analyzed in time-frequency domain. The square of the absolute value of the matrix $S = S(t, f)$ is called spectrogram. In the next step of the proposed segmentation procedure of seismic signal, for each frequency f we analyze the time series $S(:, f)$ separately. Such time series are called subsignals. For each subsignal we calculate quantiles of a given order p. Quantiles are calculated on the basis of subsignals corresponding to given frequencies. For each frequency f we obtain different quantile of order p. In the next step of our segmentation procedure, we calculate so-called binary matrix, which contains zero–one values and indicates if given observation from subsignal corresponding to frequency f exceeds the quantile $x(p, f)$. The binary matrix we define as follows:

$$S_{\text{bin}}(t, f) = \begin{cases} 1, & S(t, f) > x(p, f) \\ 0, & S(t, f) \le x(p, f) \end{cases} \tag{5}$$

Next, we calculate the vector of probabilities for each time point t, which is equal to the mean of the binary matrix at each time point. After calculating the vector of probabilities, we determine the percentage threshold k. On the basis of this threshold, we calculate the limit of normalized vector $P(t)$. This limit is calculated as follows:

$$l = k^{*}\max(P(t)) \tag{6}$$

On the basis of calculated limit we analyze a decision vector defined as follows:

$$D(t) = \begin{cases} 1, & P(t) > l \\ 0, & \text{otherwise} \end{cases} \quad (7)$$

Next, we analyze the increments of the decision vector $D(t)$: $\Delta D(t) = D(t) - D(t - 1)$.

3 Results

See Figs. 2, 3, 4, 5, 6 and 7; Tables 1 and 2.

4 Discussion

It is very important to have a visual display of the alarm signal for the detection of the precursor of the earthquake. It is also important to investigate the relationships between the received EM wave in the ELF band and the date and place of earthquake occurrence. We might also work for a system that can automatically verify the relations between the database of the received wave in the ELF band and the date and place of earthquake occurrence by the input data of the actual earthquake occurrence [8].

In this section of result analysis, we propose a simple method for signal processing of the received experimental data in order to detect and plot the seismic radiation. Here we used the tool called CWT (Continuous wavelet Transformation) to visualize the earthquake pattern. The scalogram is the absolute value of the CWT plotted as a function of time and frequency. Frequency is plotted on a logarithmic scale. The cone of influence showing where edge effects become significant is also plotted. Gray regions outside the dashed white line delineate regions where edge effects are significant [9, 10].

12–24 h before the earthquake, a crake is formed from the epicenter which is called Micro Crake which cannot be seen with the eye. Through this crake, all the radioactive elements such as radon that we have on the surface of the earth come out and take place into the atmosphere. There are always some electrons and particles moving in the atmosphere. The radioactive element is very energetic and it hits the particles and ionized them into '+' ions and '−' ions. Then those ions become energized by receiving the energy.

Some of the radioactive elements hit the electrons that move freely in the atmosphere and these electrons collect energy from those elements. Then the electrons get energized and hit other electrons and energized it too. This randomly hitting process continues in the chain process. As a result, the movement of electrons becomes random. Movement of electrons means that some noise signals are generated and then they are adapted to the main signal. Due to the energy and being light weighted, particles move into the Stratosphere. Ionosphere lies between the Stratosphere and Troposphere.

Due to the air movement and light weight, particles move to the ionosphere. There warming is created for the random movement of the particles. So, the whole process of disturbing the ionosphere by going up from the atmosphere is called "Atmosphere–Ionosphere Coupling".

In Fig. 2 we are using Morse wavelet of CWT. This one produced joint time frequency visualization of the given input data. Here the minimum and maximum scale for the analysis are determined automatically by CWT function based on the wavelet's energy spread. The magnitude of the signal is denoted by the color coded line. The gray dashed line denotes the cone of influence within this region the wavelet coefficient estimates are reliable. Looking at the plot we can see the region produced

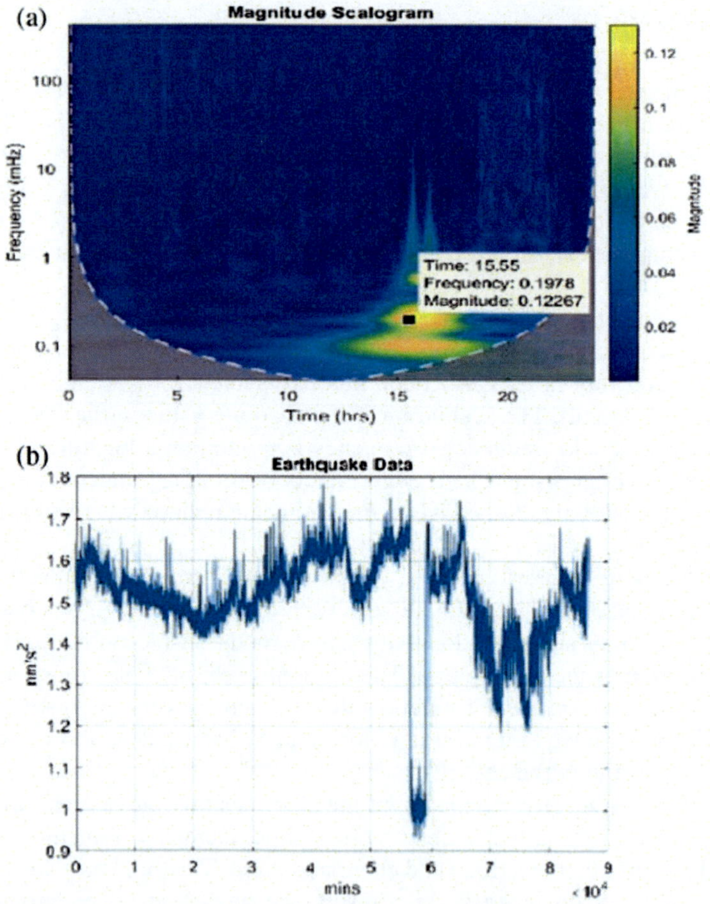

Fig. 2 **a** Scalogram representation of data; **b** wave plot of earthquake data

Fig. 3 Map showing the location of the receiver (by green rectangle) and the transmitter (by red rectangles). The transmitters are at Norway, Russia, and Australia; the receiver is located at Kolkata, respectively. The three VLF radio paths are indicated by black lines

by the earthquake in a yellowish manner. If we compare both the figure it's cleared where an abrupt change is notified the yellow spectrum also generated there.

Figure 3 shows a map where the receiver is located at Kolkata represented by a green square and the transmitters are located at three different places at Norway, Russia, and Australia represented by red squares. Thus, joining the path from the receiver to the transmitter we get the three very low frequency (VLF) radio paths represented by black straight lines. We can get the data of all the places lying on the VLF radio paths in between the transmitting stations and the receiving stations.

Table 1 is the table representation of recorded data with its magnitude and distance from earthquake region any many more key factors which have tentative effect to determine any CWT outcome from any sources.

Table 2 represents the time of occurrence of the first spectrum at different frequencies, 25 kHz, 19.8 kHz, and 16.4 kHz respectively. It's the black line (VLF radio paths) joining the receiver located at Kolkata with different transmitters placed across the world. At certain places tracing data becomes impossible as the VLF radio paths don't lie in that region.

Figures 4, 5 and 6 is the scalogram representation of the signal at different frequencies of the transmitted signal at 25 kHz, 19.8 kHz, and 16.4 kHz respectively at different time intervals.

Table 1 Date-wise occurrence of several earthquakes along with the descriptions of different connecting parameters

Date	Time (I.S.T)	Depth (km)	Magnitude (M)	Region	Distance from kokata (22.56° N, 88.5° E), *D* (km)
03.04.2013	22:05:48	10	5.5	Myanmar (17.17° N, 95.99° E)	784
04.04.2013	12:01:37	237	5.8	Hindu Kush, Afghanistan (36.89° N, 73.26° E)	2218
04.04.2013	20:46:26	40	5.4	Myanmar (17.17° N, 95.99° E)	831
05.04.2013	18:30:02	570	6.3	N.E. China Border (46.40° N, 127.61° E)	4505
06.04.2013	04:25:00	80	5.4	Hindu Kush, Afghanistan (36.89° N, 73.26° E)	2246
06.04.2013	10:12:40	90	7.1	Papua Indonesia (−3.98° N, 138.34° E)	6118
09.04.2013	17:22:55	42	6.2	Southern Iran (29.59° N, 52.56° E)	3746
11.04.2013	09:17:03	12	5.4	Myanmar (17.17° N, 95.99° E)	831
14.04.2013	07:02:24	41	6.5	Papua New Guinea (−5.68° N, 154.5° E)	7862
16.04.2013	14:04:09	10	5.0	India (Arunachal Pradesh) (27.68° N, 97.65° E)	967
16.04.2013	16:14:11	46	7.8	Pakistan-Iran Border (28.0° N, 62.1° E)	2717
17.04.2013	04:25:26	10	6.8	Papua New Guinea (−5.68° N, 154.5° E)	4519
19.04.2013	08:35:51	97	7.3	Kuril Islands (50.38° N, 155.83° E)	6112

(continued)

Table 1 (continued)

Date	Time (I.S.T)	Depth (km)	Magnitude (M)	Region	Distance from kokata (22.56° N, 88.5° E), *D* (km)
20.04.2013	01:28:40	10	6.1	Kuril Islands (50.38° N, 155.83° E)	6623
20.04.2013	05:32:48	29	6.6	Sichuan, China (30.50° N, 102.49° E)	1673
20.04.2013	18:42:47	10	6.1	Kuril Islands (50.38° N, 155.83° E)	6609
21.04.2013	08:52:16	422	6.2	Izu Islands, Japan (34.98° N, 138.95° E)	5054
24.04.2013	14:55:29	66	5.7	Hindu Kush, Afghanistan (36.89° N, 73.26° E)	2221

In Fig. 7 there are some cases in which the schematic spectrum pattern of an earthquake data where the yellow spark in between shows the high magnitude waves that pass in form of an earthquake and in some there is only a blue spectrum. The plot with only the blue spectrum Fig. 7a is the farthest from the recorder center (Kolkata) and the plot with yellow sparks Fig. 7b is the closest to the recorded center. It shows here that as the earthquake is getting closer to the recorded center (Kolkata) the intensity increases.

5 Conclusion

The continuous wavelet transform of a signal which has a finite length with a fixed scale is considered to be a linear time-invariant operator. The edge effect can be handled by using Equivalent Condition of Time-domain and Frequency domain Algorithms of CWT. 13 We have recorded the data of earthquakes of different places on different dates. With the help of the CWT algorithm, we observe that it generates some blue spectrum in which some yellow sparks are induced. Basically, the blue spectrum represented the generalized value or generalized frequencies.

Yellow sparks denoting unnatural behavior of the frequencies of nature. Here this unnatural behavior is in seismic magnitude scale which is the overall strength or size of an earthquake. These yellow sparks are also represented as an earthquake signal. These earthquake signals are evident that the earthquake has occurred at this particular time. We are jotting down those signal values from Kolkata. For this reason,

Table 2 Data-wise occurrence of the spectrum of the data

Date	Time (I.S.T)	First spectrum time (25 kHz)	First spectrum time (19.8 kHz)	First spectrum time (16.4 kHz)
03.04.2013	22:05:48	5:43 am	**	**
04.04.2013	12:01:37	4:50 am	**	5:08 am
04.04.2013	20:46:26	5:37 am	**	**
05.04.2013	21:30:02	4:50 am	**	**
06.04.2013	04:25:00	21:57 pm on 5.04.13	**	**
06.04.2013	11:12:40	5:43 am	**	5:08 am
09.04.2013	05:22:55	**	**	12:44am
11.04.2013	09:17:03	1:48 am	**	**
14.04.2013	07:02:24	**	1:58 am	**
16.04.2013	14:04:09	2:01 am	1:42 am	6:47 am
16.04.2013	16:14:11	2:01 am	**	5:52 am
17.04.2013	04:25:26	**	19:35 pm on 16.04.2013	**
19.04.2013	09:35:51	**	**	**
20.04.2013	01:28:40	**	**	**
20.04.2013	05:32:48	23:42 pm on 19.04.2013	**	**
20.04.2013	18:42:47	**	**	**
21.04.2013	08:52:16	**	**	**
24.04.2013	14:55:29	5:37 am	**	4:29 am

the signal whose distance is very large, cannot take a powerful signal. So that the intensity or the magnitude is small and the spikes are not prominent in most of the cases. The signals whose distance is small can take powerful signals. So that the intensity or the magnitude is large and thus the spikes are coming prominently.

By using a monitoring system, we will monitor that at what time, at which place, in which intensity the earthquake has occurred. We will represent it as a visual representation by using CWT Algorithm.

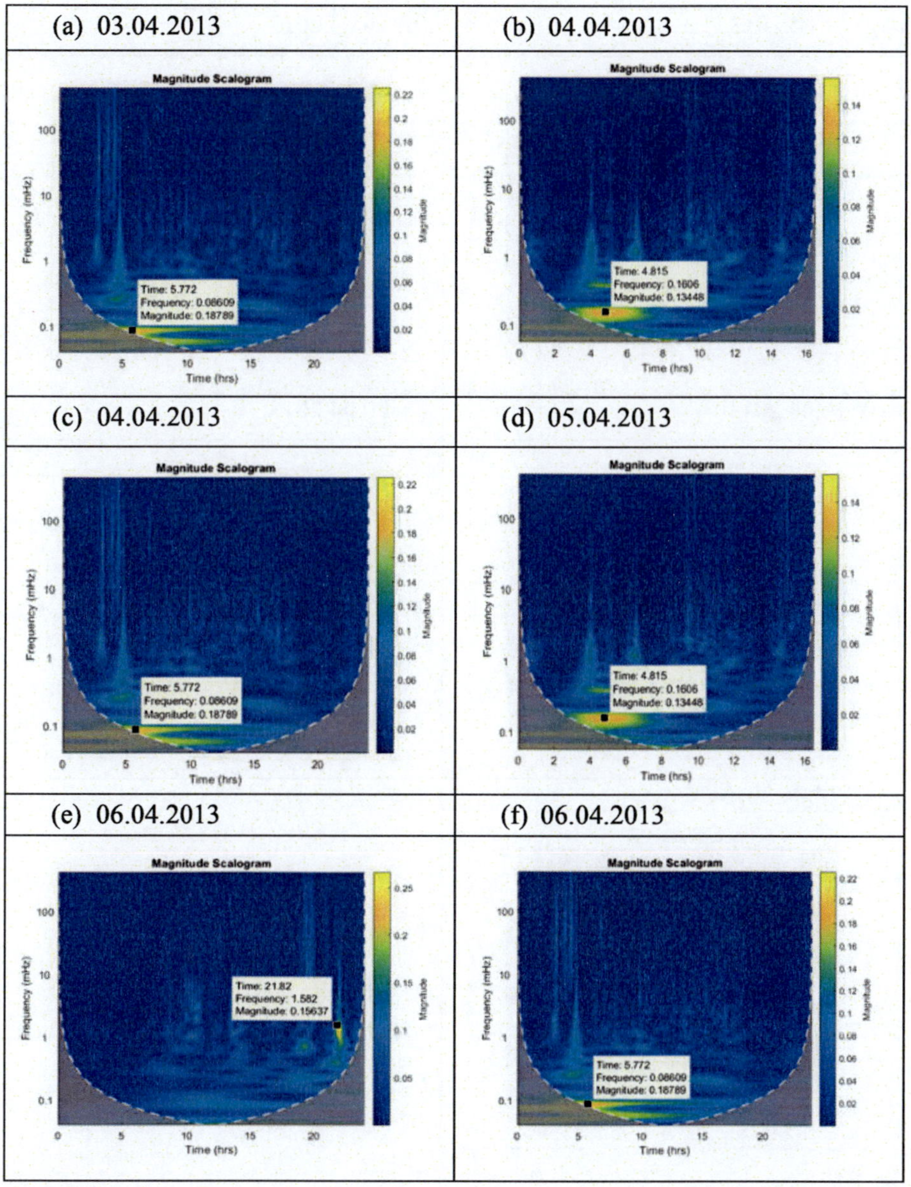

Fig. 4 Scalogram representation of signal for the earthquakes which are lie in the path of 25 kHz transmitted signal

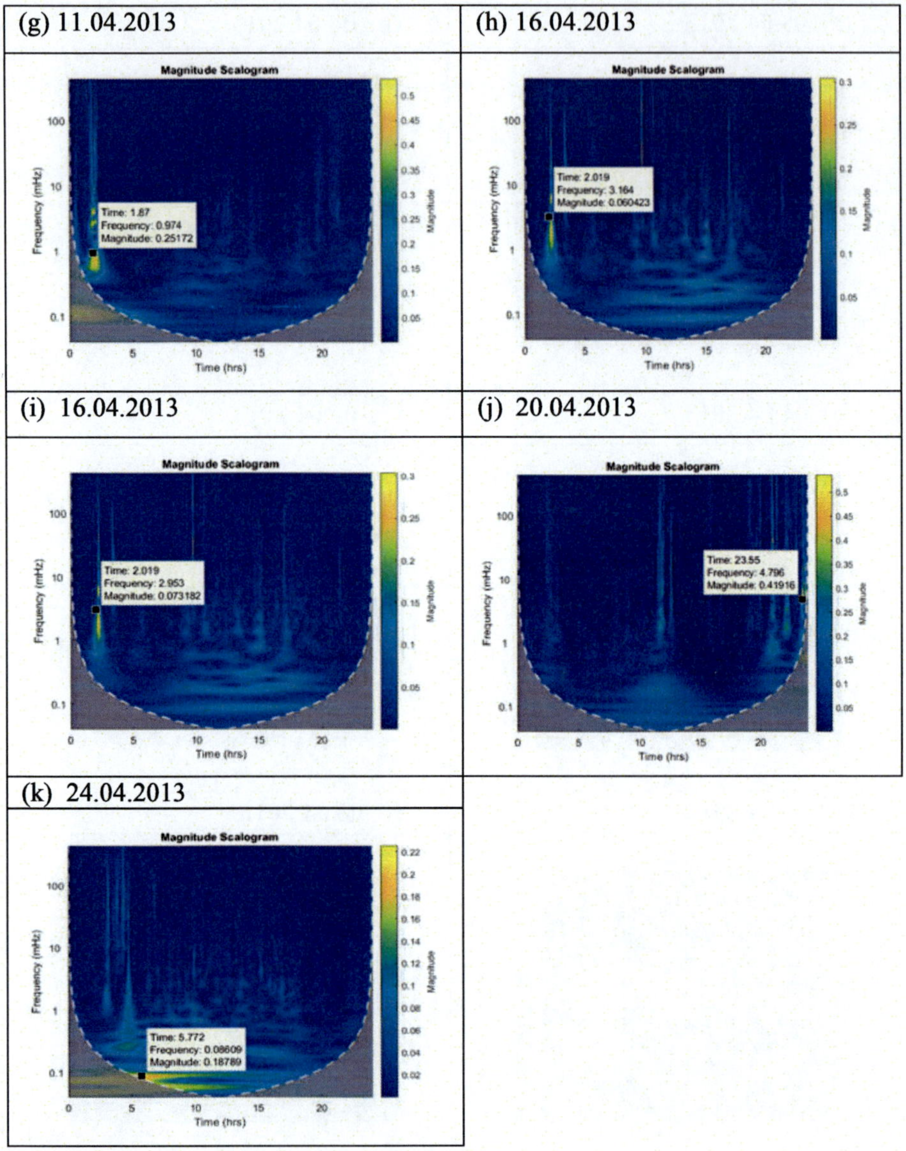

Fig. 4 (continued)

Acknowledgments The authors acknowledge with thanks the support from Narula Institute of Technology, Agarpara, and Kolkata, India for encouraging research work and giving scope in this regard. We are also thankful to the Department of Electronics and Communication Engineering of Narula Institute of Technology, for carrying out the study.

The authors gratefully acknowledge the support of the S. K. Mitra Center for Research in Space Environment, Institute of Radio Physics and Electronics, University of Calcutta, Kolkata 700,009, India, in carrying out this work as well as providing the recorded data. The authors are also thankful

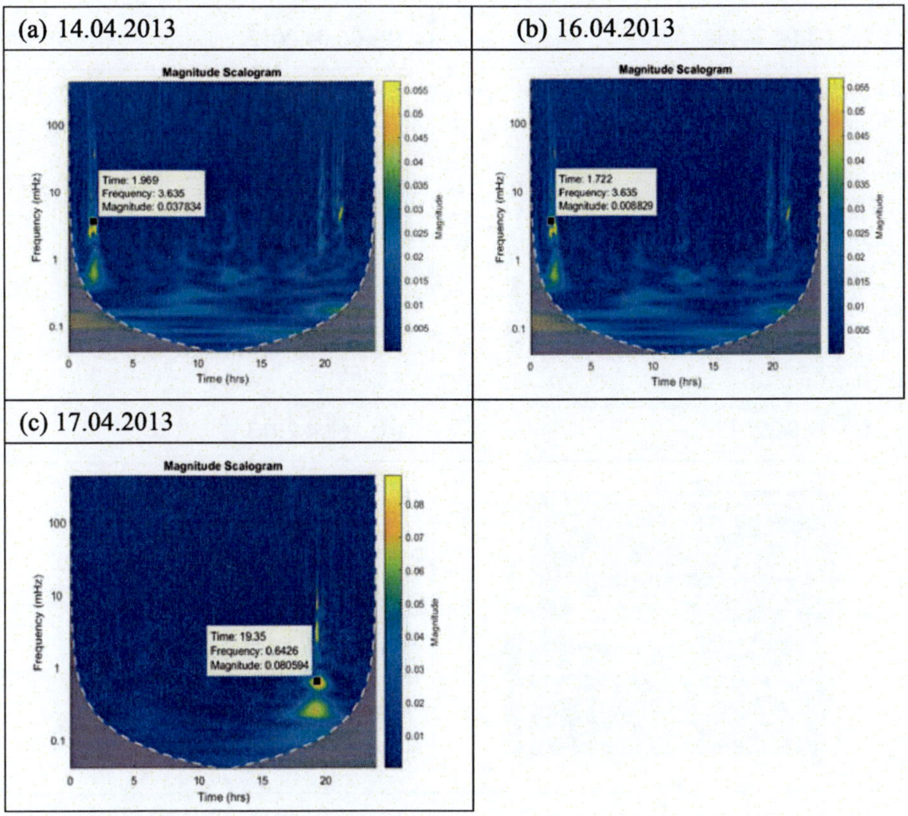

Fig. 5 Scalogram representation of signal for the earthquakes which are lie in the path of 19.8 kHz transmitted signal

to all the scholars of that lab who had helped a lot. The authors are especially grateful to Prof. S. S. De (Retd.), Institute of Radio Physics and Electronics, University of Calcutta, for his effortless guide and profound knowledge to work on this field and complete this work. Authors are thankful to the respected authors of the papers which helped greatly to complete this work.

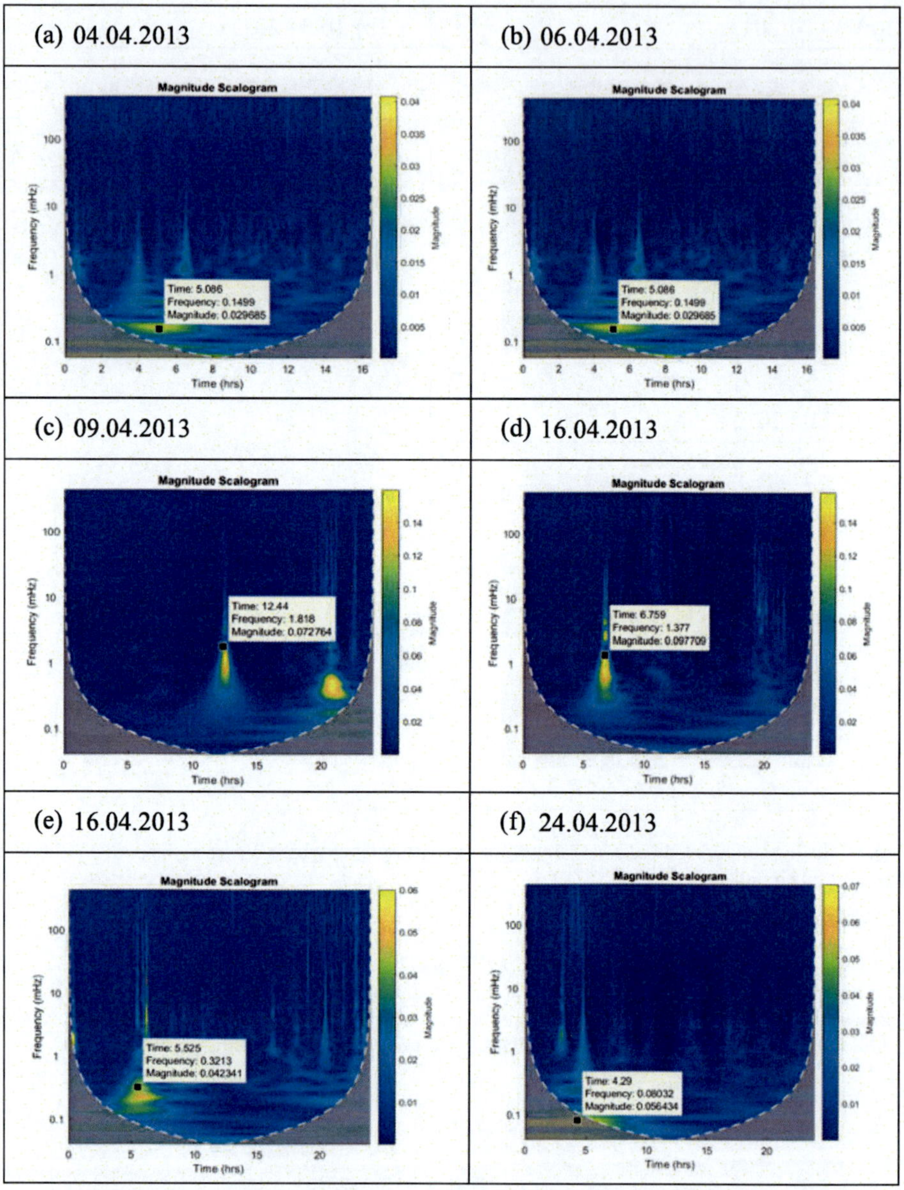

Fig. 6 Scalogram representation of signal for the earthquakes which are lie in the path of 16.4 kHz transmitted signal

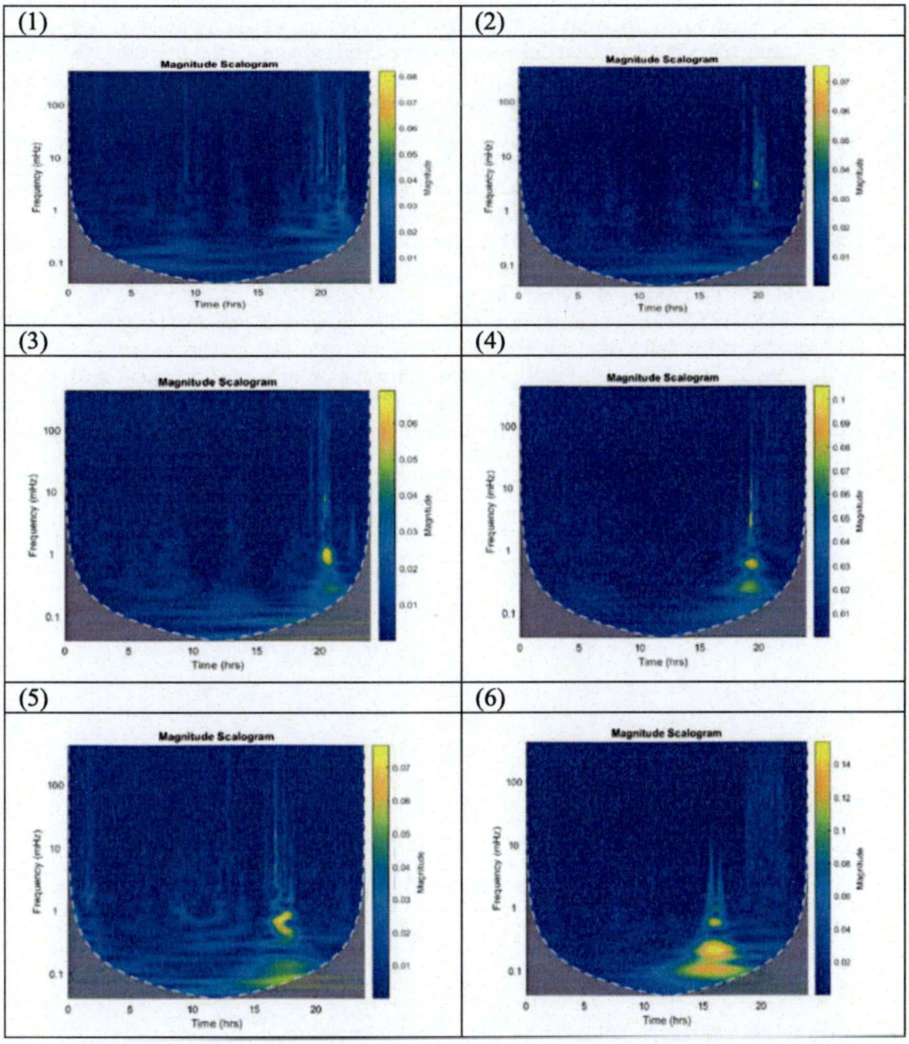

Fig. 7 Intensity increasing as earthquake is getting closer to the recorded center (Kolkata), i.e., **a** Kuril Islands (distance from Kolkata 6629), **b** Papua Indonesia (distance from Kolkata 6118), **c** Papua New Guinea (distance from Kolkata 4519), **d** 07/04/2013 (16.4 kHz), **e** Pakistan-Iran Border (distance from Kolkata 2717), **f** 08/04/2013 (16.4 kHz)

References

1. Chakraborty A, Okaya D (1995) Application of wavelet transform to seismic data. Geophysics 60
2. Ikelle LT, Amundsen L (2012) Introduction to petroleum seismology, investigation in geophysics. Series No 12
3. Liner CL (2012) Element of seismic dispersion: a somewhat practical guide to frequency-dependent phenomena

4. Ansari A, Noorzad A, Zafarani H, Vahidifard H Correction of highly noisy strong motion records using a modified wavelet denoising method. Soil Dynam Earthq Eng 30:1168–1181
5. Amezquita-Sanchez JP, Adeli H (2015) A new music-empirical wavelet transform methodology for time frequency analysis of noisy nonlinear and non-stationary signals. Digit Signal Process 45:55–68
6. Chen J, Rostami J, Tse PW, Wan X (2017) The design of a novel mother wavelet that is tailor-made for continuous wavelet transform in extracting defect-related features from reflected guided wave signals. Measurement 110:176–191
7. Lee WK, Ratnam MM, Ahmad ZA (2017) Detection of chipping in ceramic cutting inserts from workpiece profile during turning using fast fourier transform (FFT) and continuous wavelet transform (CWT). Precis Eng 47:406–423
8. Leigh GM (2012) Fast FIR algorithms for the continuous wavelet transform from constrained least squares. IEEE Trans Signal Process 61:28–37
9. Yi H, Chen Z, Cao Y (2014) High precision computation of morlet wavelet transform for multi-period analysis of climate data. J Inf Comput Sci 11:6369–6385
10. Zhao YY, Yuan X, Wei YH (2006) Realization of continuous wavelet transform of sequences by MATLAB. J Sichuan Univ 43:325–329

Effect of Cognitive Task on the Central Nervous System

Ananya Banerjee, D. K. Bhattacharya, and Anilesh Dey

Abstract Electroencephalography (EEG) signal analysis has received great acknowledgment in the domain of biomedical signal processing for the interpretation of human brain activities. There is a close bonding between the EEG signal and human brain activities. In the human brain, millions of neurons interact with one other and as a result, we obtain electrical signals by placing the electrodes on the scalp in a non-invasive way. The human behavior (polite, rude, whimsical, etc.), mood (happy, sad, anger, depressed, etc.), sensory states (movement of the eye, lip, hand, etc.), cognitive task ability (understanding, thinking, problem-solving, implementation, debugging, recalling) can be monitored, interpreted and analyzed with the exploitation of EEG signals. Moreover, to detect neurological diseases and for treatment purposes, EEG signals are countless boons in the field of biomedical signals. The central nervous system is responsible for controlling human behavior, mood, cognitive task motor, and imaginary task to some extent. To find evidence, we have focused on the effects of cognitive tasks on the central nervous system. Due to the non-linearity and non-stationarity nature of the EEG signals, we have investigated the signals using non-linear tools like the Surrogate data test and phase space plot. Moreover, we have explored the topological scalp map view to obtain the visual effects of the scalp.

Keywords Biomedical signal · EEG · Central nervous system · Surrogate data test · Phase space plot

---

A. Banerjee (✉)
Department of Computer Science and Engineering, Narula Institute of Technology, Kolkata, India
e-mail: ananya.banerjee@nit.ac.in

D. K. Bhattacharya
Department of Pure Mathematics, University of Calcutta, Kolkata, India

A. Dey
Department of Electronics and Communication Engineering, Narula Institute of Technology, Kolkata, India
e-mail: anilesh.dey@nit.ac.in

M. Mitra et al. (eds.), *Computational Advancement in Communication, Circuits and Systems*, Lecture Notes in Electrical Engineering 786,
https://doi.org/10.1007/978-981-16-4035-3_23

1 Introduction

The human brain and its association with EEG signals have magnetized a plethora of interests for researchers in the last few decades. Due to close connectivity between the human brain and EEG signal, researchers have explored a potential amount of investigation to study brain activities during a cognitive task.

Inouye et al. [1] calculated the degree of ascendance by quantifying the entropy of the power spectrum for frequency bands (alpha, beta, and theta) of EEG signal to find irregularity during rest and active state of mind performing phrenic task. They detected anomalous patterns during rest and active states in the anterior region than occipital. They additionally visualized that there was EEG desynchronization while someone performed a noetic arithmetic task.

Thalla et al. [2] observed different patterns of EEG activation patterns while subjects performed mental arithmetic tasks. The absolute power and relative power of the alpha frequency band was visually detected during recognition of sign in the parietal anti occipital zone. On the other hand, the absolute power of the delta frequency band was more dominant in the process of the apperception of symbols cognate to arithmetic than during number-crunching task execution. This prefigured that in the posterior areas of the right-hand side, the pattern of the EEG was plenarily changed while the subjects went through the process of the apperception of symbols related to arithmetic. Moreover, consequential changes in the EEG signal pattern between the mental tasks were visually found in the frontal lobes belonging to the beta frequency band. Like alpha, the higher beta frequency was observed while subjects performed the cognitive tasks rather than in the apperception of signs. There was no substantial difference visually found when subjects performed subtraction during rest conditions.

Skrandies et al. [3] studied visual information processing of subjects during the performance of arithmetic tasks for a group of salubrious and dextral males and females. Field strength and topographical view of the scalp assisted to visualize the effects of the subjects prominently. The topographical visualization also helped to understand the effect of males and females during the visualization of information processing. The amplitudes were more preponderant while the subjects went through active reading than went through reading without processing. The authors showed gender differences in terms of the event-cognate encephalon activity during the performance.

Wang et al. [4] observed that in the left side of the occipital at 3.9 Hz EEG changes were visually examined during visual encoding of words. The same effect was visible in the regions of tempo-occipital at 4.68, 5.46, and 6.24 Hz. During the apperception of the figure, at 7.8 and 17.94 Hz changes were found in front transient and prefrontal segments. At 3.12 and 7.02 Hz, no significant differentiation between assigned tasks was outwardly analyzed. Authors suggested an approach fortified by a hypothesis known as amodal semantic where the logogen and iconogen systems deported as an input system [5, 6].

Zyma et al. [7] experimented with EEG predicated noetic arithmetic task of 10 subjects and examined the precision in the amalgamation of statistical feature, GHFDS, autoregressive model (AR), and power spectrum density (PSD), methods to ameliorate the neurofeedback system of patients suffering from attention disorder quandary. They achieved a 97.87% precision level by exploiting all the features. Higher precision authoritatively mandated more enhancement and efficiency of the neurofeedback system.

Snodgrass et al. [6] experimented to amass the neurological activities of the encephalon utilizing EEG data while subjects went through a psychological undertaking and evaluated the competency of participants to accomplish the arithmetic task. The author additionally utilized the feature extraction methods such as coherence, Fourier power spectrum, and detrended fluctuation to explore the encephalon activation pattern on individual task arduousness.

In this work, EEG signals are accumulated during the relaxed state as well as the stressed state while the subjects solve the mental arithmetic tasks (serial subtraction). With the avail of global analysis, the attractors are reconstructed to presage the signal features. Most of the research papers cognate to this domain using the fractal analysis or multifractal analysis to soothsay features [7–11] whereas we have used global analysis to study the effector vicissitudes in the encephalon segments using Surrogate data test, Phase space reconstruction, and topological scalp map view. Here the aim is to ascertain which component of the encephalon is getting affected the most.

2 Method

2.1 Mental Arithmetic Task Experiment

The Neurogram EEG 23-channel system is employed for recording the EEG signals. During the build-up of information, the chloride electrodes were positioned on the scalp at frontal channels—F3, F4, Fz, F7, F8, temporal channels—T3, T4, T5, T6, central channels—C3, C4, Cz, parietal channels—P3, P4, Pz, occipital channels—O1, O2 and symmetrical anterior frontal (Fp1, Fp2) in line with the International 10/20 scheme. The subtraction of 2 numbers was the assigned task for the subjects. At first, subtraction of 4-digit numbers and 2-digit numbers were received by the participants whom they had to try verbally. The information was taken upon only those participants who had normal or rectified-to-normal sight or normal vision. For the recording of EEG data, 50 Hz transmission line notch, and 30 Hz interrupt frequency filters were utilized. For artifact removal techniques, ICA or Independent Component Analysis (ICA) was used during processing and the data was of 60 s duration. In total 66 salubrious dextral subjects (men of 19 and women of 47 strengths) were associated with the experiment. All the subjects belonged to the National University of Taras Shevchenko and were the pupils from the first to the third year of age range from eighteen to twenty-six years. The members were qualified to be selected inside the examination if they had an ordinary visual capacity, and had no clinical indications of noetic or discerning harm or verbal or non-verbal learning inabilities. Oversight models were the utilization of psychoactive prescription, medication or intoxication, and mental or neurological objections [12].

2.2 Surrogate Data Test

Theiler et al. [13, 14] proposed Surrogate data test for the evidence of non-linearity nature of the signal-dependent on time-domain and the observation of the dynamics of the experimental data. It has become a popular method to find the non-linearity existence in the underlying observed data. In this method, the null hypothesis (H0) is specified preserving linearity, and surrogate data set is generated by amplitude adjusted Fourier transform according to the null hypothesis using the method Monte Carlo. For the original time series and surrogate data set, a discriminating statistical value is considered. Comparison is made between the original output and surrogate data set under a statistical framework to order to accept or decline H0. Whenever the statistical value in the pristine time series differs from surrogate data set to a great extent, we can assure the presence of the non-linearity time series properties.

2.3 Phase Space Reconstruction Analysis of a Dynamical System

During phase space construction, we can epitomize all the probable states of a dynamical chaotic system in a high-dimensional space [15]. The evolutionary state of the system follows a trajectory path in a multi-dimensional space over time. If the authentic number of variables that rule the comportment of the dynamical system is not kenned, the phase space can be reconstructed utilizing time-delayed embedding, which is derived from Taken's theorem [16]. Fang et al. [18] verbalized that a topological view of the pristine system is identically tantamount to reconstructed phase spaces and thus can instaurate the non-linear dynamics of the system.

Let $\{X(j)\}_{j=1}^{N}$ be a discrete EEG signal. Let the three dimensional frequency-delay plot be constructed by sub-dividing this signal into three groups as x^{+}, x^{-}, x^{--} with same time delay τ, where

$$
\begin{aligned}
x^{+} &= \{|X(j)|\}_{j=1}^{N-2\tau}, x^{-} = \{|X(j)|\}_{k=1+\tau}^{N-\tau}, \\
x^{--} &= \{|X(j)|\}_{j=1+2\tau}^{N}, \quad \tau = 1, 2, \ldots, (N-1)
\end{aligned}
\tag{1}
$$

Now this coordinate system is transformed by a three dimensional rotation with the same angle $\pi/4$ with respect to X, Y and Z axis. The transform is given by

$$\begin{pmatrix} x_m \\ x_n \\ x_p \end{pmatrix} = \begin{pmatrix} \cos\frac{\pi}{4}\cos\frac{\pi}{4} & \cos\frac{\pi}{4}\sin\frac{\pi}{4}\sin\frac{\pi}{4} - \cos\frac{\pi}{4}\sin\frac{\pi}{4} & \cos\frac{\pi}{4}\cos\frac{\pi}{4}\sin\frac{\pi}{4} + \sin\frac{\pi}{4}\sin\frac{\pi}{4} \\ \cos\frac{\pi}{4}\sin\frac{\pi}{4} & \cos\frac{\pi}{4}\cos\frac{\pi}{4} + \sin\frac{\pi}{4}\sin\frac{\pi}{4}\sin\frac{\pi}{4} & -\cos\frac{\pi}{4}\sin\frac{\pi}{4} + \cos\frac{\pi}{4}\sin\frac{\pi}{4}\sin\frac{\pi}{4} \\ -\sin\frac{\pi}{4} & \cos\frac{\pi}{4}\sin\frac{\pi}{4} & \cos\frac{\pi}{4}\cos\frac{\pi}{4} \end{pmatrix} \begin{pmatrix} x^{+} \\ x^{-} \\ x^{--} \end{pmatrix}$$

$$= \frac{1}{2\sqrt{2}} \begin{pmatrix} 2\sqrt{2} & -\left(\sqrt{2}-1\right) & \left(\sqrt{2}+1\right) \\ 2\sqrt{2} & \left(\sqrt{2}+1\right) & -\left(\sqrt{2}-1\right) \\ -2 & \sqrt{2} & \sqrt{2} \end{pmatrix} \begin{pmatrix} x^{+} \\ x^{-} \\ x^{--} \end{pmatrix} \quad (2)$$

Hence,

$$x_m = \frac{1}{2}\cdot x^{+} + \left(\frac{1}{2\sqrt{2}} - \frac{1}{2}\right)\cdot x^{-} + \left(\frac{1}{2\sqrt{2}} + \frac{1}{2}\right)\cdot x^{--}$$
$$= \frac{2\sqrt{2}\cdot x^{+} - \left(\sqrt{2}-1\right)\cdot x^{-} - \left(\sqrt{2}+1\right)\cdot x^{--}}{2\sqrt{2}};$$
$$x_n = \frac{1}{2}\cdot x^{+} + \left(\frac{1}{2\sqrt{2}} + \frac{1}{2}\right)\cdot x^{-} + \left(\frac{1}{2\sqrt{2}} - \frac{1}{2}\right)\cdot x^{--}$$
$$= \frac{2\sqrt{2}\cdot x^{+} + \left(\sqrt{2}+1\right)\cdot x^{-} - \left(\sqrt{2}-1\right)\cdot x^{--}}{2\sqrt{2}};$$
$$x_p = \left(-\frac{1}{\sqrt{2}}\right)\cdot x^{+} + \frac{1}{2}\cdot x^{-} + \frac{1}{2}\cdot x^{--} = \frac{-2\cdot x^{+} + \sqrt{2}\cdot x^{-} + \sqrt{2}\cdot x^{--}}{2\sqrt{2}}. \quad (3)$$

Thus a new co-ordinate system (x_m, x_n, x_p) is formed.

Figure 1a, b show the phase space plot of subject-0's encephalon afore and during the performance of the cognitive task. As we know that the axis of the ellipse stands as a vigorous designator of the transmuting energy dynamics of the Poincare plot so we are here calculating the volume of the ellipse composed due to the 2-D Poincare plot. We made rudimentary endeavors to ascertain the distinction between the volume of the ellipse during the relaxation phase and in the period of performing a cognitive task.

2.4 Average Mutual Information

Mapping of a uni-dimensional signal is done into multi-dimensional time series to perceive its underlying dynamics as it is a fact that short term analysis of the signal may not enlighten us the felicitous nature of the signal. For this purpose, long term analysis must be done and phase space is reconstructed. The optimal delay (τ), and D, the embedding dimension parameters are required for phase space reconstruction. AMI or Average mutual information is used to compute the optimal delay τ value. On other hand, False Nearest Neighbors (FNN) function is used to compute the value of the embedding dimension parameter D [17–19].

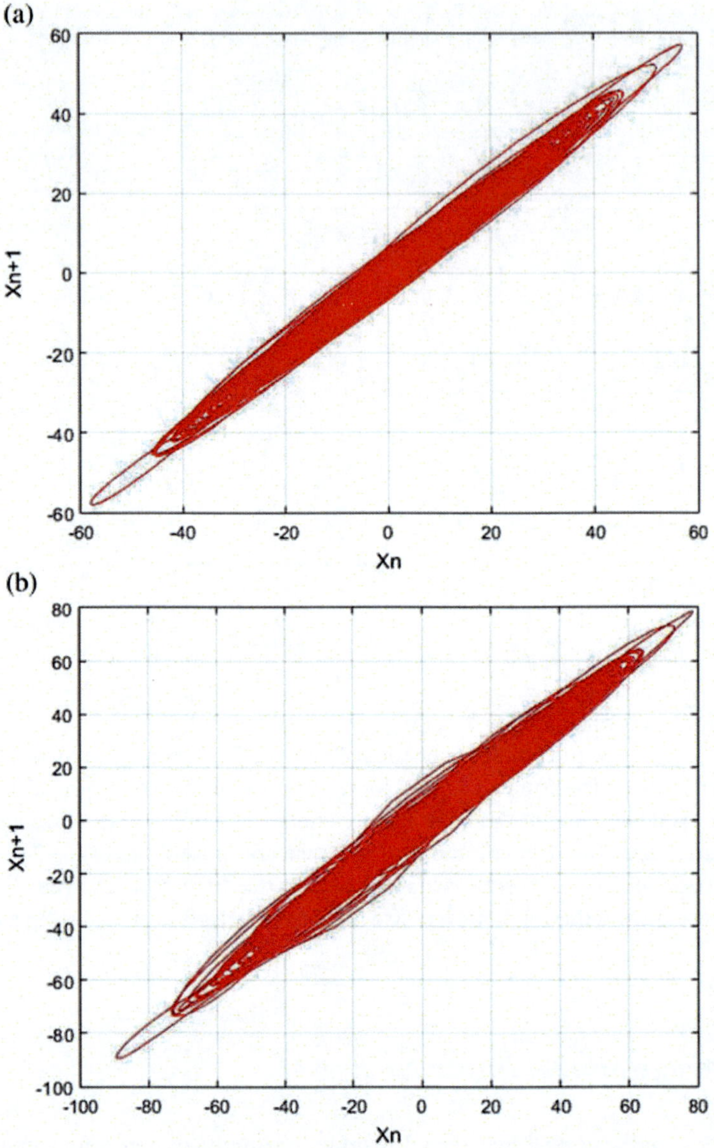

Fig. 1 **a** Phase space plot of subject's 0 brain activity before the performance of a cognitive task. **b** Phase space plot of subject's 0 brain activity during the performance of a cognitive task

For computing, the optimal time delay, Mars et al. [20] proposed a method which involves time delay coordinates which are independent of each other. We have considered $\{x(t)\}_{t=1}^{N}$, the original time series and $\{x(t+\tau)\}_{t=1}^{N}$, same time series with delay τ. They have computed $I(x(t))$, the mutual information of the experimented time series and $I(x(t+\tau))$, time-delayed version of the same time series. This method

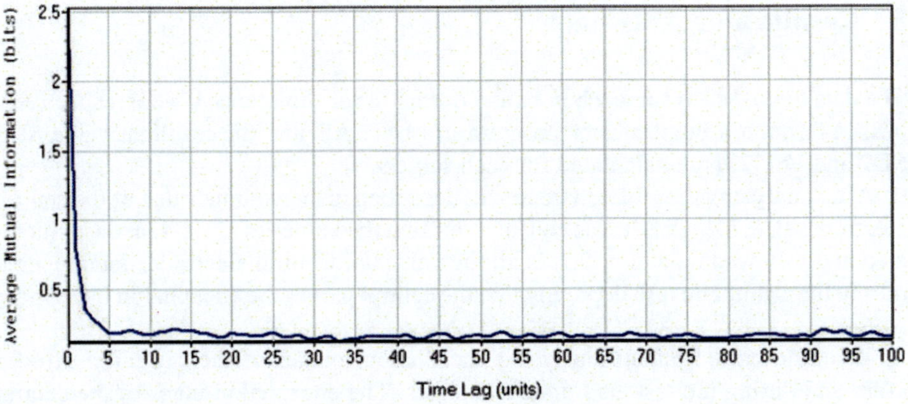

Fig. 2 $I(\tau)$ versus τ

refers to as Average or auto Mutual Information (AMI). It can be expressed in terms of non-linear autocorrelation generalization function [21, 22].

$$I(\tau) = \sum_{t=1}^{N-\tau} P[x(t), x(t+\tau)] \log\left(\frac{P[x(t), x(t+\tau)]}{P[x(t)]P[x(t+\tau)]}\right) \quad (4)$$

It was recommended in [23] that τ, the delay value, where for the 1st minimum value (at which functions level off) of $I(\tau)$ be used as the basis of the Poincaré reconstruction shown in Fig. 2.

2.5 *False Nearest Neighbors (FNN)*

To compute the embedded dimension, kennel et al. [24] proposed a method kenned as Erroneous most proximate neighbors. This method is predicated on the posit that two points proximate to each other in enough embedding dimension should remain adjacent as the dimension upsurges.

FNN proposed that consider the points in the one-dimensional time series as neighboring points which are adjacent to each other. The distance between two neighboring points is the magnitude between them.

2.6 *Topological Scalp Map*

To have a clear picture of stress level before and during cognitive task performance, the EEG lab is habituated to render images of encephalon maps utilizing the 2D and 3D topographical map features of the EEG lab. Once the 2D and 3D images are available, it is possible to visualize the difference more pellucidly [25].

3 Results and Discussion

The data files of each subject before and during the arithmetic task were respectively passed through a point cluster program in MATLAB and the resultant value SD1 SD2 and Vol [23] are calculated for each subject.

After analyzing the tables, a trend was described in the volume value of the central electrodes (C3, C4, Cz). It was conspicuous that the values of C3, C4, and Cz fluctuated and followed a pattern, that is all the values of Central electrodes incremented during the arithmetic task depicting that the central part of the encephalon faces more stress.

When described punctiliously the encephalon images of the central electrodes afore and during the task had drastic changes. The encephalon maps of the central

Table 1 Data table of the subject-0's brain's electrodes before the cognitive task and during the task

Subject_0						
	Before			During		
	SD1	SD2	VOL	SD1	SD2	VOL
Fp1	0.911747	12.95953	37.12049	1.170691	22.78237	83.78981
Fp2	1.088629	15.51992	53.07861	1.10415	17.51566	60.75815
F3	1.16371	16.46751	60.20361	1.137178	18.26749	65.2615
Fz	1.170693	16.97289	62.42358	1.124624	17.48145	61.76389
F4	1.177513	17.13305	63.37969	1.135215	17.38368	61.99683
C3	1.142732	16.91499	60.72477	1.069905	16.46239	55.33348
Cz	1.214558	19.07731	72.7923	1.163568	18.80935	68.75679
C4	1.166597	16.48481	60.41636	1.085293	16.2103	55.26982

Table 2 Data table of the subject-1's brain's electrodes before the cognitive task and during the task

Subject_1						
	Before			During		
	SD1	SD2	VOL	SD1	SD2	VOL
Fp1	0.792568	11.28539	28.09979	0.964557	19.68141	59.63948
Fp2	0.841156	12.14234	32.087	0.940372	14.91191	44.05376
F3	1.008345	14.26884	45.20099	0.899045	16.97616	47.94803
Fz	0.968933	14.74667	44.88875	0.962346	18.47951	55.86913
F4	0.933143	15.17596	44.48916	0.927386	15.69906	45.73871
C3	1.123476	15.14804	53.46504	0.922323	15.45825	44.79124
Cz	1.003048	16.13217	50.83521	0.924461	15.8681	46.0854
C4	0.952377	14.71225	44.01878	0.912664	16.3486	46.875

electrodes from the 'during' dataset were red which designates more stress was put in the central part of the encephalon when a subject was verbalized to solve an arithmetic quandary. The following Figs. 3 and 4 will make it more pellucid.

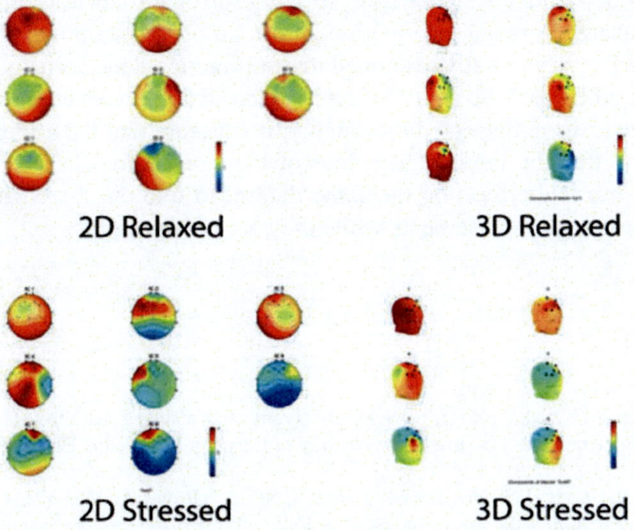

Fig. 3 Topological scalp maps of the subject-0's brain

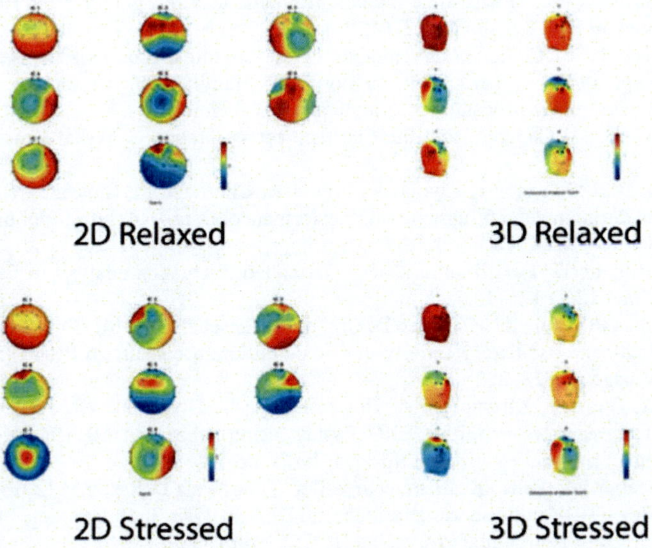

Fig. 4 Topological scalp maps of the subject-1's brain

4 Conclusion

We can conclude that some part of the encephalon is stressed while doing any kind of cognitive task. This work is a unique one because no one else has worked with the datasets provided from Physionet afore utilizing surrogate data tests, Phase space reconstruction, average mutual information, and topological scalp map. While the result was studied, it was visually examined that subtraction done verbally results in the incremented volume of the central electron recorded in a conventional pattern. After all the images were inspected thoroughly, it was sure that the arithmetic task had a prominent effect on the central sulcus of the encephalon. In the future, we will incorporate musical effects on the same dataset to find out the cognitive task efficiency amelioration of the subjects while they are under stress.

References

1. Inouye T, Shinosaki K, Sakamoto H, Toi S, Ukai S, Iyama A, Katsuda Y, Hirano M (1991) Quantification of EEG irregularity by use of the entropy of the power spectrum. Electroencephalogr Clin Neurophysiol 79(3):204–210
2. Fernfindez T, Harmony T, Rodriguez M, Bernal J, Silva J, Reyes A, Marosi E (1995) EEG activation patterns during the performance of tasks involving different components of mental calculation. Electroencephalogr Clin Neurophysiol 94:175–182
3. Skrandies W, Reik P, Kunze C (1999) Topography of evoked brain activity during mental arithmetic and language tasks: sex differences. Neuropsychologia 37:421–430
4. Wang Q, Sourina O (2013) Real-time mental arithmetic task recognition from EEG signals. IEEE Trans Neural Syst Rehabil Eng 21(2)
5. Seymour PHK (1976) Contemporary models of the cognitive processes: II. Retrieval and comparison operations in permanent memory. In: Hamilton V, Vernon PV (eds) The development of cognitive processes. Academic Press, London
6. Snodgrass JG (1984) Concepts and their surface representation. J Verbal Learning Behav 23:3–22
7. Zyma I, Tukaev S, Seleznov I, Kiyono K, Popov A, Chernykh M, Shpenkov O Electroencephalograms during mental arithmetic task performance. Data 4:14. https://doi.org/10.3390/data4010014
8. Marton LF, Brassai ST, Bako L et al (2014) Detrended fluctuation analysis of EEG signals. Procedia Technol 12(1):125–132
9. Lally N, Mullins PG, Roberts MV, Price D, Gruber T, Henschel C (2014) Glutamatergic correlates of gamma-band oscillatory activity during cognition: a concurrent ER-MRS and EEG study. Neuroimage 85:823–833
10. Mazumder A, Ghosh P, Khasnobish A, Bhattacharyya S, Tibarewala DN (2015) Selection of relevant features from cognitive EEG signals using relief and MRMR algorithm. In: Advancements of medical electronics. Springer India, pp 125–136
11. Dey A, Chakraborty S, Palit SK, Bhattacharya DK, Tibarewala DN, Roy M (2014) Study the effect of music on HRV impulse using multifractal DFA analysis. In: International Conference On Communications And Signal Processing (ICCSP), pp 1338–1342
12. Dey S, Bhimani R, Mazumdar A, Ghosh P, Tibarewala DN (2015) Multifractal detrended fluctuation analysis used to analyse EEG signals originating from different lobes of the brain. In: IEEE 2nd International conference on Recent Trends in Information Systems (ReTIS), pp 422–427

13. Zyma I, Tukaev S, Seleznov I, Kiyono K, Popov A, Chernykh M, Shpenkov O (2019) Electroencephalograms during mental arithmetic task performance. Data 4(1):14. https://doi.org/10.3390/data4010014
14. Theiler J, Eubank S, Longtin A, Galdrikian B, Farmer JD (1992) Testing for nonlinearity in time series: the method of surrogate data. Physica D 77–94
15. Theiler J, Prichard D (1996) Constrained realization Monte Carlo method for hypothesis testing. Physica 221–235
16. Henry B, Lovell N, Camachos F (2012) Nonlinear dynamics time series analysis. In: Nonlinear biomedical signal processing, dynamic analysis and modeling, vol 2, pp 1–39
17. Jaeger J, Monk N (2014) Bioattractors: dynamical systems theory and the evolution of regulatory processes. J Physiol 592(11):2267–2281
18. Fang Y, Chen M, Zheng X (2015) Extracting features from phase space of EEG signals in brain-computer interfaces. Neurocomputing 151:1477–1485
19. Kim HS, Eykholt R, Salas JD (1999) Nonlinear dynamics, delay times, and embedding windows. Physica D 127(1–2):48–60
20. Mars N, Arragon G Van (1981) Time delay estimation in nonlinear systems. IEEE Trans Acoust Speech Signal Process 29(3):619–621
21. Swinney AM, Harry L (1986) Independent coordinates for strange attractors from mutual information. Phys Rev A 33(2):1134–1140
22. Rakshit A, Banerjee A, Mazumder A, Ghosh P, Dey A, Tibarewala DN (2017) Fractal analysis of EEG signals for studying the effect of cognitive stress on the brain. Int J Biomed Eng Technol 25(2/3/4)
23. Dey A, Palit SK, Bhattacharya DK, Tibarewala DN, Das D (2013) Study of the effect of music on central nervous system through long term analysis of EEG signal in time domain. Int J Eng Sci Emerg Technol 5(1):59–67
24. Kennel MB, Brown R, Abarbanel HDI (1992) Determining embedding dimension for phase-space reconstruction using a geometrical construction. Phys Rev A 45(6):3403–3411
25. Lin WC, Chiu HW, Hsu CY (2005) Discovering EEG signals response to musical signal stimuli by time-frequency analysis and independent component analysis. In: Proceedings of the 2005 IEEE engineering in medicine and biology 27th annual conference, Shanghai, China

Microcontroller-Based Heart Rate Monitor

Aniket Saha, Subhojit Saha, Pritam Mandal, Priyanka Bawaly, and Moupali Roy

Abstract This project is used to measure heartbeat rate by using an embedded technology. This project can measure and monitor the patient's condition simultaneously. This project described the design of a simple, low-cost wireless patient monitoring system. Heartbeat rate of the patient is measured through fingertip using infrared device sensor. The pulse counting sensor is used to check whether the heart rate is normal or not. In case of abnormal condition, a SMS is sent to the mobile number using GSM module. The heart rate can be measured by monitoring one's pulse using medical devices such as an electrocardiograph [ECG], portable device. The heartbeat monitoring systems is the wrist strap watch or any other commercial heart rate monitors.

Keywords Microcontroller · Heart rate · Sensor

1 Introduction

This project is beneficial in medical applications and offers cost-effective services with size less than that of electrocardiogram (ECG). In the case of emergency, for those peoples who are suffering with heart diseases, continuous monitoring of the patient is required which is sometimes not possible in the hospital due to several circumstances [1]. In such serious case, this template circuit is useful to measure the heart rate and the other required information is transmitted to the medical counsellor for the preliminary precautions so as to get a glimpse of that patient and to take necessary actions if needed. The whole situation is to prevent the patient from having serious situation before reaching to the hospital [2].

A. Saha · S. Saha · P. Mandal · P. Bawaly · M. Roy (✉)
ECE, Narula Institute of Technology, Kolkata, India

M. Mitra et al. (eds.), *Computational Advancement in Communication, Circuits and Systems*, Lecture Notes in Electrical Engineering 786,
https://doi.org/10.1007/978-981-16-4035-3_24

Fig. 1 Heart rate measurement through fingertip

2 Photoplethysmography

Photoplethysmography is a process where it optically estimates the volumetric measurement of an organ. Heart rate monitoring, pulse oximetry, respiration monitoring, cardiovascular monitoring, etc., are some of the common applications of photoplethysmography. When the heart contracts, the volume of blood inside the fingertip decreases else when the heart expands the volume of blood inside the fingertip increases. The final pulsing of blood volume inside fingertip is equal to the heart rate [3, 4]. Here an IR transmitter/receiver pair is placed in close contact with the fingertip. When heart beats, the volume of blood cells under the sensor increases which then reflects more IR waves to the sensor and when there is no beat then the intensity of this reflected beam decreases (Fig. 1).

3 Working Principle

The device can sense the heart rate from the fingertip by using IR reflection method and will display it on three-digit seven segment display in beats per minute. The circuit has an accuracy of 4 beats per minute, and it can be used very easily. In medical use, the technique that is used for sensing heart rate is called photoplethysmography [5] (Fig. 2).

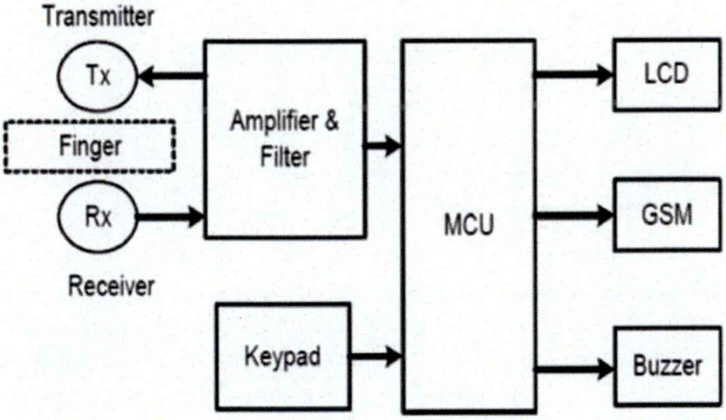

Fig. 2 Working principle

3.1 *Circuit Diagram*

This is a simple LCD version of the heart rate monitor. LCD displays are very popular; nowadays, in one most of the embedded system, designers prefer them over multiplexed seven segment LED displays. Using LCD displays, we can display text, custom characters, graphics, and a lot of stuff, and it is a great advantage over the LED counterparts. JHD162 is the LCD display used here. It is a 16 × 2 LCD display based on the HD44780 driver IC, interfacing LCD display to 8051 [6, 7] (Fig. 3).

3.2 *Expected Outcome of the Project*

This project is used to alert the family members about patient's heartbeat via SMS. The objective of the project is to detect and monitor the patient's heartbeat rate using PPG technique, interfaced with the GSM modem, and then it sends alert to the family and medical experts via SMS [8] (Figs. 4, 5, 6, 7, 8 and 9) .

The connection between the microcontroller and the hyperterminal is successfully established. At the moment of ongoing test, it can send alert directly from PIC circuit to mobile phone to get a stable system. It can be modified to very light, portable, smart, and elegant, e.g., like a watch or embed with i-POD [9, 10]. By using the value of heart rate monitor, we can also know the ages, oxygen contents in human body, and the patient's weight also.

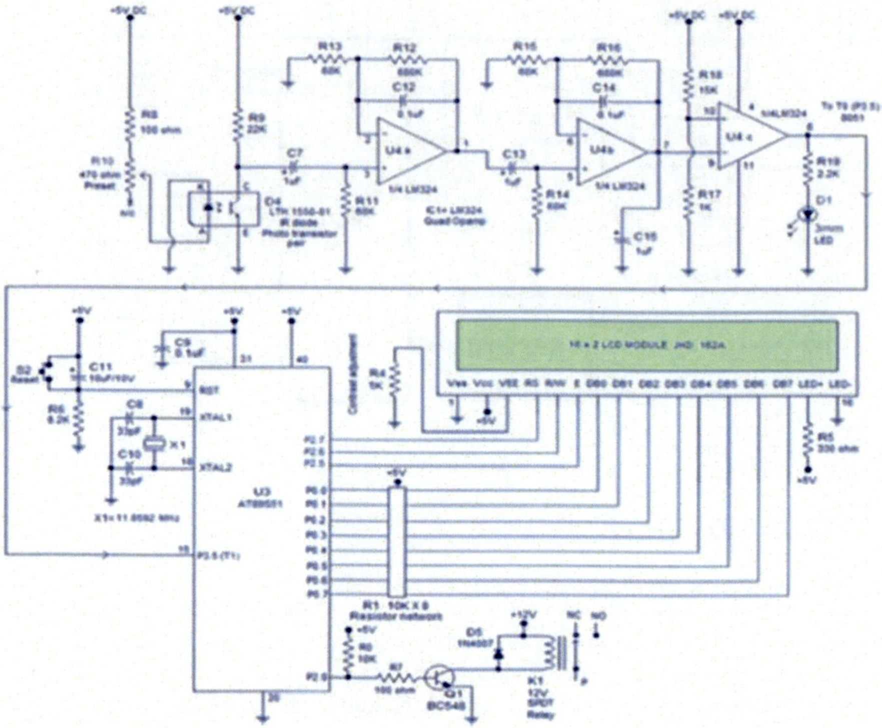

Fig. 3 Circuit diagram

3.3 Application of the Project

The healthcare industry is greeting to the increasing popularity and availability of technological transformations, such as tablets, smartphone, and ultramodern pieces of equipment. Successfully utilizing smartphones and employing it in the field of health care and medicine is helpful as it eases the operation of medical devices to enable lightly trained individuals whether patients or medical advisory to reliably collect medical data for diagnosis and other health-related stuffs [11, 12]. This phone accessory heart rate monitor is therefore a part of a wide viewed project that proposes to make health monitors used in clinical practices compact and available to the public with straightforwardness and crystal-clear views to use and no consideration of time or place. Having the data on the phone, it can be used as a health monitor or sent to a medical advisory elsewhere for remote evaluation so the patient can seek advice from his/her counsellor without having to book for an appointment [13].

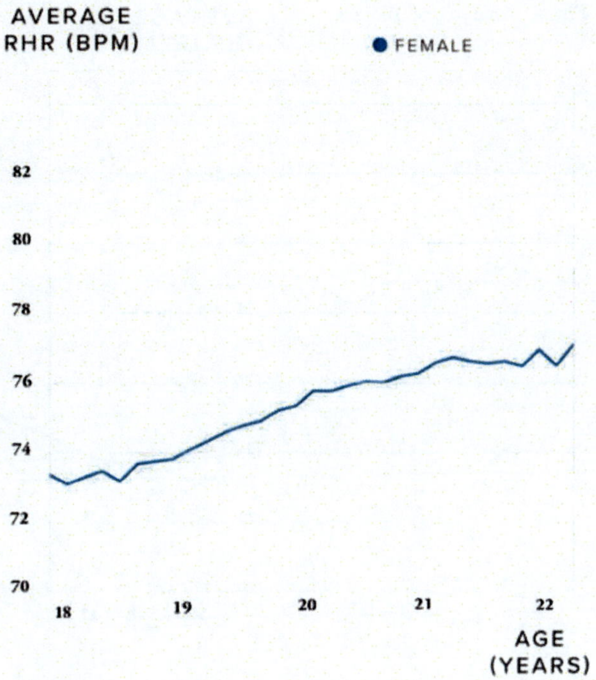

Fig. 4 Analysis of Table 1

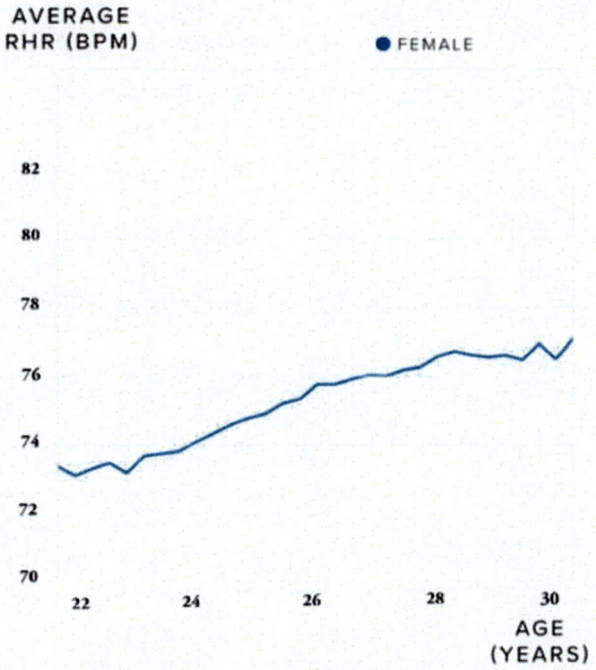

Fig. 5 Analysis of Table 2

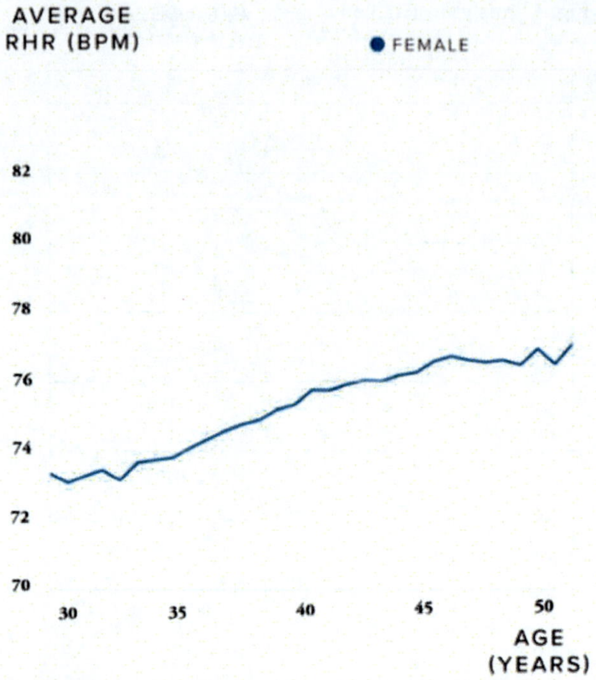

Fig. 6 Analysis of Table 3

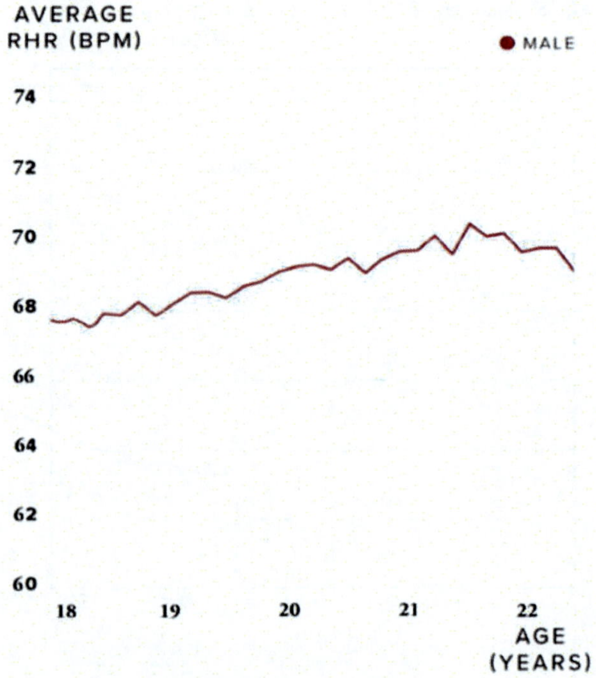

Fig. 7 Analysis of Table 4

Fig. 8 Analysis of Table 5

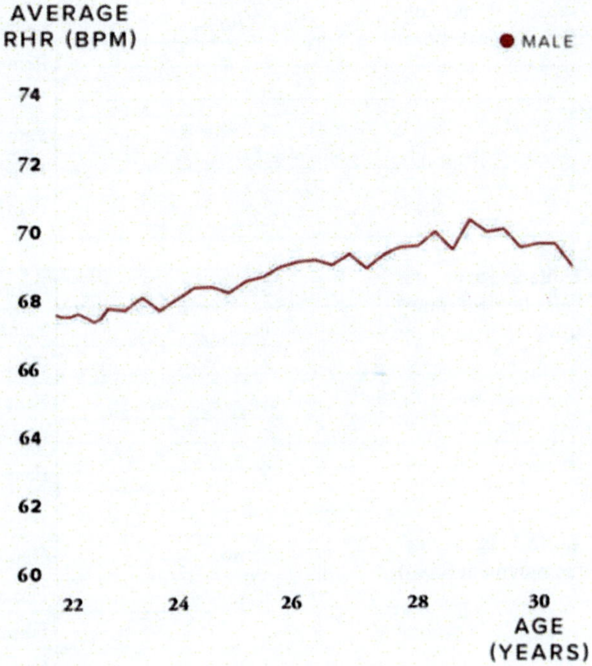

Fig. 9 Analysis of Table 6

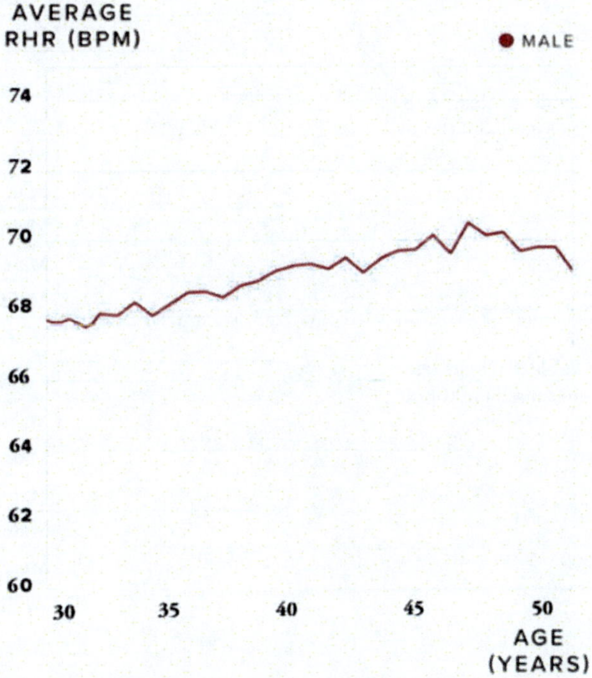

Table 1 Heart rate measurement (bpm)

S. no.	Gender	Age	HR
1	Female	18	73
2	Female	19	74
3	Female	20	75
4	Female	21	77
5	Female	22	77

Table 2 Heart rate measurement (bpm)

S. no.	Gender	Age	HR
1	Female	22	73
2	Female	24	74
3	Female	26	75
4	Female	28	77
5	Female	30	77

Table 3 Heart rate measurement (bpm)

S. no.	Gender	Age	HR
1	Female	30	73
2	Female	35	74
3	Female	40	75
4	Female	45	77
5	Female	50	77

Table 4 Heart rate measurement (bpm)

S. no.	Gender	Age	HR
1	Male	18	67
2	Male	19	68
3	Male	20	69
4	Male	21	70
5	Male	22	71

Table 5 Heart rate measurement (bpm)

S. no.	Gender	Age	HR
1	Male	22	67
2	Male	24	68
3	Male	26	69
4	Male	28	70
5	Male	30	71

Table 6 Heart rate measurement (bpm)

S. no.	Gender	Age	HR
1	Male	30	67
2	Male	35	68
3	Male	40	69
4	Male	45	70
5	Male	50	71

4 Conclusion

The instrument has a very simple structure, reliable operation, high accuracy, low power consumption, and good portability full featured function. The real-time monitoring system for cardiac patient physical state is based on the wireless transreceiver module technology. It can be taken by patient, and it keeps the patient moment intact because it is miniature and portable. The system can monitor and record the physical states and moment parameters in real time. With the intelligent transreceiver module, the sign of acute disease for patient can be found early, and then the patient can be helped in time and the sudden death of patient can be avoided. The wireless transreceiver module technology can be suited for short distance communication, and the transmission distance is limited only in about 10 m, and then it can be made suitable for in-patient monitoring. This system is important to be applied to patient care.

References

1. Prasath J (2013) Wireless monitoring of heart rate using microcontroller. Int J Adv Res Comput Sci Electron Eng 2:214–219
2. Babiker SF (2014) Microcontroller based heart rate monitor using fingertip sensors. Khartoum Univ Eng J 1
3. Wei YC (2013) Design of a microcontroller based real-time heart rate variability measurement system using a lowcomplexity R-peak detection algorithm. Instrum Sci Technol 41:274–289
4. Proulx J, Clifford R, Sorensen S, Lee DJ, Archibald J (2009) Development and evaluation of a Bluetooth EKG monitoring sensor. In: Proceedings of the IEEE international symposium on Computer-Based Medical Systems (CBMS)
5. Hashem MMA, Shams R, Abdul Kader M, Abu Sayed M (2010) Design and development of a heart rate measuring device using fingertip. Department of Computer Science and Engineering, Khulna University of Engineering and Technology (KUET), Khulna 9203, Bangladesh
6. Ibrahim D, Buruncuk K (2014) Heart rate measurement from the finger using a low cost microcontroller. Near East University, Faculty of Engineering, TRN
7. Kara S et al (2006) Low-cost compact ECG with graphic LCD and phonocardiogram system design. J Med Syst 30:205–209
8. Abenstein J, Tompking W (1982) A new data reduction algorithm for real time ECG analysis. IEEE Trans Biomed Eng 29(1):43–47
9. Arif C (2004) Embedded cardiac rhythm analysis and wireless transmission (Wi-CARE). MS Thesis, School of Computing and Software Engineering, Southern Polytechnic State University, Marietta, Georgia, USA

10. Ayang-ang C, Sison L (2001) Electrocardiograph pre-filtering, QRS detection, and palm display programming for biomedical applications. In: Proceedings of the ECE conference, University of St. Tomas, Manila
11. Choi W, Kim H, Min B (1996) A new automatic cardiac output control algorithm for moving actuator total artificial heart by motor current waveform analysis. Int J Artif Organs 19(3):189–197
12. Leffler C, Saul J, Cohen R (1993) Rate-related and autonomic effects on atrioventricular conduction assessed through beat-to-beat PR interval and cycle length variability. J Cardiovasc Electrophysiol 5(1):2–15
13. https://worldwidescience.org/topicpages/m/microprocessor+based+pulse.html

A QCA-Based Improvised TRNG Design for the Implementation of Secured Nano Communication Protocol in ATM Services

Arindam Sadhu, Kunal Das, Debashis De, Maitreyi Ray Kanjilal, and Pritam Bhattacharjee

Abstract In this paper, an attempt has been made for quantum-dot cellular automata (QCA) based design of TRNG (True Random Number Generator) to support the implementation of developed nano communication protocol targeting more secured operation over automated teller machine (ATM). TRNG is an ingenious design can generate non-deterministic and distinctive stream digital bit, and a major aspirant for any secured cryptography process. Here, the mode of design is using QCA technology due to its advantageous aspects of consuming low-design area and ultra-low power during high-frequency operations. Overall the functional verification of our proposed setup is carried out using QCA Designer 2.0.3 where from its efficacy is rightly depicted.

Keywords QCA · TRNG · Crossed loop circuit · Seed input · Majority Voter

1 Introduction

Quantum-dot cellular automata (QCA) is one of the more-than-moore design paradigms [1, 2] which happen to be a promising technology in constructing secured

A. Sadhu
Greater Kolkata College of Engineering and Management, Dudhnai, Ramnagar, Baruipur, Kolkata, West Bengal 743387, India

K. Das
Department of Computer Science, Acharya Prafulla Chandra College, Kolkata 700131, India

A. Sadhu · D. De
Maulana Abul Kalam Azad University of Technology, Kolkata, West Bengal 700064, India

M. R. Kanjilal
Narula Institute of Technology, Agarpara, Kolkata, India

P. Bhattacharjee (✉)
Department of Computer Science and Engineering, Amrita Vishwa Vidyapeetham, Clappana, Amritapuri, India
e-mail: pritambhattacharjee@am.amrita.edu

M. Mitra et al. (eds.), *Computational Advancement in Communication, Circuits and Systems*, Lecture Notes in Electrical Engineering 786,
https://doi.org/10.1007/978-981-16-4035-3_25

communication systems. Here, the information is transmitted from one QCA cell to another in terms of electrostatic charge based on columbic interaction (i.e., no physical charge movement). This makes QCA (which was pioneered by C.S. Lent and his group back in 2005) a good alternative to the present CMOS technology. It's attracting features like having low covering area, controlled power dissipation, while operating in high frequency zone (in the range of THz) is very intriguing for circuit and system designers. Though there were initial troubles, but by now, the physical design of a QCA cellis also possible which is operable around ~ 293°K [3]. This achievement has given researchers an impetus for further work in QCA-based architecture designs.

On the other hand, we also see how important it is to have secure devices/gadgets in our daily life. Consequently, there are several kinds of research being pursued on security issues, and some maybe the development of new algorithms or new hardware design supporting secured operation in the modern communication systems [4]. Several researches on the accomplishment of contemporary attack and the novel countermeasures of each finding have been presented in different scientific literatures. However, vital aspects of hardware structure for a secured arrangement remain untouched to drape different sticky security situation in various levels of secured communication. The algorithm generated secret key could be recovered by using some physical quantities indirectly correlated with the key. Herein, the hardware generated secret key generator or TRNG is competent to breed a distinctive digital bit array, and non-deterministic that can be accumulated to construct secured hardware communication structure. Our intention is to design a hardware cryptographic architecture using the QCA framework as an imperative application in nano communication domain. In our proposed design, the non-deterministic bits are extorted from a crossed loop majority voter (MV) based symmetrical TRNG array, and the seed inputs are used for enhancing the unpredictability of generated number chain in addition to the deviation of the random numbers. We realize a bi-decryption archetype for a secured ATM (automated teller machine) system in this paper as an application of TRNG using QCA.

2 Preliminaries

2.1 Random Number Generator

Random number generator (RNG) is a vital component used in designing safe and secure systems. Unpredictability of successive numbers and non-deterministic nature are the strength of a RNG device [5]. RNG can be broadly categorized in two ways viz., HRNG (Hardware Random Number Generator) and SRNG (Software Random Number Generator). SRNG on the other hand is also identified as PRNG (Pseudo Random Number Generator) because it follows a deterministic algorithm to generate random number, but it is not suitable for high-secured device [6]. The reason is the generated number is quite predictable to anyone who has idea of the used algorithm.

However, no algorithm is needed to generate random numbers using HRNG. Here, output sequence is unpredictable even if the internal circuit design of HRNG is somehow known to others. The HRNG structure is more relevant to the modern design of TRNGs (True Random Number Generator) [7]. Till date, the popularity has been with the design of analog TRNGs. But due to the presence of noise, jitter and radioactive decay as the seed input of TRNG device, and people are now more inclined toward design of digital TRNGs. The quantum aspects of logic design for TRNG can be brought in using QCA which is our objective in this paper. As a matter of fact, very few works on QCA-based TRNG design is reported for example, in [8–11]. Considering these works as point of reference, it is intuitive to us that the TRNG designs can be improvised and used in the development of some nano communication protocol supporting the ATM dual security.

2.2 *QCA-Based TRNG Seed Circuit*

In QCA circuit design, a specific arrangement of QCA cells can turn out to be more unpredictable and irregular. This issue was addressed in [8–11] with certain solutions as depicted in Fig. 1a–c which happened to be mainly the typo of TRNG seed generator circuit.

Apart from the seed circuit, TRNG also has output generator circuits viz., cross-oriented where these kind of improvisations are also required when QCA-based design is taken in account.

3 Proposed MV Based TRNG in QCA

For the purpose of generating unpredictable and non-deterministic number sequence, two closed loop MV based QCA circuit has been proposed in this fragment.

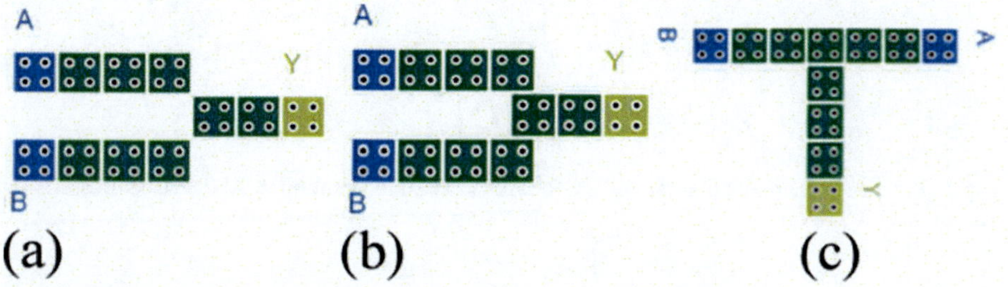

Fig. 1 QCA layout of **a** SEED1 [8] **b** SEED2 [10, 11] **c** SEED3 [10, 11]

3.1 Crossed Loop MV Circuit

Two asymmetric type crossed loop QCA circuit are proposed in this subsection: one is OR based logic and another is AND based logic depicted in Fig. 2a, b, respectively. The corresponding QCA layouts are individually presented in Fig. 3a, b. Both these circuits are composed of 43 QCA cell with 0.032 μm^2covering area and 1QCA clock cycle latency.

Now if two asymmetric circuits are connected in row wise with single QCA clock pulse then it will start to generate two bit random number. This schematic diagram and corresponding QCA layout are shown in Fig. 2c and 3c, respectively.

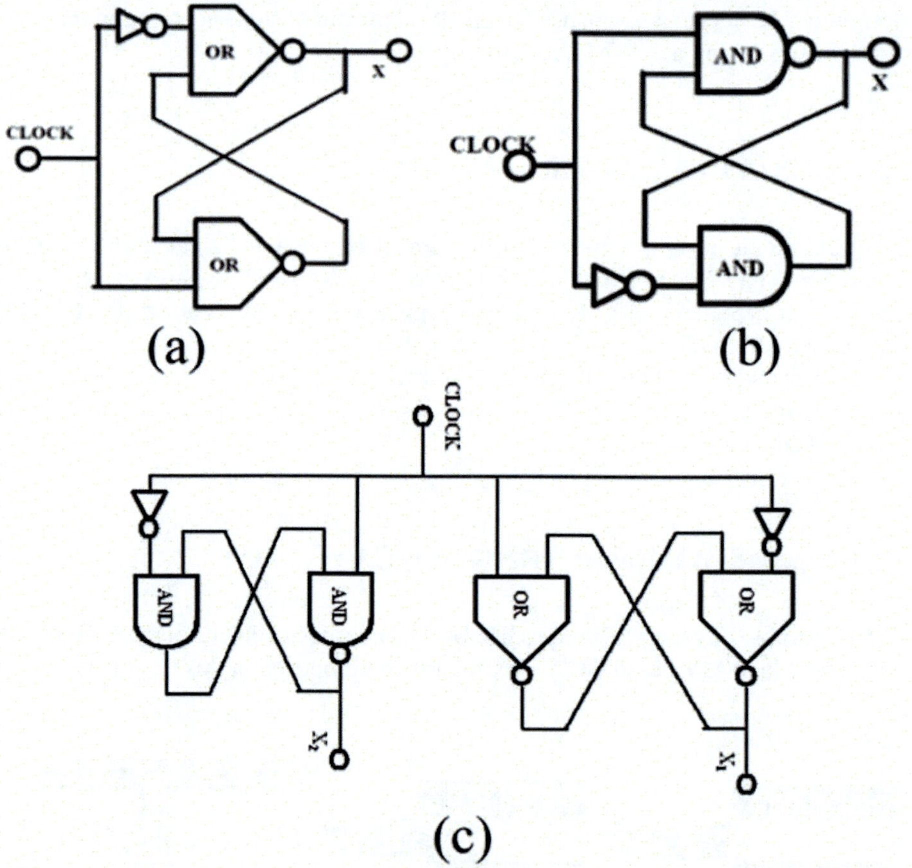

Fig.2 Proposed schematic diagram of **a** OR based crossed loop circuit, **b** AND based crossed loop circuit, **c** 2 bit TRNG

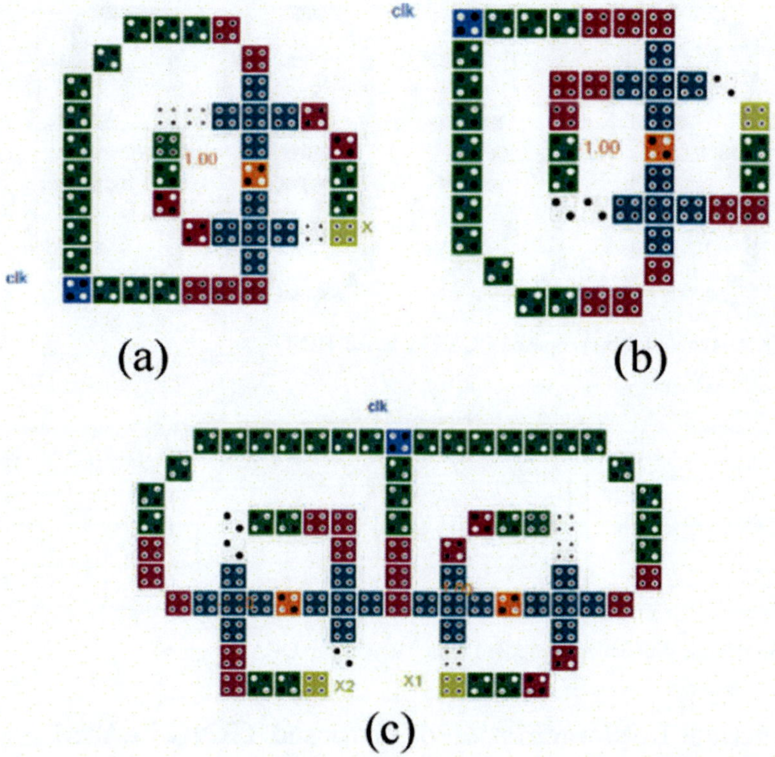

Fig. 3 Proposed QCA layout of **a** OR based crossed loop circuit **b** AND based crossed loop circuit **c** 2 bit TRNG

3.2 Majority Voter Based QCA 8-bit TRNG Design

In this subsection, straight forward 8-bit QCA layout is illustrated with its simulated outcome. The straight forward 8-bit TRNG is realized by connecting the 2-bit asymmetric crossed loop QCA circuit in column wise as depicted in Fig. 4. The feed forward structural design is useful to generate more non-deterministic state. The Seed 3 design [10, 11] is used to connect consecutive asymmetric QCA structure. Herein, we have used QCA Designer [12] to simulate the outcome of our 8-bit QCA circuit and portrayed the report in Fig. 5.

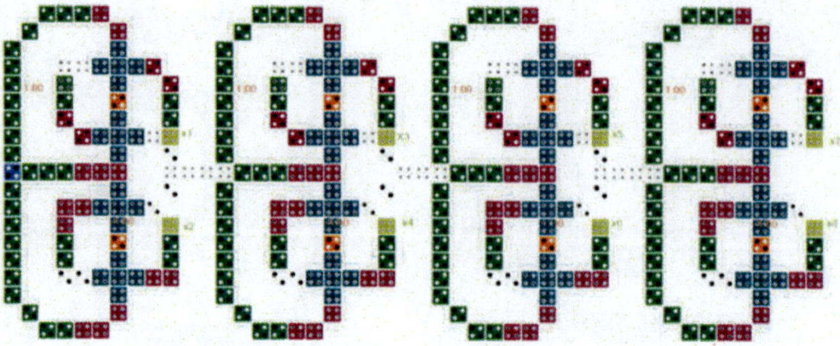

Fig. 4 QCA layout of 8-bit proposed QCA MV based TRNG

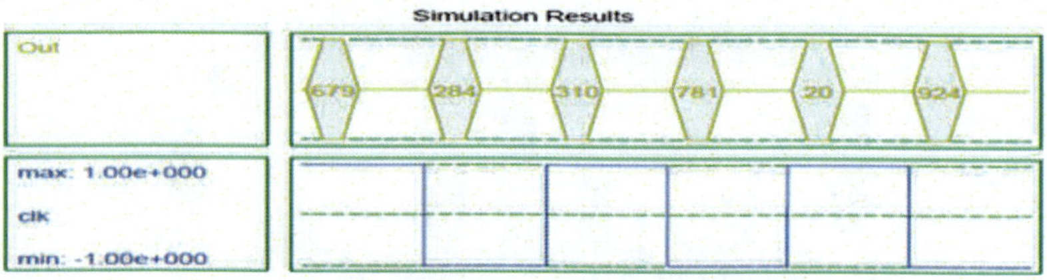

Fig. 5 Simulation outcome of proposed 8-bit TRNG from QCA Designer

4 Practical Implementation of Proposed TRNG in ATM Machine

In this section, multi-level authentication process is proposed in ATM machine to enhance security in the process of money laundering. The flow diagram of proposed technique is depicted in Fig. 6.

To start the process, punch the ATM card inside the machine. After the card being inserted properly, a request is generated to input the user's password. If the input is not paired with the actual password the process will restart, else it will proceed to next step. As the password match, the TRNG generates a six bit code, one is stored in the processor of the ATM machine through the SRAM cell, and same code is sent to the user's registered number through the microwave link. The code from the user's number is requested to insert as input through the manual keyboard on the machine. After insertion of the code, the hamming distance is detected. If the value returns 0, user can proceed to the further operations; else it begins the process all over again.

The QCA layout of this idea is depicted in Fig. 7. It consists of control block, microwave block, 6-bit MV based TRNG, QCA SRAM block, input block for user, read enable block and hamming distance checker circuit with output port. There are 3 input ports in control block, i.e., control signal, seed input and read–write signal ($WR/\overline{R}$). Control signal and $WR/\overline{R}$ signal will set to "1" if and only if password matches with customer pressed key. At the same time, the auto-generated seed input incites the TRNG block.TRNG generates a 6-bit code, a copy of which is stored in

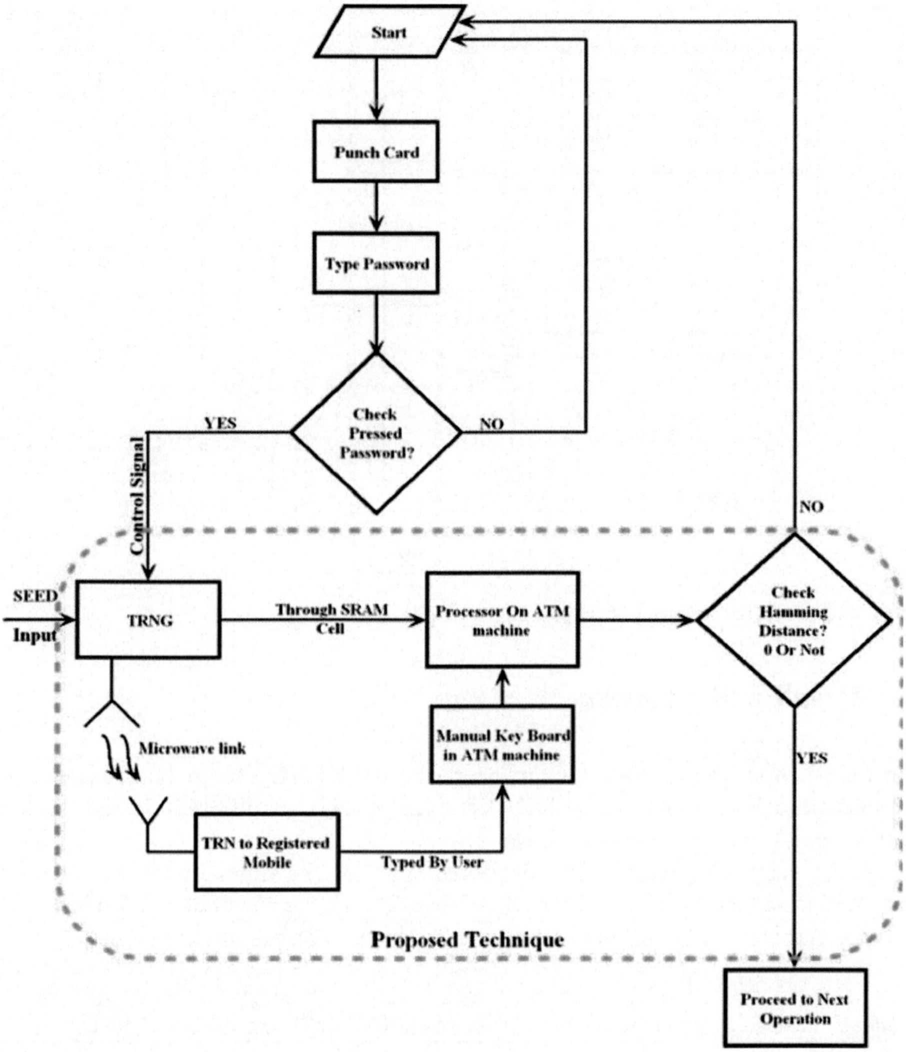

Fig. 6 Flow diagram of proposed ATM application of TRNG

the SRAM cell, as $WR/\overline{R}$ is already set at "1" and the same code is sent to the user's registered number through the microwave block. Read enable block has a vital role in this footprint. Read enable signal will set to high only when user presses any number, otherwise it remain fixed at "0". As it is generally stable at "0", so the output port will set to high state. Hence, ATM will not proceed to the next operation until "0" is generated as output. This "0" will be produced from the output port once the user had typed the proper number and high-read enable signal. The ATM service thus achieves double-security based on this topology, i.e., by unique user pin and TRNG generated number.

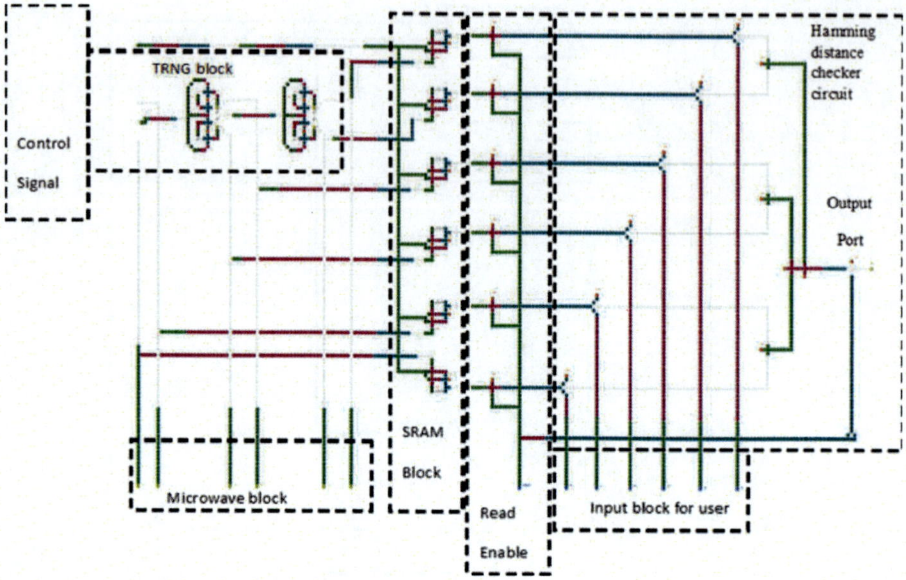

Fig. 7 QCA layout footprint of proposed ATM application of TRNG

5 Result and Analysis

In this section, the performance of proposed MVTRNG design is assessed with respect to earlier reported QCA TRNG designs [8–11] over different parameters like area, delay, power and quantum cost.

These parametric readings are extracted based on the simulation setup executed in QCA Designer 2.0.3 as well as QCAPro [13], and it is illustrated in Table 1. It is observed that there is decent amount of drop down in all parametric sectors.

Table 1 Parametric comparison of our proposed 8-bit MVTRNG w.r.t., reported TRNGs

Parameter	[8]	[9]	[10]	[11]	Proposed TRNG	% Benefit (w.r.t., range)
Complexity (No. of QCA cell)	498	463	388	339	334	32.93% to 2%
Area(μm^2)	0.52	0.46	0.36	0.32	0.29	44.23% to 10%
Avg. Power/bit (meV)	62.355	57.735	57.06	52.175	47.126	25% to 10%
Quantum Cost	288	288	584	277	280	52% to -1%

6 Conclusion

In the present time, many researchers are involved in solving the security related issues and reported many relevant architectures. However, this paper is the first study of an inventive QCA MV-based TRNG design which facilitates a secured communication protocol for ATM operation. Here, the intended hardware cryptographic communication architecture demonstrates its double-secured protocol and verify using QCA-based nano architecture. The framework of our proposed architecture is based on uncertain polarization in the output of a QCA cell with the use of crossed loop of a QCA based "OR" as well as "AND gate" in the non-deterministic seed circuit. The presented TRNG of 8-bit comprises of 334 QCA cells covering an area of 0.29 μm^2. The latency observed is minimal since the circuit generates any random number in a single QCA clock cycle, despite of that it also runs at a high range of frequency. Average energy dissipation is observed, exhibits 47.126 meV energy per bit and 280 quantum cost. The design strenuously fulfills the desired standard of industry concerning NIST standard. Thus, the proposed TRNG and double-secured communication protocol can be interpreted as a potential design for the next generation quantum cryptography.

References

1. Lent, CS, Douglas Tougaw P, Porod W, Gary Bernstein H (1993) Quantum cellular automata. Nanotechnology 4(1):49
2. Bhattacharjee P, Das K, Dey A, De D, Chakraborty SK (2017) Estimation of power dissipation in ternary quantum dot cellular automata cell. J Low Power Electronics 13(2):231–239.
3. DiLabio, GA, Wolkow RA, Pitters JL, Piva PG (2014) Atomistic quantum dot. U.S. Patent 8,816,479, issued August 26, 2014
4. Krishnan P, Najeem J (2017) A multi plane network monitoring and defense framework for sdn operational security. International Conference on Operating System Security (ICOSS 2017)
5. Devi DI, Chithra S, Sethumadhavan M (2019) Hardware random number generator using FPGA. J Cyber Security Mobility 8(4):409–418.
6. Mohan, AK, Devi N, Sethumadhavan M, Santhya R (2018) A selective generation of hybrid random numbers via android smart phones. Int J Pure Appl Math 118(8):311–317
7. Akhileswar Y, Raghul S, Meghana C, Mohankumar N (2020) Hardware-assisted QR code generation using fault-tolerant TRNG. In: International conference on communication, computing and electronics systems, pp 203–212. Springer, Singapore
8. Abutaleb MM (2018) A novel true random number generator based on QCA nanocomputing. Nano Commun Networks 17:14–20
9. Pain, P, Das K, Sadhu A, Kanjilal MR, De D (2019) Power analysis attack resistable hardware cryptographical circuit design using reversible logic gate in quantum cellular automata. Microsystem Technologies, pp. 1–13.
10. Sadhu, A, Das K, De D, Kanjilal MR (2020) SSTRNG: self-starved feedback SRAM based true random number generator using quantum cellular automata. Microsystem Technol 26(7):2203–2215.
11. Sadhu A, Das K, De D, Kanjilal MR (2020) MVTRNG: majority voter-based crossed loop quantum true random number generator in QCA nanotechnology. In: Computational advancement in communication circuits and systems, pp 241–253. Springer, Singapore

12. Walus K, Dysart TJ, Jullien GA, Budiman AR (2004) QCA designer: a rapid design and simulation tool for quantum-dot cellular automata. IEEE Trans Nanotechnology 3(1):26–31
13. Srivastava S, Asthana A, Bhanja S, Sarkar S (2011) QCAPro-an error-power estimation tool for QCA circuit design. In 2011 IEEE international symposium of circuits and systems (ISCAS), pp 2377–2380. IEEE

Diagnoses of Melanoma Lesion Using YOLOv3

Shubhendu Banerjee, Sumit Kumar Singh, Atanu Das, and Rajib Bag

Abstract The most modifiable risk factor for skin cancer is ultraviolet radiation (UVR) exposure. Melanoma or malignant melanoma is the rarest but at the same time deadliest form of skin cancer. While prevention of melanoma is possible to some extent by educating masses to involve in safe sun practices as avoiding sun exposure during peak radiation hours, using protective clothing, applying sunscreen and distancing oneself from artificial sources of UV light, early detection and accurate treatment of the disease may curtail the fatality of the deadly disease. If statistics are to be believed, the lifetime risk of developing melanoma in the year 1935 was 1 in 1500 as compared to 1 in 50 in 2010, indicating its dramatic increase in the last century. While effective and timely treatment of melanoma has been a subject of prime importance for researchers and the medical fraternity alike, several invasive and non-invasive techniques have come to the fore from time to time for diagnosis of melanoma. Analysis of the several methods developed during the years suggests that easier access to skin examinations increase the chances of accurate and well-timed detection of melanoma and computer-aided diagnosis (CAD) has played a major role in fulfilling the same. This work proposes a novel CAD approach which includes preprocessing of the dermoscopic images by Dull Razor algorithm followed by classification by deep learning-based algorithm 'You Only Look Once' (YOLO) and finally segmentation of the identified image by a self-designed algorithm. The experiments have been conducted on three publicly available datasets—PH2, ISBI 2017 and ISIC 2016. The combination of the total methodology offers a Jac score of 86.12% and Dic of 92.55% which is way superior to results of contemporary works in the area.

S. Banerjee (✉)
Department of CSE, Narula Institute of Technology, Kolkata 700109, India

S. K. Singh
Department of CSE, University of Essex, Colchester CO43SQ, UK

A. Das
Department of MCA, Netaji Subhash Engineering College, Kolkata 700152, India

R. Bag
Principal, Indas Mahavidyalaya, Bankura 722205, India

M. Mitra et al. (eds.), *Computational Advancement in Communication, Circuits and Systems*, Lecture Notes in Electrical Engineering 786,
https://doi.org/10.1007/978-981-16-4035-3_26

Keywords Skin cancer · Melanoma · Skin lesion segmentation · YOLO · Deep learning

1 Introduction

With the advent of global warming, the harmful ultraviolet rays from the sun have depleted the ozone layer and has affected the largest sense organ of human body. Human anatomy is covered with three protective layers of epithelial tissues namely: dermis, epidermis and hypodermis. Prolonged exposure to the strong and injurious UV radiation increase the production of melanin in the epidermis of the skin, which leads to abnormal growth of melanocytes, thereby causing skin cancer [1]. According to World Health Organization and the international agency for research on cancer, skin cancer is the most widely spread cancerous cell in the world and accounts for about one-third of total cancer cases [2]. In 2019, the annual report of the American cancer society estimated approximately 96,480 new cases of melanoma with 7230 death due to skin cancer alone [3]. According to a report, five million cases are reported every year in the USA, and in 2015, 60,000 deaths were reported out of a total of 350,000 melanoma cases. Skin cancer can be ordered chiefly into three different types—squamous cell carcinoma (SCC), basal cell carcinoma (BCC) and malignant melanoma (MM).

According to dermatologists, melanoma lesion is an irregular or amoeboid shaped mole of greater than 5 mm in diameter which can evolve on a preexisting mole or can form as a new mole on the skin. Dermatologist also states that a melanoma lesion may or may not show any regular sign on the skin and can also occur in the area which is never exposed to direct radiation of the sun. As compared to a normal mole, a melanoma lesion is normally smaller in size and it is in the shades of brown, deep gray and black which is typically different from that of regular mole [4]. A malignant melanoma is an irregular surfaced, rapidly growing tumor which is asymmetrical in shape and shows different shades of pigmentation ranging from brown to deep gray and black. The colors also form a contour of different color pigmented as a cluster within the lesion. As the lesion develops, it may show inflammation, itching, ulceration or sometimes even bleeding from the area where the lesion developed. Beginning of twentieth century marks the diagnoses of melanoma lesion manually by a dermatologist, who physically examines the lesion by naked eyes. Later, the manual diagnosis process was replaced by immunotherapy and radiation treatment along with invasive methods. Anatomical-pathological examination determines the possibilities of a lesion being a malignant tumor. However, this method of detection is an aciurgy method and is not likely to be practiced by the clinical patients. Moreover, the manual detection requires a well-trained specialist and it often suffers variation of report for a single lesion due to inter-observer variations. Thus, the lesion detection accuracy lies between 75 and 84% if diagnosed by an inexperienced dermatologist [5]. Although 95% of the early stage malignant tumor is curable if detection and removal of the tumor is executed in time, only 15% of melanoma is curable at advance stage. If it is not removed in time, the disease might metastases as liver or lung tumor and then the death chances increase to 80% even post-surgery. The timely detection and removal of malignant tumor defines the survival rate of the patients; thus, our fundamental aim should be fast, easy and accurate detection of melanoma in its very early stage. Therefore, a non-invasive computer-aided diagnosis (CAD) system can be brought into regular practice which can help the patient easily detect melanoma thereby helping them not to undertake unnecessary invasive method for detection of malignant tumors [6]. Computer-aided diagnosis (CAD) system of detection of malignant tumor will not only ease the process of detection but also increase the accuracy and efficiency of pathologists. The dermoscopy technologies developed in recent times are non-invasive in nature and the illuminated and magnified images of skin region increase the clarity of the spot thereby visually enhancing the skin lesion and deducting any surface reflection on the skin lesion. Dermoscopy has improved the diagnoses performance to a significant extent. Modern methods and algorithms of detection of melanoma includes most commonly used visual inspection methods like ABCD (asymmetry, border, color, diameter) rule, Menzies method, seven-point checklist and other pattern analysis methods. The CAD system can be chronologically framed as; preprocessing of the dermoscopy image, segmentation of lesion from the image which is indeed most challenging and important step in the diagnoses process, followed by feature extraction, and classification of the lesion [7–10].

It is a laborious task of detection of melanoma lesion from dermoscopy images, as it includes several digital imaging factors like difficulty in segmentation of lesion area with regular skin region due to low contrast of images which is a result of poor lighting conditions, variable skin condition like presence of hair or blood vessels over melanoma lesion, dark skin tone which might visually appear indistinguishable from the lesion color, air bubbles caused due to immersion fluid which is applied on the lesion during medical imaging, different texture and pattern of melanoma lesion among different clinical patients and most importantly high degree of visual similarities between melanoma and non-melanoma or benign lesions which might result in faulty detection of a lesion even by a trained dermatologist. Gazing into the stated diagnosis problems, this paper further reshapes the conventional CAD system; the preprocessing step involves Dull Razor algorithm [11] which is a digital image hair removal technique, further the dermoscopy image is pictographically enhanced and location of the lesion is detected using YOLO v3 as the classifier. YOLO is an open-source convolutional neural network used for deep learning and to execute other computational algorithms [12]. Various segmentation methods and algorithms have already been proposed in recent years; particularly, CNN-based skin lesion segmentation has achieved remarkable results. Although most of the CNN models tend to decrease the resolution of the image thereby demining the visual properties of the lesion, YOLO, a deep learning classifier which works on the principal of You Look Only Once, tends to be faster and more accurate than any other neural network model. Furthermore, color-based segmentation is performed on the preprocessed dermoscopy images, thereby extracting the original lesion from the image with the average AUC of 0.99.The training and test images used in the proposed algorithms are from the open-source and free available datasets: PH2 dataset, International Skin

Imaging Collaboration (ISIC2016) and International Skin Imaging Collaboration (ISIC2017). The entire work is summarized in three subparts—proposed method, result analysis and conclusion, where each subpart deals minutely with the concerned area.

2 YOLO Architecture and Implementation

Detection of melanoma lesion before classification or segmentation of lesion leads to increase in accuracy of the segmentation. Several neural network algorithms like decision trees, *k*-nearest neighbor (kNN) and support vector machine (SVM) have proved to be efficient classifiers. However, we have adapted in our work a comparatively new and faster convolution neural network model that is You Look Only Once or simply YOLO. It is a regression-based algorithm which contains a set of convolutions and max-pooling layers which scans the whole image and makes presumptions to identify, localize and classify objects inside the image. Unlike most of the classifiers, it only depends on single neural network to run sundry components; thus, it is easy to optimize algorithms. It yields the detection result as fast as 45 frames per second with an accuracy of above 90% in most cases. It is not only useful and fast for research and collaboration purpose but also efficient for real time medical vision.

Advancement in the emerging technologies of medical vision has led to many commendable and efficient approaches of dealing with the various setbacks to be faced by these algorithms. In the arena of computer vision, object detection algorithms encompass some of the ingenious ideologies where a person works on pointing out the objects and the locations of those multiple objects inside a delineated image. Contrarily, classification algorithms signalize the contents of the prescribed image or more precisely checks the existence of an image beforehand, in the given image. Classification algorithms deal with the passing of the CNN classifiers in the previously designated boxes, i.e., those boxes to be created at the very beginning of the procedure after which all the anomalies of the processes are removed pertaining to the deceleration of the method as well as strenuous upsurge of every individual constituent of the stated image (see Fig. 1). On the other hand, object detection algorithm imparts regressive approach of deep learning which concludes to a one step process of the envision of those bounding boxes in the whole image and not by focusing on its discrete enticing portions, henceforth making the process much faster compared to the former one.

A mastermind in the field of computer vision and deep learning would be the YOLO or You Only Look Once. The algorithm marks itself as a well-planned approach mainly applicable to real time object detections. The most vital factor present in this algorithm infers that researchers focus on the passing of classifiers on the whole image, instead of its individual compositions, performing the overall job in 45 frames per second. YOLO comprises of the consistent CNN which includes twisting and amalgamating layers with the involvement of two completely analogous CNN layers. YOLO serves as the fastest technique regarding real time object

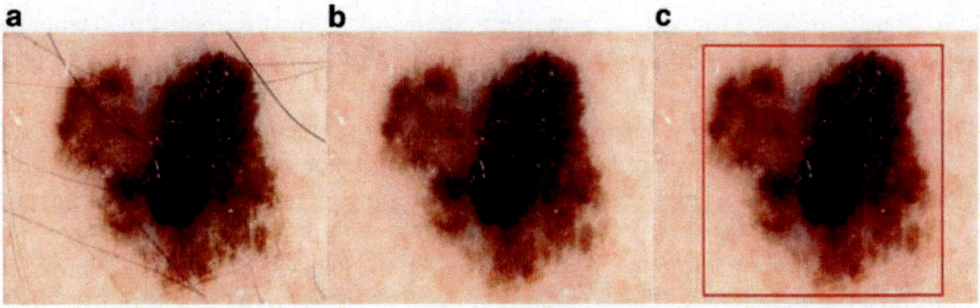

Fig. 1 **a** Input images (ISIC_0000161), **b** after hair removal by Dull Razor method, **c** location detection

detection due to its usage of the regression as a primary step of learning, henceforth using a sole neural network in its approach.

The algorithm advances to the knowledge of its prognosis tensor, i.e., $(S \times S) \times B \times (5 + C)$ where $(S \times S)$ represents the non-intersecting grid cells after cleaving the specified image. Here, B indicates the bounding boxes whose predictions and confidence score calculations depends fully on the grid cells. Confidence score finds the presence or absence of an object beforehand within the bounding box and considers itself as zero regarding the absence of the existing object.

$$\text{Confidence score} = \Pr(\text{Object}) \times \text{IOU}(\text{ground truth predicted})$$

The architecture of YOLO as per Google-Net possesses 24 convolution layers for providing researchers with a tensor (7, 7, 1024) with shape as the final result succeeded by 2 fully connected layers providing $7 \times 7 \times 30$ parameters and finally molding it as (7, 7, 30) means 2 boundary box predictions for every location. YOLO finds the highest IOU for calculating the forfeiture of the numerous bounding boxes for every grid cell and moreover moves on to the sum squared error approach for enumerating the loss function comprising the classification loss, localization loss and the confidence loss.

3 Proposed Methodology

3.1 Training YOLOv3 with PH2, ISIC 2016 and ISBI 2017 Dataset

Skin cancer detection being the most vital area of research in the field of medical vision, where training and testing of proposed algorithms requires appropriate datasets with pertinent dermoscopic images. The studies have been conducted from the publicly accessible datasets—PH2, ISIC 2016 and ISIC 2017, where the latter two datasets are used for both training the system and testing the algorithms, whereas the former class of dataset is only used for testing purpose thereby forming a holdout dataset for the proposed algorithm. These dermoscopic images are collected by expert dermatologist after several dermoscopic analyses on pigmentation, streaks, dots and color regression patterns. These datasets are developed for specialized research collations on classification, segmentation and skin lesion analysis. There are 900 dermoscopic images available in ISIC 2016 for training and addition of 379 dermoscopic images for testing melanoma lesion are present in the dataset. These images are of 8-bit RGB with the range of image resolution from 722×542 to 2272×1704. Whereas in ISIC 2017 dataset, a total of 2000 dermoscopic images are present for training the system; along with it, 600 images are for testing and 150 for validation. All the images in this dataset are of 8-bit RGB with image resolution ranging from 540 × 722 to 4499 × 6748 pixels. In addition to the rigorous testing process, 200 images of PH2 dataset are also being used; it consists of 40 melanoma instances and 80 dermoscopic images of atypical and normal nevi each, with the resolution of 768 × 560 pixels and captured with a magnification of 20x. PH2 dermoscopic images are acquired from Department of Dermatology of Pedro Hispano hospital. All the images are resized to a uniform resolution of 512 × 512 pixel before processing the images to training the classifier. The resized images are being used to train the classifier, YOLOv3 under the training parameters of batch size equal to 32, subdivision of 16, momentum of 0.9, decay rate of 0.0005, learning rate of 0.001, and it is trained through 50,000 epochs where weight is saved after every 10000th epoch which results in best fitting of the location detection results.

3.2 *Preprocessing and Segmentation*

The next step includes segmentation of the dermoscopic image after enhancement using Dull Razor method. The image in question is preprocessed for augmentation by subjecting it to Dull Razor method for removal of artifacts. The preprocessed image thus derived is a color image formed on RGB plane.

This image (M × N) can be considered as a matrix where pixel location can be defined as.

The process is determined on the R, G and B planes and we present the instance on the red plane, where ri,j be the intensity value of the pixel location. Simultaneously, a (M × N) flag image is generated taking all the values of the pixels fr(i,j) = 0.

A 5 × 5 window will be created taking center pixel taking where.

For a particular window having center, the pixel values are rk-2,l-2;…;rk,l;…rk + 2,l + 2; If(|rk-2,l-2-rk,l|≥T) where T = |10| then fr(k-2,l-2) = 1 else fr(k-2,l-2) = 0.

…… Proceeding in this way……

If(|rk + 2,l + 2-rk,l|≥T) then fr(k + 2,l + 2) = 1 else fr(k + 2,l + 2) = 0.

A new flag image will be generated where the pixel values will be either 1 or 0. In a similar manner, flag images with pixel values 1 or 0 will be generated on Green (fg) and Blue (fb) planes as well. After getting all the flag images, the following operation is performed and the final flag image (ff) is generated.

If (fr(x,y) && fg(x,y) && fb(x,y)) = = 1), then ff(x,y) = 1 and a border line is created in the preprocessed (M × N) image else ff(x,y) = 0 where.

4 Result and Discussion

4.1 Parameters for Evaluations

The proposed method adopts location detection of the lesion on dermoscopic image using YOLOv3 classifier model and further applies segmentation algorithms to precisely segment the lesion from dermoscopic image. To classify the precision of the proposed algorithms, they are tested on predefined parameters like: Sensitivity (Sen), Specificity (Spe), Accuracy (Acc), Jaccard score (Jac) and Dice coefficient (Dic). Sensitivity is the measure of total number of correct detections of melanoma lesion over total number of melanoma lesion in the dataset, whereas specificity is measure of correct detection of non-melanoma lesion over total non-melanoma dermoscopic images in the dataset. Accuracy is the calculation of overall segmentation performance. The assessment of metric over achieved segmentation result and ground truths is termed as Jaccard score. Dice coefficient quantifies segmentation performance and explains connection with the ground truth. The formulas for the above stated evaluation metrics are stated below:

$$\mathrm{IOU} = \frac{\text{Area of Overlap}}{\text{Area of Union}}$$

$$\mathrm{Sen} = \frac{\mathrm{TP}}{\mathrm{TP} + \mathrm{FN}}$$

$$\mathrm{Spe} = \frac{\mathrm{TN}}{\mathrm{TN} + \mathrm{FP}}$$

$$\mathrm{Dic} = \frac{2 \times \mathrm{TP}}{(2 \times \mathrm{TP}) + \mathrm{FP} + \mathrm{FN}}$$

$$\mathrm{Jac} = \frac{\mathrm{TP}}{\mathrm{TP} + \mathrm{FN} + \mathrm{FP}}$$

$$\mathrm{Acc} = \frac{\mathrm{TP} + \mathrm{TN}}{\mathrm{TP} + \mathrm{TN} + \mathrm{FN} + \mathrm{FP}}$$

True positive, false positive, true negative and false negative are represented as TP, FP, TN, FN, respectively. If the melanoma lesion is detected or segmented as they melanoma, it is regarded as true positive image, otherwise considered to be a

Table 1 Location detection analysis of skin lesion using YOLOv3

Dataset	Sensitivity	Specificity
PH2	97.50	97.50
ISIC 2016	97.33	97.05
ISBI 2017	97.97	97.02

false negative dermoscopic image. In case of non-melanoma lesion, the dermoscopic images which are detected or segmented as non-melanoma lesion are stated as true negative pixels, else they are stated as false positive pixel.

4.2 Result Analysis

This subpart deals with the analysis of algorithms and methods proposed in this paper. The result is analyzed on the basis of location detection and segmentation performance on the dermoscopic images of the lesion from the following publicly accessible datasets: PH2, ISIC 2016 and ISBI 2017. Entire operation was computed on a personal computer with configuration of i7 processor, RAM of 16 GB and on Windows 10 environment. Moreover Python 3.5 along with OpenCV framework was used for developing the proposed algorithms.

Performance of the algorithm and classifier was calculated on the basis of Location detection using two fundamental metrics—sensitivity and specificity, where the PH2 dataset results to 97.5% sensitivity rate and 97.5% specificity. While the sensitivity and specificity result to 97.33% and 97.05%, respectively, on ISIC 2016 dataset. Both the metrics results to a bit higher rates on ISBI 2017 dataset, with the sensitivity and specificity results to 97.97% and 97.02%, respectively. Table 1 briefly expresses the location determination of dermoscopic images using YOLOv3 classifier, and Fig. 2 outlines the overall process involve in the proposed model.

Segmentation of the lesion is the very next process after location detection, and it is measured on the following parameters of: sensitivity, specificity, accuracy, Jac and Dic scores, where the accuracy score for PH2, ISIC 2016 and ISBI 2017 results to be 94%, 95.26% and 95.17%, respectively. Table 2 illustrates overall segmentation performance on different parameters for PH2, ISIC 2016 and ISBI 2017 datasets.

4.3 Discussion

Segmentation performance plays a vital role in accurate detection of melanoma lesion from a dermoscopic image. In recent times, research on medical vision has accelerated, and particularly in the field of melanoma lesion detection, scholars across the globe have redefined the segmentation process. The proposed method was accessed

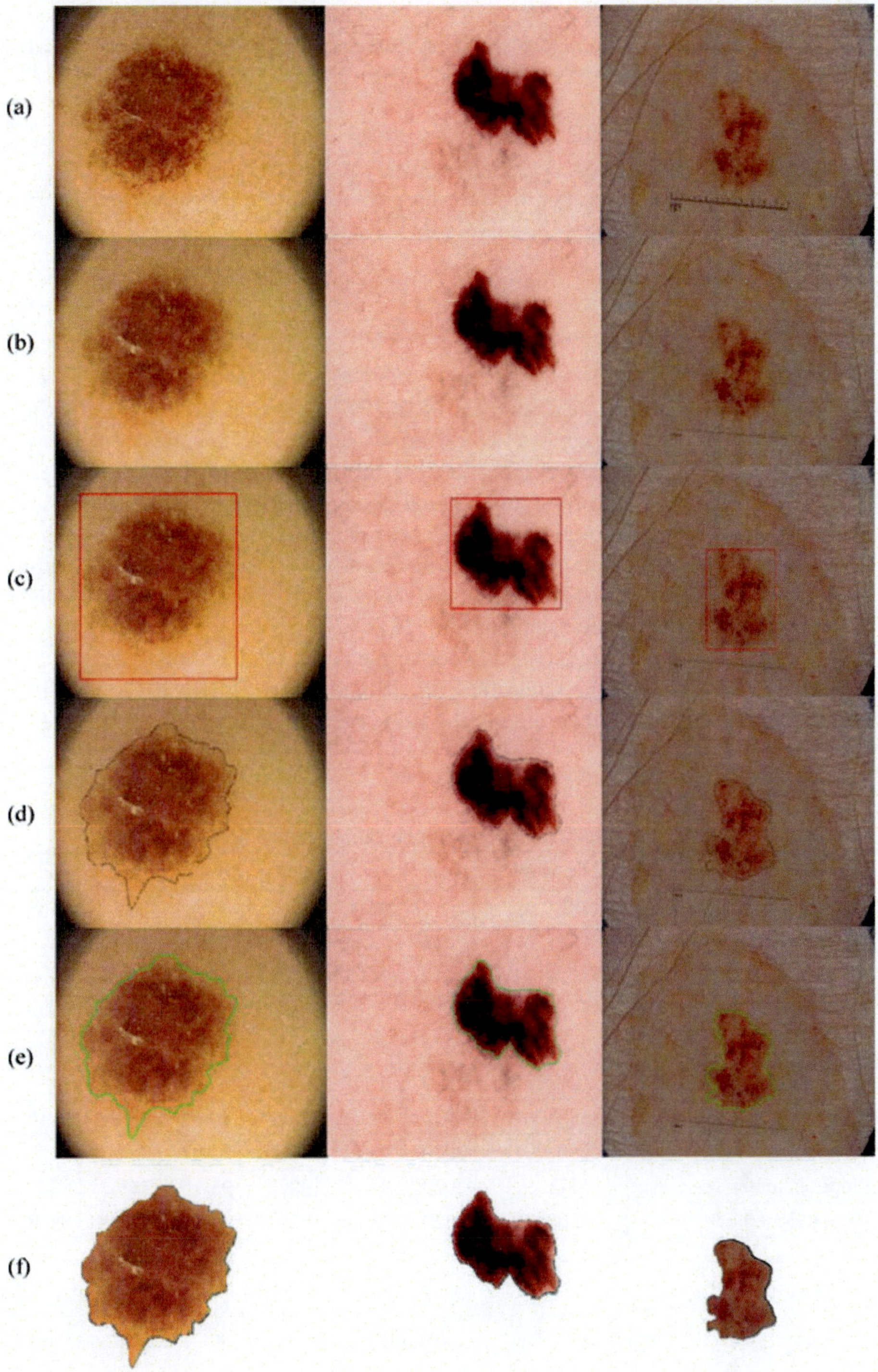
(a)
(b)
(c)
(d)
(e)
(f)

◂**Fig. 2** **a** Input images (IMD022, ISIC_0000165 and ISIC_0013615), **b** after hair removal, **c** location detection, **d** segmented area detection with dotted line, **e** final segmented area detection, **f** final segmented output

Table 2 Location detection analysis of skin lesion using YOLOv3

Dataset	Sen	Spe	Acc	Jac	Dic
PH2	90	95	94	75	85.71
ISIC 2016	92	96.06	95.26	79.31	88.46
ISBI 2017	91.37	97.02	95.17	86.12	92.55

on publicly accessible datasets such as: PH2, ISIC 2016 and ISBI 2017. Segmentation performance of the proposed algorithm was compared against deep convolutional neural network framework (DCNN)[13], multistage fully convolutional neural network(FCN)[14] and a fully convolutional residual network(FCRN) which is an advancement of fully convolutional neural network architecture[15–17]. Besides, we have also drawn a segmentation analysis with few advanced segmentation algorithms like deep neural network of convolutional–deconvolutional neural network (CDNN) which do perform automatic segmentation [18].

Table 3 illustrates the segmentation performance on the basis of Acc, Sen, Spe, Jac and Dic for all the three datasets: PH2, ISIC 2016 and ISBI 2017. The listed parameters are calculated according to the value of true positive, true negative, false positive and false negative.

Table 3 Comparative analysis of the proposed segmentation method

References	Dataset	Year	Sen	Spe	Acc	Jac	Dic
Yuan et al. (CDNN) [18]	ISBI 2017	2017	93.4	82.5	97.5	76.5	84.9
Li et al. [15]	ISBI2017	2018	93.2	82	97.8	76.2	84.7
Bi et al. (ResNets) [14]	ISBI2017	2017	93.4	80.2	98.5	76	84.4
Al-Masni et. al. [18]	ISBI2017	2018	94.03	85.4	96.69	77.11	87.08
Unver et al [12]	ISBI2017	2019	93.39	90.82	92.68	74.81	84.26
Saba et al. [13]	ISBI2017	2019	94.78	*	*	*	*
Al-Masni et al. [17]	ISBI2017	2020	81.57	75.67	80.62	*	*
Proposed method	ISBI2017	2020	95.17	91.37	97.02	86.12	92.55
Proposed method	PH2	2020	94	90	95	75	85.71
Proposed method	ISIC 2016	2020	95.26	92	96.06	79.31	88.46

5 Conclusion

Though the last decade has seen an upsurge in the cases of occurrence of melanoma, the recent study by the American Cancer Society presents a gleam of hope in the treatment of melanoma as the death rate of the disease seem to have decreased in recent years. Reports show an average of 7% per year decline in deaths for adults younger than 50 and an average of 5.7% per year decline in deaths for adults older than 50. Innovative ideas and state-of-the-art methods have contributed to the advancement in the field of diagnosis and treatment of melanoma. Our work attempts to make a humble contribution toward its timely and accurate detection by proposing a self-designed algorithm for segmentation purpose. The application of the deep learning-based YOLO classifier increases the rapidity of the process without compromising on the validity of the output. Testing and training of images were drawn on three distinguished publicly available datasets—PH2, ISBI 2016 and ISIC 2017. The work is evaluated against distinguished research works in today's time and fares moderately better in comparison to the rest on most defined parameters. Though we have seen reducing mortality rate among the melanoma victims, more advanced research is still the need of the hour for the benefit all patients irrespective of the patients' color, sex, ethnicity or age.

References

1. Feng J, Isern NG, Burton SD, Hu JZ (2013) Studies of secondary melanoma on C57BL/6J mouse liver using 1H NMR metabolomics. Metabolites **3**:1011–1035
2. Abuzaghleh O, Faezipour M, Barkana BD (2014) SKINcure: an innovative smartphone- based application to assist in Melanoma early detection and prevention. Signal Image Process Int J **15**:1–13
3. Jemal A, Siegel R, Ward E, Hao Y, Xu J, Thun MJ (2019) Cancer statistics, 2019. CA Cancer J Clin **69**:7–34
4. Mayer JE, Swetter SM, Fu T, Geller AC (2014) Screening, early detection, education, and trends for melanoma: current status (2007–2013) and future directions Part II. Screening, education, and future directions. J Am Acad Dermatol **71**:e1–e611
5. Rigel DS, Russak J, Friedman R (2010) The evolution of melanoma diagnosis: 25 years beyond the ABCDs. CA Cancer J Clin 60:301–316
6. Banerjee S, Singh SK, Chakraborty A, Das A, Bag R (2020) Melanoma diagnosis using deep learning and fuzzy logic. Diagnostics 10:577
7. Togawa Y (2017) Dermoscopy for the diagnosis of Melanoma: an overview. Austin J Dermatol 4:1080
8. Kroemer S, Frühauf J, Campbell TM, Massone C, Schwantzer G, Soyer HP, Hofmann-Wellenhof R (2011) Mobile teledermatology for skin tumour screening: diagnostic accuracy of clinical and dermoscopic image tele-evaluation using cellular phones. Br J Dermatol 164:973–979
9. Harrington E, Clyne B, Wesseling N, Sandhu H, Armstrong L, Bennett H, Fahey T (2017) Diagnosing malignant melanoma in ambulatory care: a systematic review of clinical prediction rules. BMJ Open **7**:e014096
10. Robinson JK, Turrisi R (2006) Skills training to learn discrimination of ABCDE criteria by those at risk of developing melanoma. Arch Dermatol 142:447–452

11. Lee T, Ng V, Gallagher R, Coldman A, McLean D (1997) Dullrazor®: a software approach to hair removal from images. Comput. Boil. Med. 27:533–543
12. Ünver HM, Ayan E (2019) Skin lesion segmentation in dermoscopic images with combination of YOLO and GrabCut algorithm. Diagnostics 9:72
13. Saba T, Khan MA, Rehman A (2019) Region extraction and classification of skin cancer: a heterogeneous framework of deep CNN features fusion and reduction. J Med Syst 43:2–19
14. Bi L, Kim J, Ahn E, Feng D (2017) Automatic skin lesion analysis using large-scale dermoscopy images anddeep residual networks. arXiv, arXiv:1703.04197
15. Li Y, Shen L (2018) Skin lesion analysis towards melanoma detection using deep learning network. Sensors 18:556
16. Al-masni MA, Al-antari MA, Choi MT, Han SM (2018) Skin lesion segmentation in dermoscopy images via deep full resolution convolutional networks. Comput Methods Programs Biomed 162:221–231
17. Al-masni AM, Kim D, Kim T (2020) Multiple skin lesions diagnostics via integrated deep convolutional networks for segmentation and classification. Comput Methods Programs Biomed 190:1–12
18. Yuan Y (2017) Automatic skin lesion segmentation with fully convolutional-deconvolutional networks. arXiv, arXiv:1703.05165

Detection of COVID-19 Using Deep Transfer Learning-Based Approach from X-Ray and Computed Tomography(CT) Images

Kumar Kalpadiptya Roy, Ipsita Mazumder, Arijit Das, and Subhram Das

Abstract Clinical authorities need technological support aided with artificial intelligence for early diagnosis and slowing the spread of pandemic diseases. The outbreak of COVID-19 disease caused by the newly discovered SARS-CoV-2 virus was reported by the officials in Wuhan City, China, in December 2019. Since then the virus had a disrupting impact on the health of people accompanied by psychological, financial, and social distress. In this paper, a deep learning-based approach for early detection of COVID-19 has been proposed. Five deep neural network architectures have been trained through transfer learning based on the available X-ray and computed tomography image dataset. The chosen architectures have given quite promising results in terms of accuracy. Thus, the proposed experiment provides an efficient tool for the early detection of COVID-19.

Keywords Chest X-ray · COVID-19 · Deep learning · Transfer learning

1 Introduction

Coronavirus belongs to the group of RNA viruses that cause infection in both humans and animals. The first human case of COVID-19 disease caused by the newly discovered coronavirus and afterward named as SARS-Cov-2 was reported by the officials in Wuhan City, China, in December 2019. Investigations by Chinese authorities have identified that the earliest known cases were linked to a wholesale food market in Wuhan. Environmental samples taken from that market in December 2019 tested positive for SARS-CoV-2, suggesting that it played a major role in the initial amplification of the outbreak. The WHO declared the virus a pandemic on March 11 which has almost collapsed the healthcare and economic system of many countries. The virus that causes COVID-19 is mainly transmitted through droplets generated when an infected person coughs, sneezes, or exhales. These droplets stick to the floors or surfaces and hence infect other people who come in contact with these. People

K. K. Roy (✉) · I. Mazumder · A. Das · S. Das
Computer Science and Engineering, Narula Institute of Technology, Kolkata, India

M. Mitra et al. (eds.), *Computational Advancement in Communication, Circuits and Systems*, Lecture Notes in Electrical Engineering 786,
https://doi.org/10.1007/978-981-16-4035-3_27

who severely fall sick with COVID-19 experience pneumonia and multiple organ failure which can lead to death. The incubation period of COVID-19 is between 1 and 14 days. It is contagious before symptoms appear, which is why many people get infected. X-rays which are cheaper, faster and expose the patient to less radiation provide an accuracy rate of 60–70% and computed tomography images additionally provide 50–75% accuracy, helping in the early diagnosis and treatment of COVID-19 disease. However, a specialist is needed to examine in detail the nodular ground glass opacities accompanied with reticulation in the X-rays and computed tomography images respectively [1]. This is because the ground glass opacities are synonymous to many other non-COVID lung conditions [2]. The alarming rate at which the virus is spreading has impacted the medical personnel both in terms of time and shortage of resources. In this context, artificial intelligence-based solutions (human independent) can provide faster, cheaper, and accurate diagnosis for COVID-19 and other types of pneumonia. Artificial intelligence chest radiography models based on deep learning have given researchers results as effective as any experienced radiologists [3, 4]. Since the publication of COVID-19 datasets [5], many researchers have trained their models by classifying those X-ray images which are gathered from all around the world. We have studied few related works in this field like the 'Deep-COVID: Predicting COVID-19 from chest X-ray images using deep transfer learning,' [6] where the cohen et al dataset along with a subset of images from the CheXpert dataset to produce a total of 5000 images was used. ResNet18, ResNet-50, SqueezeNet, and DenseNet-161 were then applied, and the best results were obtained from SqueezeNet with a sensitivity rate and specificity rate of almost 98 and 92.7%, respectively. Another work is the 'Early diagnosis of COVID-19-affected patients based on X-ray and computed tomography images using deep learning algorithm' where the COVID-19 X-ray dataset [7] was used to train VGG-16 and InceptionV2. An accuracy of 100% was found after training VGG-16 without over-fitting for normal versus COVID-19 X-ray images. For pneumonia versus COVID-19 classification, they have achieved an accuracy of 94% again without over-fitting. From the InceptionV2 model, they have achieved a precision of 80 and 76% for COVID-19 negative and positive patients, respectively [8]. There are few other studies also where transfer learning has been used efficiently [9, 10]. In this paper, we have presented a logical scheme based on our experimentation and analysis through mathematical results obtained, to guide the choices that might be taken for COVID-19 detection using currently available datasets and the deep learning algorithms. We have employed VGG-16 [11], DenseNet-121 [15], AlexNet [14], ResNet-34 [12], InceptionNet-V3 [16] on our dataset collected from the IEEE COVID-19 available as open source [13]. These architectures have given us quite promising results in terms of accuracy on the limited dataset available. DenseNet-121 has given the best result among all with an accuracy of 99.5%. Thus, our classification guide on the dataset and deep learning models comes out to be a reasonable choice for further experimentation and research.

2 Data Collection

We have collected the data mainly from two sources.

2.1 *COVID-19 Image Data*

It is an open-source repository of COVID-19 images [13]. At the time of writing this article, it had 9224 number of frontal X-ray and CT scan images of potentially COVID-19-affected people.

2.2 *Chest X-Ray Images (Pneumonia)*

This dataset [13] contains 5857 chest X-ray images of normal patients (1583), patients affected by bacterial pneumonia (2780), and affected by viral pneumonia (1493).

From the above-mentioned sources, we have collected total 535 COVID-19 chest X-ray images and 390 images that consist of both normal chest X-ray images and Pneumonia patient's chest X-ray images. We have used this dataset for all the experiments. In Fig. 1, few samples of the dataset have been shown.

3 Data Preprocessing

First of all we have made three datasets from the entire data available to us. Dataset A consists of images of only CT scans of both COVID patients and non-COVID patients. Dataset B consists of images of only chest X-rays of both COVID and non-COVID patients. Dataset C is the combination of Datasets A and B and consists of all the images in those sets, respectively. We have reshaped the images of our datasets into 224×224 dimension, and then we have applied random horizontal

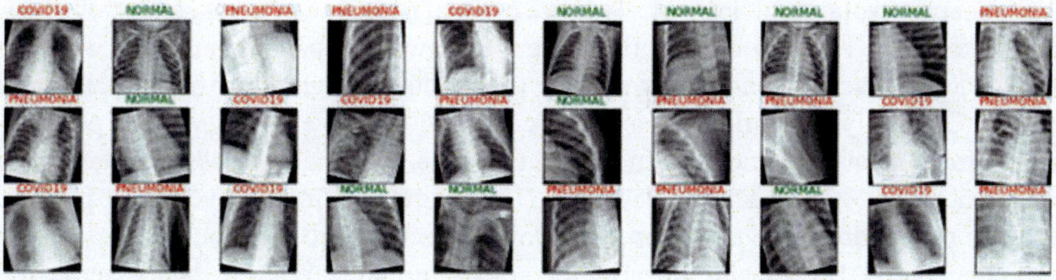

Fig. 1 Sample image dataset

flip. We have also normalized the image pixels by conventional mean values and standard deviation values of RGB images, such that data augmentation is achieved over this dataset and we can almost feed twice the number of images to our model during training. During testing of the model over a test datasets of 60 images each for COVID-19 and non-COVID-19 category, we have only used the technique of normalization as augmentation.

4 Methods

We have used various kinds of very popular deep learning architectures, namely deep learning residual networks [12] better known as ResNet, VGG convolutional neural network [11], AlexNet, DenseNet-121, and InceptionNet-V3 and trained it on the same dataset.

4.1 Residual Networks

Residual networks [12] are a family of convoluted neural network architectures that have shortcut connections among its layers. In these kinds of networks, we implement the concept of passing the residue left after passage through the previous convolutional layer to one of the succeeding layers. This shortcut identity can be directly used if the output size of the previous layer and the input size of the succeeding layer is of the same size. But if the output size of the previous layer and the input size of the succeeding layer differs, then we use a method called 'padding' to equalize both the dimensions, so that the shortcut link can be implemented between the layers. In 'padding,' we generally add extra zero(s) to the output size which is generally smaller to make it equivalent to the input size.

4.2 Very Deep Convolutional Networks

Very deep convolutional networks [11] are plain in nature and do not consist of shortcut links as the residual convolutional neural networks do. The networks work on the idea of extracting features by passing the inputs through filters of convolutional layers. Generally, the dimensions of the convolutional layers may vary as per configurations when implemented. In the general case, the smallest possible filter has a dimension of 3×3 and a stride of 1 pixel. The spatial padding is done with 1 pixel so that the output dimensions of a previous layer matches with the input dimensions of the next layer. After passing through the convolutional layers, the data is subjected to spatial pooling which has 5 max pooling layers. Max pooling in these networks is performed with stride = 2 and over a window of 2×2 dimensions.

4.3 *AlexNet*

AlexNet [14] is famous because this deep learning model won the 2012 ImageNet LSVRC-2012 competition by a large margin. It contains 5 convolutional layers and 3 fully connected layers. ReLU is applied after very convolutional and fully connected layer. Dropout is applied before the first and the second fully connected year.

4.4 *DenseNet*

DenseNet [15] is the family of neural networks which are deep but tends to solve the problem vanishing gradient while keeping the number of parameters low at the same time. DenseNet architectures exploit the potential of traditional neural networks by reuse of features. Moreover, in a comparison it can be seen that ResNet layers have a large number of parameters because each layer of the network has to learn its own weight, whereas in DenseNet architectures the layers are narrow due to less number of parameters and with each passing layer only a few new features are added to the network.

4.5 *InceptionNet-V3*

Inception networks [16] came into existence originally to solve the problem of large variations of features distributed over a wide area within an image. To maintain a check on the computational expenses and at the same time maintain a good accuracy of predictions, the concept of multiple-sized kernels for filters within a convolutional neural network was introduced which forms the basis of InceptionNet model. We have employed the third version of this network, where it is equipped with improvements in RMSProp optimizers, 7×7 convolutions, batch normalization in the auxiliary classifiers and label smoothing (for preventing over-fitting).

5 Results

We have implemented two varieties of convolutional neural networks, namely VGG-16 [11], AlexNet [14], ResNet-34 [12], DenseNet-121 [15], and InceptionNet-V3 [16] on our dataset. During training, we have fed the models with a batch size of 10 images at a time. After the model trains itself on those dataset, we have stored the output in the memory and then we have passed the output to an optimizer. The optimizer here is nothing but the gradient descent mechanism. Out of the various available gradient descent techniques we have used the 'Adam' for residual network models and stochastic gradient descent (SGD) for VGG models in our training procedure.

5.1 AlexNet Architecture

Our training consists of 794 training batches in each epoch for dataset A, 942 training batches in each epoch for dataset B and 1734 batches for the dataset C. After running the algorithm for 2 epochs, we have got an accuracy of 98% for dataset A, an accuracy of 100% for dataset B, and an accuracy of 99% for dataset C, which is a very promising result from such limited dataset. The training and validation curves have been plotted in Fig. 2a–c.

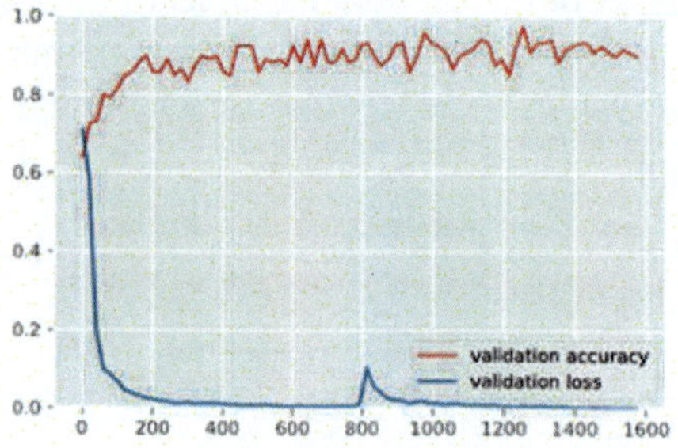

(a) Validation plot for Alexnet on Dataset A

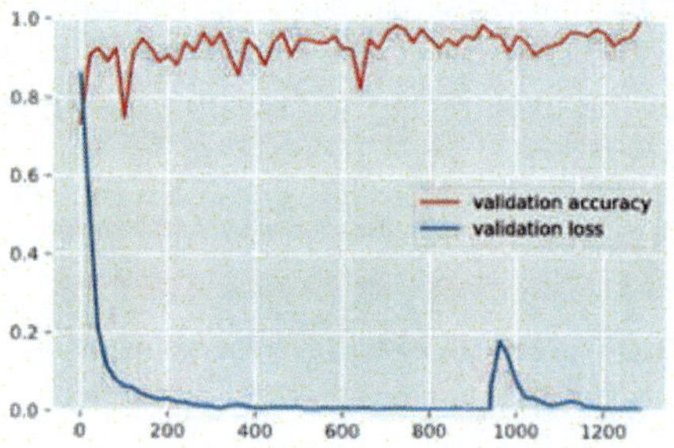

(b) Validation plot for Alexnet on Dataset B

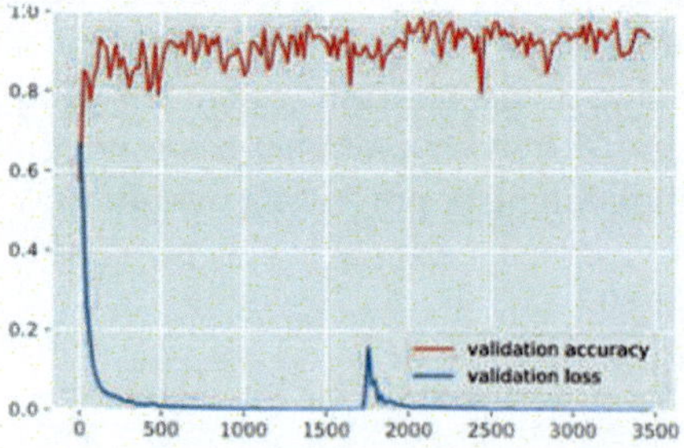

(c) Validation plot for Alexnet on Dataset C

Fig. 2 Validation analyses for AlexNet

5.2 *ResNet-34 Architecture*

Here we have got an accuracy of 96.32% for dataset A, an accuracy of 100% for dataset B, and an accuracy of 99% for dataset C which is significant because of the less training time of this model. The training and validation curves have been plotted in Fig. 3a–c.

5.3 *DenseNet-121 Architecture*

In this architecture, we have achieved an accuracy of 100% for all the three datasets A, B, and C, respectively, which highlights a strong evidence for deployment of this model in practical and real-life scenarios for COVID-19 detection. The training and validation curves have been plotted in Fig. 4a–c.

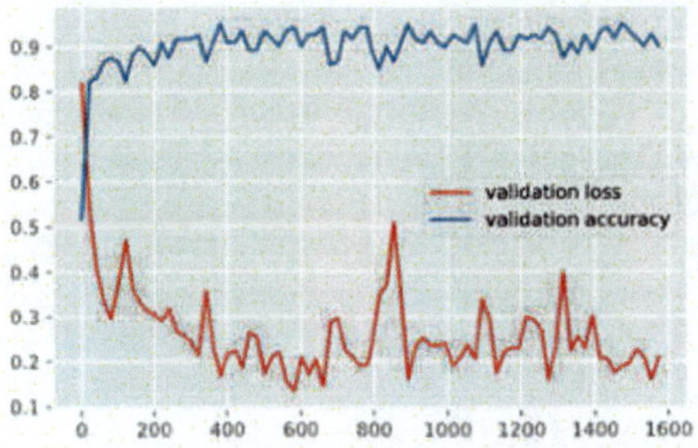

(a) Validation plot for ResNet-34 on Dataset A

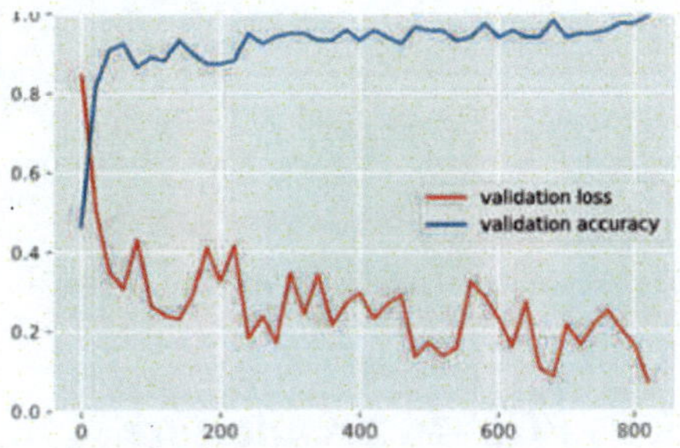

(b) Validation plot for ResNet-34 on Dataset B

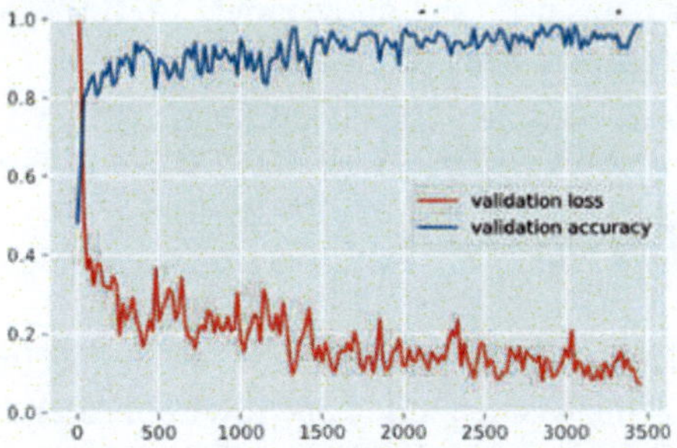

(c) Validation plot for ResNet-34 on Dataset C

Fig. 3 Validation analyses for ResNet-34

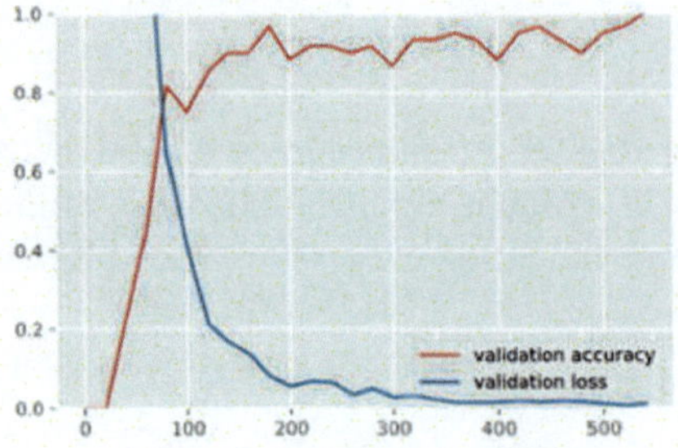

(a) Validation plot for DenseNet-121 on Dataset A

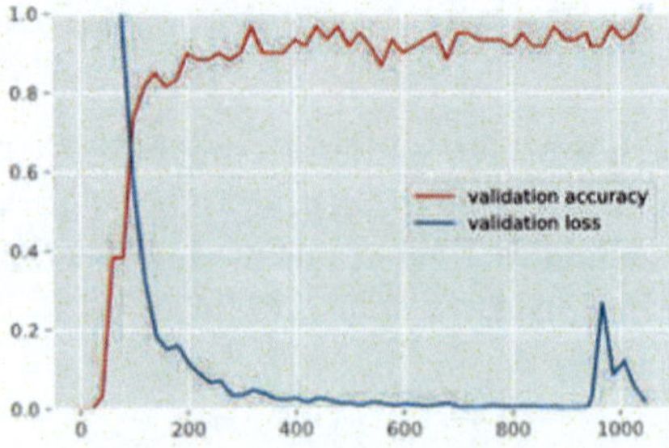

(b) Validation plot for DenseNet-121 on Dataset B

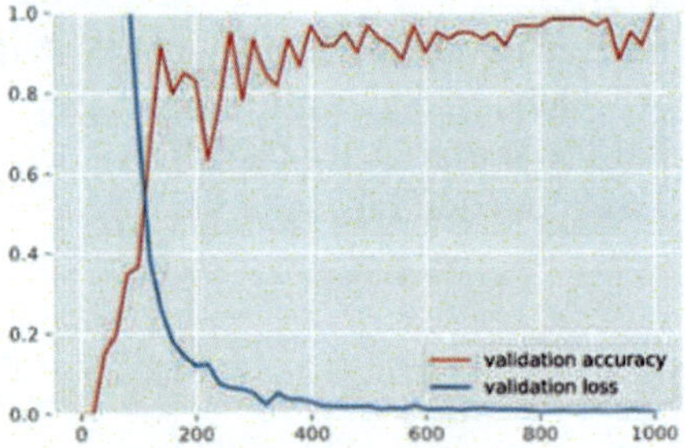

(c) Validation plot for DenseNet-121 on Dataset C

Fig. 4 Validation analyses for DenseNet-121

5.4 VGG-16 Architecture

During the training of this model, we have recorded an accuracy of 97% for dataset A, an accuracy of 99% for dataset B, and an accuracy of 100% on dataset C. The training and validation curves have been plotted in Fig. 5a–c.

5.5 InceptionNet-V3 Architecture

Here we have got an accuracy of 99% for dataset A and an accuracy of 100% for dataset B and dataset C, which is again a milestone if we consider the less number of training steps in which we have achieved it. The training and validation curves have been plotted in Fig. 6a–c.

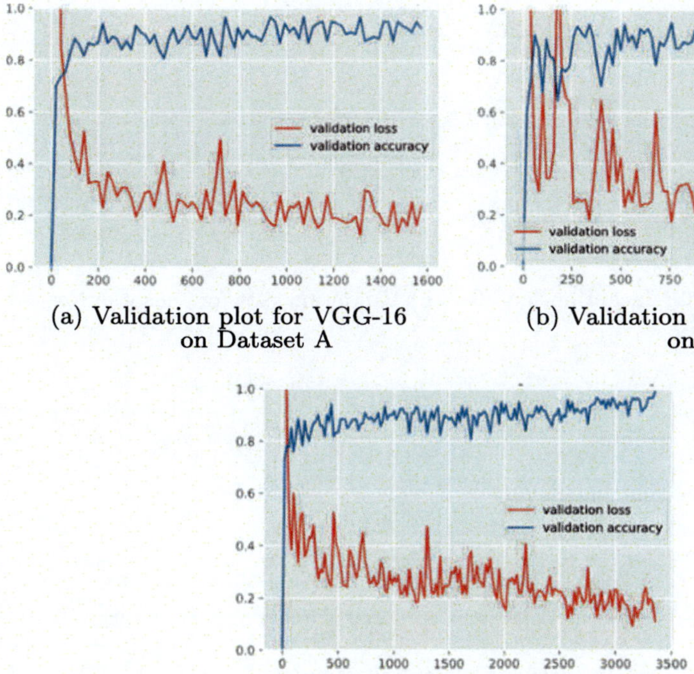

(a) Validation plot for VGG-16 on Dataset A

(b) Validation plot for VGG-16 on Dataset B

(c) Validation plot for VGG-16 on Dataset C

Fig. 5 Validation analyses for VGG-16

6 Conclusion

X-ray and computed tomography images are very important diagnostic tools for clinical management of COVID-19. Artificial intelligence has the ability to gauge these clinical tools and differentiate COVID-19 from other lung diseases. Keeping this in mind, we have proposed a deep transfer learning-based approach from X-ray and CT images for rapid and early detection of COVID-19 virus. We have collected our dataset from two open-source repositories and run them on five CNN models. Eventually we have got the highest accuracy of 100% with the DenseNet-121 architecture for all the three datasets, respectively. Thus, our classification guide with DenseNet-121 architecture on top comes out to be a reasonable choice for further experimentation and research.

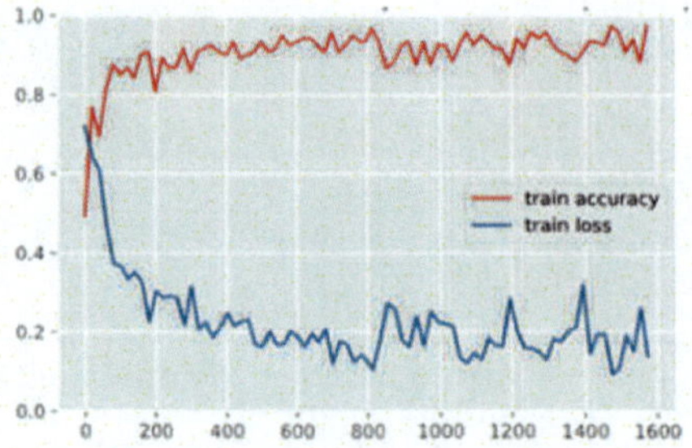

(a) Validation plot for InceptionNet-V3 on Dataset A

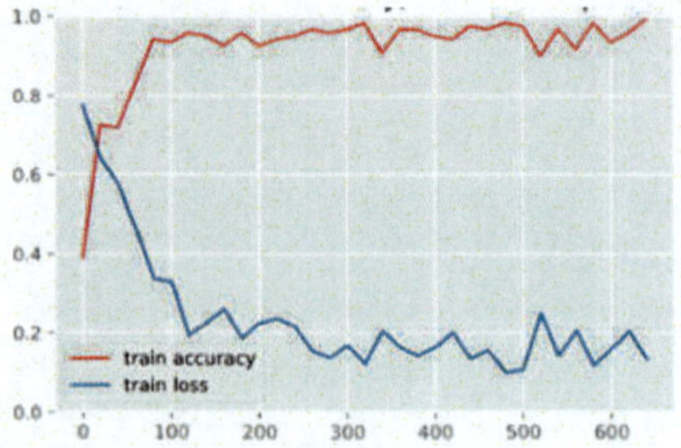

(b) Validation plot for InceptionNet-V3 on Dataset B

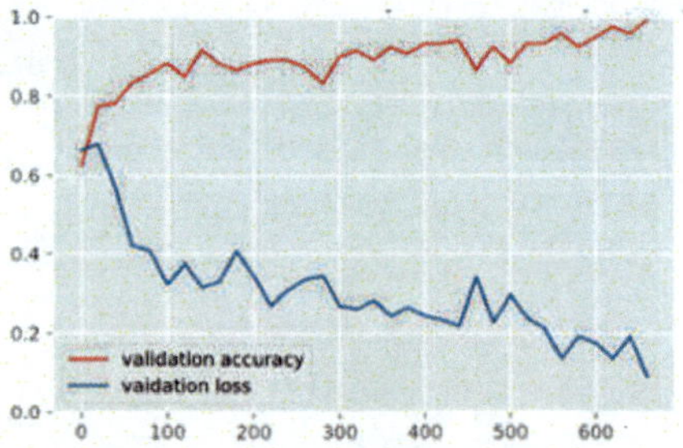

(c) Validation plot for InceptionNet-V3 on Dataset C

Fig. 6 Validation analyses for InceptionNet-V3

References

1. Schmitt W, Marchiori E (2020) Covid-19: round and oval areas of ground-glass opacity. Pulmonology. https://doi.org/10.1016/j.pulmoe.2020.04.001
2. Mossa-Basha M et al (2020) Radiology department preparedness for COVID-19: radiology scientific expert panel. Radiology 200988. https://doi.org/10.1148/radiol.2020200988
3. Gu X et al (2018) Classification of bacterial and viral childhood pneumonia using deep learning in chest radiography. https://doi.org/10.1145/3195588.3195597
4. Sivaramakrishnan R et al (2018) Comparing deep learning models for population screening using chest radiography. Medical imaging 2018: computer-aided diagnosis, vol 10575. International Society for Optics and Photonics. https://doi.org/10.1117/12.2293140
5. Maguolo G, Nanni L (2020) A critic evaluation of methods for covid-19 automatic detection from x-ray images. arXiv preprint arXiv:2004.12823, https://arxiv.org/abs/2004.12823
6. Minaee S et al (2020) Deep-covid: predicting covid-19 from chest x-ray images using deep transfer learning. https://arxiv.org/abs/2004.09363
7. https://www.kaggle.com/plameneduardo/sarscov2-ctscan-dataset
8. Dansana D et al (2020) Early diagnosis of COVID-19-affected patients based on X-ray and computed tomography images using deep learning algorithm. Soft Comput 1–9. https://link.springer.com/article/10.1007/s00500-020-05275-y
9. Kumar SP, Kumari BS (2020) Detection of coronavirus disease (covid-19) based on deep features. Preprints 2020030300:2020
10. Apostolopoulos, ID, Mpesiana TA (2020) Covid-19: automatic detection from x-ray images utilizing transfer learning with convolutional neural networks. Phys Eng Sci Med 1

11. Simonyan K, Zisserman A (2014) Very deep convolutional networks for large-scale image recognition
12. He K, Zhang X, Ren S, Sun J (2016) Deep residual learning for image recognition
13. https://github.com/ieee8023/covid-chestxray-dataset
14. Alex K, Ilya S, Geoffrey H (2012) ImageNet classification with deep convolutional neural networks. Neural Inf Process Syst 25. https://doi.org/10.1145/3065386
15. Huang G, Liu Z, Van Der Maaten L, Weinberger KQ (2017) Densely connected convolutional networks. In: 2017 IEEE conference on computer vision and pattern recognition (CVPR), Honolulu, HI, USA, pp 2261–2269. https://doi.org/10.1109/CVPR.2017.243
16. Szegedy C, Vanhoucke V, Ioffe S, Shlens J, Wojna Z (2016) Rethinking the inception architecture for computer vision. https://doi.org/10.1109/CVPR.2016.308

Quantum Random Number Generators for Cryptography: Design and Evaluation

Puspak Pain, Arindam Sadhu, Kunal Das, and Maitreyi Ray Kanjilal

Abstract In this article, quantum circuit-based secure communication architecture has been envisioned. Herein, all the proposed circuits and architecture are verified by IBM Qiskit and established on the quantum nanostructure. The salient goal of this hardware-based cryptographic structure in the quantum domain is to attain a diversified invulnerable quantum communication arrangement through a commendable post-CMOS technology. This architecture consists of a novel quantum random number generator (QRNG) and swap gate-based quantum shuffler. In our intended framework for cryptographically secured communication model as well as its implementation through a novel quantum encryption–decryption prototype by the random bits extracted from quantum Hadamard gates, rotation gate is (R_z) postulated on QRNG.

Keywords Quantum cryptography · QRNG · Hadamard gate · Swap gate · CNOT · IBM Qiskit

1 Introduction

Bennett and Brassard's research keystone goes out in front of the budding field in quantum cryptography for modern day's secure communication systems. Afterward, proposed sundry quantum protocols make communication more exuberant and promising in a given security and cryptographic consideration. The Shor's factoring

P. Pain (✉) · A. Sadhu · M. R. Kanjilal
Narula Institute of Technology, Kolkata, West Bengal 700109, India
e-mail: puspak.pain@nit.ac.in

A. Sadhu
Greater Kolkata College of Engineering and Management, Baruipur, West Bengal, India

Maulana Abul Kalam Azad Universities of Technology, Kolkata, West Bengal, India

K. Das
Department of Computer Science, Acharya Prafulla Chandra College (Govt. Sponsored), New Barrackpur, Kolkata 700131, India

M. Mitra et al. (eds.), *Computational Advancement in Communication, Circuits and Systems*, Lecture Notes in Electrical Engineering 786,
https://doi.org/10.1007/978-981-16-4035-3_28

algorithm [1], quantum teleportation [2–7], superdense coding [8], quantum secure direct communication [9–11], quantum secret sharing [12] and quantum key agreement [13] are illustrious in today's rapidly growing field of quantum communication. Moreover, Yao's (1982) Millionaires' problem [14] has been pertained to secure multiparty computation (SMC) and envisioned for the practical state of affairs [15, 16]. However, to enhance the security and precision of Yang and Wen [17] proposed quantum private comparison (QPC) protocol, a quantum gate-based semi-honest third-party (TP) aspect is also validated in modern literature [18]. In this point of view, Lo [19] insecurity and its inaccessible correspondence in two-party quantum computations are also remarkable.

Most of the previous works with security essence had established on the reconciliation of its actualized functionality where the resources desired either settled classical networks or realizing in the quantum setup.

But the quantum random number generators (QRNG)-based secured communication proposal had deserted in the literature. This proposal is an attempt to quest those secure communication models in a quantum cryptographic frame of mind.

The rest of this article is organized as follows. Section 2 introduces the overview of quantum cryptography with the preliminaries of QRNG, CNOT, quantum Hadamard, R_z, shuffler and QSWAP gates. In Sect. 3, our proposed framework and its implementation in IBM Qiskit that verifies the correctness of our proposed scheme have been described. A conclusive future aspect is drawn in Sect. 4.

2 An Overview of Quantum Cryptography

Mostly, cryptography is the challenge in communication and computation for the transmittal of secret information while concerned parties may have not faith in another. Among the several aspects of classical and quantum cryptographic protocols, the ultimate contrast lies between the private and public—key cryptosystems are the key and its distribution. So the exceedingly noted implementation of quantum potencies in security and cryptography is the key distribution [20]. Lots of recent researchers are interested in the use of quantum properties for key distribution between two parts like "Alice" and "Bob" through the BB84 protocol that uses the non-cloning theorem and the collapse of probability waves [21].

The dubious effect of entanglement states [22] also brings about many approaches in the quantum era by one more key distribution protocol is called E91 [23]. Furthermore, an acclaimed entanglement-based teleporting circuit has also diversified cryptographic functions and applications [24].

A further vicinity of physics and its quantum mechanical model exposes the cryptographic application of random number generation. Quantum random number generators (QRNG) are one of the prime factors for portraying a QKD approach to obtain pure random bit streams. In quantum cellular automata, majority voter and self-starved feedback SRAM-based true random number generators are already described in the recent literature [25, 26]. Furthermore, quantum repeaters give a

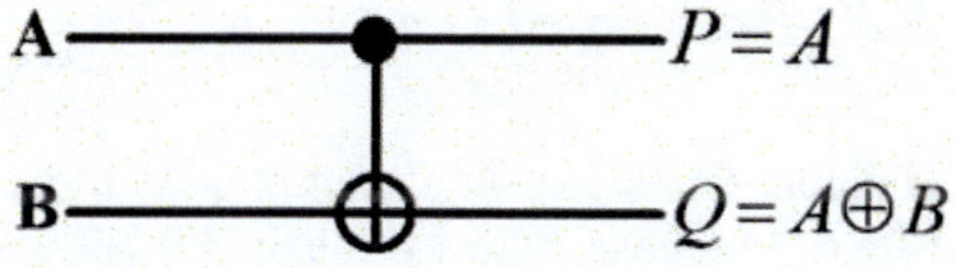

Fig. 1 Controlled not gate

vital breakthrough for elongated secure communication accomplished by the QKD in real life [27].

2.1 Basics of QRNG

In today's probabilistic age of scientific research, industrial testing, labeling, numerical simulations, computing and in hazard playoffs, random number generators (RNG) are a significant resource. RNG are having more of the essence to facilitate an arbitrary set of numbers that should not unveil any definite ideas in their generation and continuation for both classical and quantum cryptography further to security approaches.

The precedent of the legendary BB84 quantum key distribution protocol [28] is evidence that, as per Kerhoff's principle [29], a random number generator has to effectuate unpredictable bits for success in cryptography attainment. Among the ideas of random number generators, algorithmic software-based pseudorandom RNG and non-deterministic hardware RNG schemes are prevalent [30, 31].

2.2 CNOT Gate

The physical approach to quantum akin familiar electronic digital gates where unitary operators U deed on the quantum states and provide a unitary evolution is called a quantum gate. This is a basic quantum 2 × 2 reversible gate [32]. It is also known as Feynman gate [32–35] as depicted in Fig. 1.

2.3 Swap Gate

This gate is two-qubit gate. It swaps the state of the two involved input qubits.

Matrix representation of SWAP gate [34] is following.

$$\begin{pmatrix} 1\,0\,0\,0 \\ 0\,0\,1\,0 \\ 0\,1\,0\,0 \\ 0\,0\,0\,1 \end{pmatrix}$$

2.4 Hadamard Gate

Hadamard gate is a single-qubit quantum gate. This gate creates an equal superposition of two base states. It means $|0\rangle$ is converted to $\frac{|0\rangle+|1\rangle}{\sqrt{2}}$ and $|1\rangle$ to $\frac{|0\rangle-|1\rangle}{\sqrt{2}}$. Matrix representation of Hadamard gate [31–35] is following.

$$H = \frac{1}{\sqrt{2}}\begin{bmatrix} 1 & 1 \\ 1 & -1 \end{bmatrix}$$

2.5 R_Z Gate

The R_z gate is a single-qubit diagonal gate, rotation operators. It can be arranged for hardware by frame changes where rotation through angle θ (radians) around the z-axis [35–39].

$$R_Z(\theta) = \begin{pmatrix} e^{\frac{-i}{2}} & 0 \\ 0 & e^{\frac{i}{2}} \end{pmatrix}$$

3 The Proposed Scheme

Our proposed model has achieved two stages of security. Quantum true random number generator takes a crucial part of the proposed cryptographic architecture at the initial phase of communication Fig. 2.

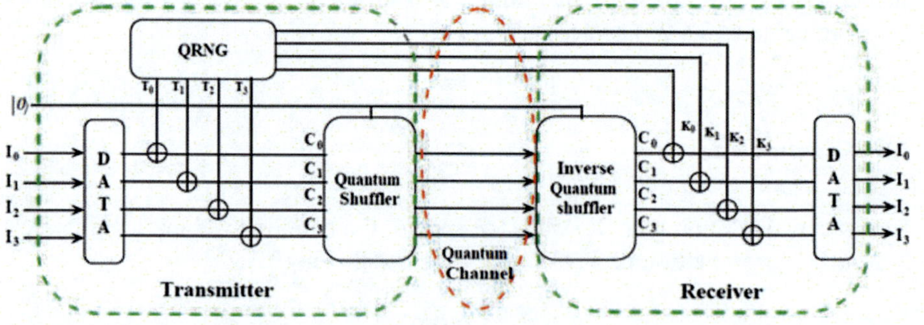

Fig. 2 Proposed quantum cryptographic secure communication scheme

3.1 *Implementation and Evaluation of the Proposal in IBM Qiskit*

All proposed designs and their functionalities are verifiable by the IBM Qiskit [31] tool. In our proposal, the quantum CNOT gate has been used for encryption of the plaintext or data stream (I_0–I_3). Again, a quantum true random number generator has been designed to produce exclusively unpredictable sequences (T_0–T_4). A quantum shuffler block intended to shamble the outputs of the CNOT operated plaintext data streams (I_0–I_3) and the QRNG's random bit streams. The cipher texts (C_0–C_3) are transferred from the sender end via quantum shuffler through a quantum channel to the decryption end. Over the same quantum channel, QRNG generated random data streams K_0–K_3 directly transmit at the same time, shuffled cipher texts are accepted at the receiver end. In the decryption process, a quantum inverse shuffler has been incorporated to rearrange the cipher text after a relevant circuital delay. At the final stage, the consequent ciphers and the random bits (K_i) are operated through the relevant decryption method of CNOT transformation to retrieve the original plain texts. Therefore, five modules: a QRNG block; two quantum shuffler blocks; two CNOT blocks are solely designed to implement our proposed quantum state of affairs.

IBM Qiskit simulated outcomes of our proposed architecture are illustrated in Fig. 3a–c. Therefore, our proposed QRNG-based hardware cryptographic architecture using quantum technology shown in Fig. 2 can be considered as a building block for secure communication which also reduces the design complexity.

4 Conclusion and Future Aspect

In this paper, we proposed a quantum circuit-based cryptographic scheme to implement a two-state quantum secure communication model. The intended communication architecture and its IBM Qiskit simulation results conclude that only using parallel Hadamard (H) gate, the probability of occurring 00–11 is 6.25%, while using R_z gate with H gate, the possibility of occurrence 00–11 varies from 5.396% to 6.921%. A comprehensive simulation of QRNG circuit reports randomness 4.688–7.6% for quantum state 0000–1111 that implies the randomness of QRNG output will improve and more secured quantum cryptographic aspect will be achieved. An improved post-quantum cryptographic architecture with quantum private comparison can be incorporated in the future to enhance the authentication of sender information.

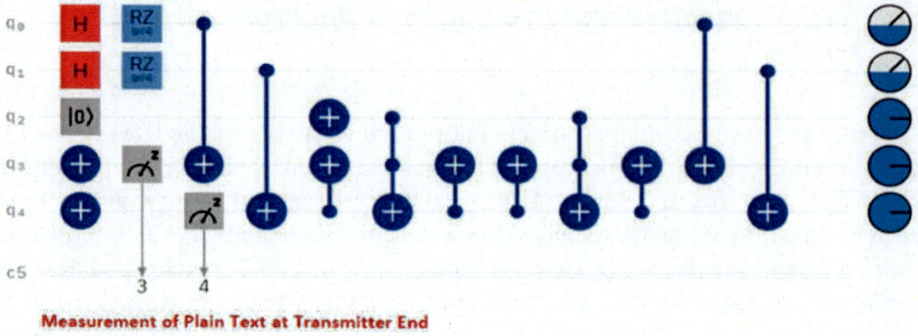

(a)

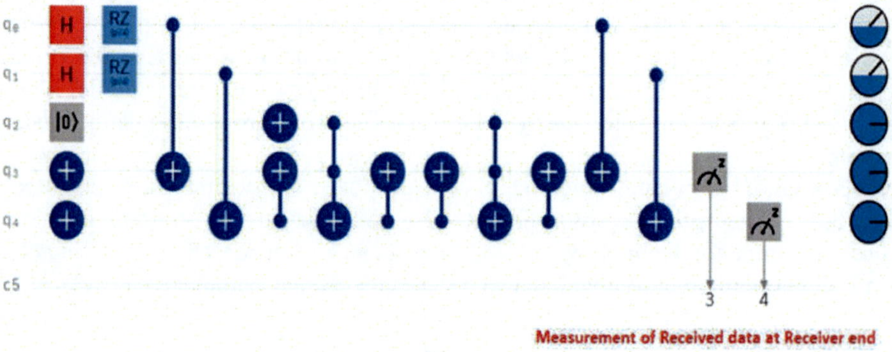

(b)

(c)

Fig. 3 Implementation of the proposed architecture in IBM Qiskit **a** measurement of transmitter end architecture in IBM-QISkit, **b** measurement of receiver end architecture in IBM Qiskit, **c** histogram output of proposed architecture shows "11" plain text is securely decrypted at receiver end

References

1. Shor PW (1995) Scheme for reducing decoherence in quantum computer memory. Phys Rev A 52:R2493
2. Martin-Lopez E et al (2012) Experimental realization of Shor's quantum factoring algorithm using qubit recycling. Nat Photo 6:773
3. Jin X-M et al (2010) Experimental free-space quantum teleportation Nat Photo 4:376
4. Cacciapuoti AS, Caleffi M, Van Meter R, Hanzo L (2020) When entanglement meets classical communications: quantum teleportation for the quantum Internet. IEEE Trans Commun 68(6):3808–3833
5. Yin J et al (2012) Quantum teleportation and entanglement distribution over 100-kilometre free-space channels. Nature 488:185
6. Zhang Q et al (2006) Experimental quantum teleportation of a two-qubit composite system. Nat Phys 2:678
7. Huang Y-F, Ren X-F, Zhang Y-S, Duan L-M, Guo G-C (2004) Experimental teleportation of a quantum controlled-NOT gate. Phys Rev Lett 93:240501
8. Ren J-G et al (2017) Ground-to-satellite quantum teleportation. Nature 549:70
9. Bennett CH, Wiesner SJ (1992) Communication via one-and two-particle operators on Einstein-Podolsky-Rosen states. Phys Rev Lett 69:2881
10. Sun Z, Song L, Huang Q, Yin L, Long G, Lu J, Hanzo L (2020) Toward practical quantum secure direct communication: a quantum-memory-free protocol and code design. IEEE Trans Commun 68(9):5778–5792
11. Deng F-G, Long GL, Liu X-S (2003) Two-step quantum direct communication protocol using the Einstein-Podolsky-Rosen pair
12. Farouk A, Zakaria M, Megahed A, Omara FA (2015) A generalized architecture of quantum secure direct communication for N disjointed users with authentication. Sci Reports 5:16080
13. Dintomon J, Sabir M, Behera BK, Panigrahi PK (2020) Implementation of quantum secret sharing and quantum binary voting protocol in the IBM quantum computer. Quantum Inf Process 19(1):33
14. Liu H-N, Liang X-Q, Jiang D-H, Zhang Y-H, Xu G-B (2019) Multi-party quantum key agreement protocol with bell states and single particles. Int J Theor Phys 58(5):1659–1666
15. Yao AC (1982) In: 23rd annual symposium on foundations of computer science (SFCS'08). IEEE, pp 160–164
16. Goldwasser S (1997) Multi party computations: past and present. In: Proceedings of the sixteenth annual ACM symposium on Principles of distributed computing, pp 1–6
17. Boudot F, Schoenmakers B, Traore J (2001) A fair and efficient solution to the socialist millionaires' problem. Discrete Appl Math 111:23–36
18. Yang YG, Cao WF, Wen QY (2009) Secure quantum private comparison. Physica Scripta 80(6):065002
19. Lang Y-F (2020) Quantum gate-based quantum private comparison. Int J Theor Phys 59(3):833–840
20. Lo HK (1997) Insecurity of quantum secure computations. Phys Rev A 56(2):1154–1162
21. Ltkenhaus N (2006) Theory of quantum key distribution (QKD). In: Lectures on quantum information. Weinheim, Germany: Wiley-VCH Verlag GmbH, pp 271–284
22. Ortigoso J (201) Twelve years before the quantum no-cloning theorem
23. Rosen N (1979) Can quantum-mechanical description of physical reality be considered complete? in Albert Einstein. Vieweg+Teubner Verlag, Wiesbaden, pp 57–67
24. Naik DS, Peterson CG, White AG, Berglund AJ, Kwiat PG (2000) Entangled state quantum cryptography: Eavesdropping on the ekert protocol. Phys Rev Lett 84(20):4733–4736
25. Spiller TP (1996) Quantum information processing: cryptography, computation, and teleportation. Proc IEEE 84(12):1719–1746
26. Sadhu A, Das K, De D, Kanjilal MR (2020) SSTRNG: self starved feedback SRAM based true random number generator using quantum cellular automata. Microsyst Technol 26(7):2203–2215

27. Sadhu A, Das K, De D, Kanjilal MR (2020) MVTRNG: majority voter-based crossed loop quantum true random number generator in QCA nanotechnology. In: Computational advancement in communication circuits and systems. Springer, Singapore, pp 241–253
28. Munro WJ, Azuma K, Tamaki K, Nemoto K (2015) Inside quantum repeaters. IEEE J Sel Top Quantum Electron 21(3):78–90
29. Bennett CH, Brassard G (2014) Quantum cryptography: public key distribution and coin tossing. Theor Comput Sci 560:7–11
30. Petitcolas FAP (2011) Kerckhoffs' principle, p 675
31. Pain P, Das K, Sadhu A, Kanjilal MR, De D (2019) Power analysis attack resistable hardware cryptographical circuit design using reversible logic gate in quantum cellular automata. Microsyst Technol:1–13
32. Qiskit IBM (2018) Quantum information science kit. https://qiskit.org/
33. McMahon D (2007) Quantum computing explained. Wiley
34. Qiskit Tutorials. https://nbviewer.jupyter.org/github/Qiskit/qiskit-tutorial/blob/master/index.ipynb
35. Aleksandrowicz G, Alexander T, Barkoutsos P, Bello L, Ben-Haim Y, Bucher D, Cabrera-Hernández FJ et al (2019) Qiskit: an open-source framework for quantum computing. Accessed 16 Mar 2019
36. Sergienko AV (ed) (2018) Quantum communications and cryptography. CRC Press
37. Pain P, Das K, Sadhu A, Kanjilal MR, De D (2019) Novel true random number generator based hardware cryptographic architecture using quantum-dot cellular automata. Int J Theor Phys 58(9):3118–3137
38. Stipčević M (2011) Quantum random number generators and their use in cryptography. In: 2011 Proceedings of the 34th international convention MIPRO. IEEE, pp 1474–1479
39. Basak A, Sadhu A, Das K, Sharma KK (2019) Cost optimization technique for quantum circuits. Int J Theor Phys 58(9):3158–3179

Performance of 60 GHz Signal as a mm Wave Access Link for 5G eMBB Access Points

Ardhendu Shekhar Biswas, Sanjib Sil, Rabindranath Bera, and Monojit Mitra

Abstract The enhanced mobile broadband is one of the prime features of 3GPP specified 5G NR technology which defines a maximum broadband speed of 20Gbps. It is to be mentioned that data traffic is expected to grow to the tune of 590% for conventional applications, 770% for mobile video, 950% for mobile virtual reality (VR), and 1320% for M2M/IoT by year 2021 (CISCO Whitepaper, "CISCO Visual Networking Index: Global Mobile Data Traffic Forecast Update, 2016–2021," Feb. 2017.). To facilitate the need and consumption of high speed data by mobile users for various user applications, mm wave carrier is the only options to be exploited due to their huge available bandwidth. A feasibility study of 60 GHz mm wave link is conducted to be used typically as an access link for 5G eMBB APs in small cell deployment scenario like airports, shopping malls, stadiums, etc. Small cells like femto cells with a radius of 50 m are generally considered for the above mentioned scenarios. A 5G NR link, suitable for femto cell network, is simulated and the performance of the link is evaluated in terms of data throughput and BER in varying conditions of SNR. The results obtained are very promising. The post-decoder BER and throughput at varying received signal power are measured. With carrier aggregation (CA) of five component carriers (CC) and bandwidth 1950.703 MHz (~2 GHz), the combined throughput achieved is 7.17 Gbps. The BER obtained above SNR of 20 dB and received signal power of −60 dBm is almost zero. Our results suggest the potential use of 60 GHz signal as an access link for 5G eMBB APs or could be used as a back haul link in small cell networks.

A. S. Biswas (✉)
Department of Electronics and Communication Engineering, TINT, New Town, Kolkata, India

S. Sil
Department of Electronics and Communication Engineering, CIEM, Kolkata, India

R. Bera
Department of Electronics and Communication Engineering, SMIT, Rangpo, Sikkim, India

M. Mitra
Department of Electronics and Telecommunication Engineering, IIEST, Shibpur, Howrah, India

M. Mitra et al. (eds.), *Computational Advancement in Communication, Circuits and Systems*, Lecture Notes in Electrical Engineering 786,
https://doi.org/10.1007/978-981-16-4035-3_29

Keywords Enhanced mobile broadband (eMBB) · New radio (NR) · Access points (AP) · Carrier aggregation (CA) · Component carrier (CC)

1 Introduction

From the past few decades, mobile wireless technologies have experienced a number of technology revolutions and evolutions from 1 to 5G. It has been understood that 5G wireless systems technology will extend mobile communication services into new application domains, namely the so-called vertical domains including the smart home, smart city, smart factory, smart vehicles, etc. Further, it is the challenge of this decade to provide very high speed wireless communication at a very small cost. Demand of data traffic is doubling every year. One way of achieving very high throughput is to increase bandwidth (BW) of transmission for accommodating more and more resource blocks. Another possible method may be implementing huge network densification using frequency reuse. The most suitable band is at 60 GHz because of its available bandwidth and scope of frequency reusability. A large bandwidth of 7 GHz is available at the carrier of 60 GHz [2]. At this frequency, the oxygen absorption is very high, and it restricts signal to travel a large distance. Oxygen loss is 15 dB/km which leads to very strong attenuation over large distance [3]. This gives rise the opportunity of frequency reusability and small cell deployment for network densification. The bandwidth expansion is possible using carrier aggregation (CA) technique. According to 5G specification by 3GPP, flexible bandwidths are offered for different services of demand. So instead of using single large bandwidth, several smaller bandwidths are preferable. In fact, bandwidth parts (BWP) are allowed in 5G wireless standards.

The present paper is organized as follows: Sect. 1 is the introduction part, Sect. 2 discusses about the related works based on literature survey, Sect. 3 discusses the methodology adopted and working idea, Sect. 4 details the simulation of the scheme, Sect. 5 thoroughly investigates the results of the simulation, and finally, Sect. 6 concludes the paper with the proposed suggestions based on the results and outcome of the simulation.

2 Related Works

A mathematical queuing-theoretic framework is suggested in [4] to provide the solution for using licensed and unlicensed mm wave frequency band for 5G NR. It is shown that a proportional sharing of two mm wave bands gives better performance in a massive augmented/virtual reality environment. It is carrier aggregation of licensed and unlicensed mm wave band for enhanced performance. A cross-layer network scheduling scheme is proposed [5] for non-contiguous or contiguous carrier aggregation with Markov decision process-based algorithm and adaptive packet scheduling algorithm. Its numerical results give very good performance in term of network reward and packet failure rate. A multi-stream carrier aggregation technique is proposed in [6] for energy minimization and capacity maximization in multi-layer HetNet 5G systems. The performance is evaluated for energy optimization and trade-off between the energy and capacity is also analyzed. A dual carrier modulation scheme is used for carrier aggregation in WLAN to achieve higher gain than non-dual carrier modulation scheme at 60 GHz [7]. A combination of coordinated multi-point and carrier aggregation is used to enhance the user equipment availability for successful communication in [8] for HetNet environment. A link level aggregation of LTE and WiFi is proposed in [9]. The two key components are mode selection and bearer-split scheduling. Results show better performance than transport level aggregation mechanism. The radio resource allocation plays an important role in LTE-A transmission. The component carrier allocation is rescheduled at every transmission interval in [10]. Then a greedy-based method is also proposed to maximize the system throughput maintaining a proportional resource allocation for all user equipments. A load-aware model of carrier aggregation is proposed [11] in multi-band heterogeneous networks. The results show that it enhances the spectral efficiency and peak data rate in small cell deployment scenario. The channel estimation at 60 GHz is very important because of very high penetration loss from outdoor to indoor communication. An efficient scheme of channel estimation is proposed sending demodulation reference signals over 60 GHz band, and it achieves channel estimation performance of the level of 3GPP release 15 standard with higher power efficiency [12]. The filter bank transceiver architecture is proposed for massive non-contiguous carrier aggregation with very small complexity increase [13]. The performance of the proposed scheme is evaluated analytically for OFDM systems. Several 5G mm wave deployment examples are demonstrated with system architecture and hardware prototype in [14]. A deployment scenario of 60 GHz for indoor access at Tokyo-Narita airport is analyzed. Mobile edge computing is used to achieve ultra-high speed download with small latency.

3 Methodology and Working Idea

A novel scheme of carrier aggregation for 5G NR down link(DL) propagation is proposed using five component carriers (CC) around 60 GHz carrier frequency because of its huge BW availability and scope of frequency reuse as shown in Fig. 1. Five different 5G new radio (NR) sources are used for CA with five different CCs. The scheme is actually applicable for five-user scenario over a combined large BW. The carrier aggregated signal is sent over a 3D channel model in the simulation environment. The signal is received through five corresponding modulator for all the CCs with five 5G NR receiver.

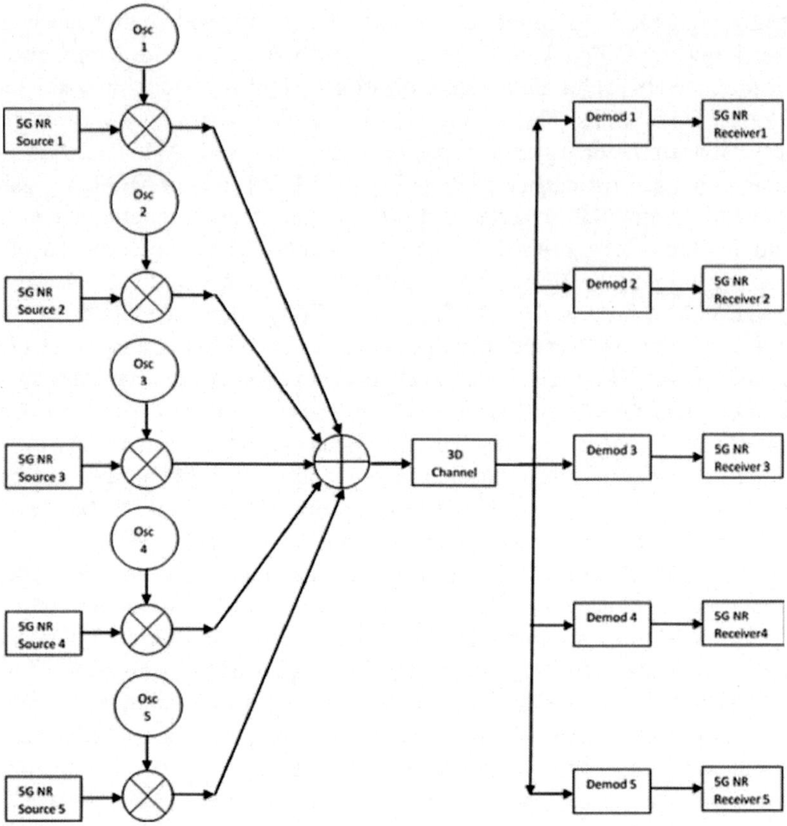

Fig. 1 Carrier aggregation scheme

4 Simulation

The scheme of carrier aggregation is simulated in Keysight SystemVue software environment. The 5G NR sources are adjusted for successful communication. The parameters are chosen as given in Table 1. The BW range of each component carrier (CC) is set at 400 MHz. This is the maximum allowable BW in one CC according to 3GPP technical specification [3]. The actual BW will depend on the number of resource blocks (RB) allocated and should be measured from the spectrum of the transmitted signal after simulation. Oversampling ratio is chosen 4 in view of the fact that the system should accommodate 4 CCs, and it is related to the overall BW. Numerology is taken 3, i.e., subcarrier spacing (SCS) is 120 kHz. At numerology 3, the number of RBs maximum specified is 264 which leads to the capacity of the system. Due to large value of SCS, slot duration reduces to 125 μs, and there are 8 such slots in a subframe of 1 ms and it leads to 80 slots finally in a frame of 10 ms.

It creates a huge scope of increasing data capacity and hence the throughput enhancement. The number of RBs in the system is chosen is 264 in one bandwidth part

Table 1 Parameters of 5G NR source

S. No.	Parameters	Values
1	BW range	400 MHz
2	Oversampling ratio	4
3	Numerology	3: 120 kHz
4	Cyclic prefix	Normal
5	System RB offset	0
6	System RB number	264
7	Number of BWP	1
8	SSB numerology	3: 120 kHz
9	Modulation	256 QAM
10	TBS	200,000
11	Allocated slots	80
12	PDSCH RB offset	24
13	PDSCH RB number	240
14	Number of OFDM symbol	13
15	DMRS port	1000
16	Bandwidth	379.92 MHz (measured)
17	Number of CC	5
18	Overall BW	1950.703 MHz (measured)
19	Gap between successive CC	14.297 MHz (measured)
20	CC frequencies	59.21, 59.61, 60, 60.39 and 60.79 GHz (measured)

(BWP). According to 3GPP specification, multiple BWPs are allowed for different services demanding different bandwidths. In this simulation, only one BWP is chosen because here the aim is to achieve more and more throughput in a certain BW. Modulation scheme is 256 QAM. TBS is set at 200,000 to maintain a code rate of 0.789236 which is quite high. Maximum number of allocated slots are used, i.e., it is set at 80. Though the number of maximum RBs is 264, RBs allocated for physical downlink shared channel (PDSCH) are 240. An offset of 24 RBs is taken to accommodate sync support block (SSB) signal. It is because of the fact that without SSB signal, synchronization is not possible, and hence, decoding will not be possible for successful communication. The number of OFDM symbol is 13. Only one port of each of the 5G NR sources is used and the port number is 1000. There are five component carriers which are aggregated to achieve the overall bandwidth. The BW of each CC is 379.92 MHz. A gap band of 14.279 MHz is maintained between successive CCs. The overall BW is 1950.703 MHz. These bandwidth parameters are measured from the simulated spectrum and some amount of measurement error may be there because it is measured manually from the visual spectrum.

A periodic sequence is considered as a data source and it is used to generate the 5G NR frames. Since the BW range selected is 400 MHz and carrier frequency will

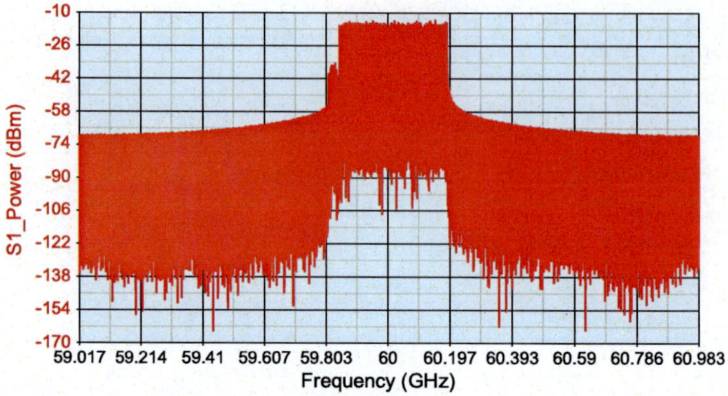

Fig. 2 Spectrum of the signal of one component carrier

be 60 GHz, definitely it should go for FR2 specifications. The SSB signal is coming out from port 4000 and the PDSCH is coming out from port 1000 and hence they are taken on a single RF chain. The spectrum of the signal for one component carrier is shown Fig. 2. Clearly two parts of the spectrum are visible. The first tiny part is the SSB signal and the wide part is the BWP for PDSCH, i.e., for data signal. A small gap between the two parts of the signal spectrum is visible. This is due to the offset maintained in the resource block allocation in PDSCH. An offset of 24 RBs is set to ensure that SSB and PDSCH do not overlap. Overlapping between SSB and data spectrum leads to inaccurate decoding at the receiver.

The 5G NR signal from all five generators are first RF modulated to the corresponding component carrier frequencies 59.21, 59.61, 60, 60.39, 60.79 GHz, respectively. Then they are added to generate carrier aggregated signal. The carrier aggregated signal spectrum is shown in Fig. 3. The five CCs are clearly visible around 60 GHz central carrier with each having SSB and data signal parts.

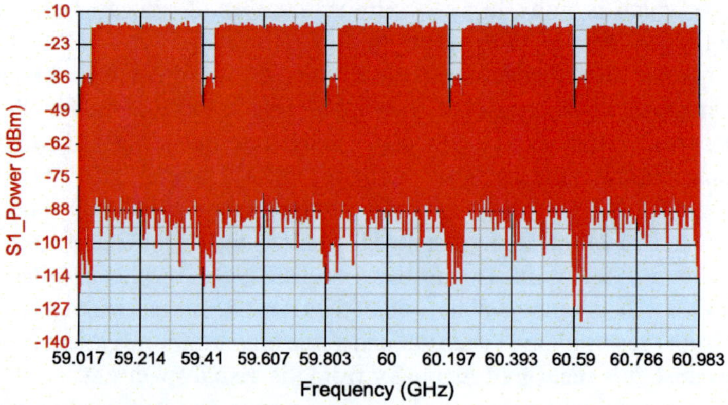

Fig. 3 Spectrum of carrier aggregated signal

Table 2 Parameters of 3D channel model

S. No.	Parameters	Values
1	Environment	Downlink
2	Scenario type	UMi
3	Indoor or outdoor	Outdoor UT
4	Carrier frequency	60 GHz
5	Sampling rate	1.966 GHz
6	LOS theta	90°
7	LOS phi	0°
8	UT speed	0
9	BS height	10 m
10	UT height	1.5 m
11	Distance 2D	50 to 300 m variable
12	Number of transmitting antenna	1
13	Number of receiving antenna	1
14	Transmitting antenna pattern type	Isotropic
15	Receiving antenna pattern type	Isotropic
16	Propagation condition	LOS
17	Path loss	Specified
18	Oxygen loss	Used
19	Penetration loss	0
20	Shadow fading	0
21	Noise density	−174 dBm/Hz

A 3D channel model is used to study the performance characteristics of the proposed design. The channel model parameters are shown in Table 2. The 3D channel block in the Keysight SystemVue simulation environment offers almost all necessary aspects of the transmission channel. In this simulation work, the downlink environment is used for the scenario of urban micro (UMi) channel model. The UMi scenario is of special interest because of the scope of 5G and 6G wireless standard deployment in small cells or femto cells. Deployment of carrier frequencies above 52 GHz is actually progress toward 6G. Specifications at such higher frequencies are proposed in technical reports of 3GPP release 17. It is advised to proceed with the same kind of frames above 52 GHz. This simulation work is carried forward in view of the future 5G/6G deployment scenario that these designs will be hopefully implemented in small cells. In the future, base stations (BS) will be at every lamp post which are used for street lighting. The user terminal (UT) is placed in outdoor location at a certain distance from the base station. The carrier frequency should be set at 60 GHz because the signal transmission will be in 60 GHz for this carrier aggregation scheme. The sampling rate f_s calculated on the basis of BW of individual CCs and number of CCs in carrier aggregation is 1.996 GHz.5G NR signal is basically an orthogonal frequency division multiplexed (OFDM) signal and the sampling rate f_s is given by

$$f_s = N.\Delta f.R_o \tag{1}$$

where N is the FFT size, Δf is the subcarrier spacing, R_o is the over sampling ratio. It is also related to the BW of the OFDM signal. In this simulation, line of sight (LOS) communication between the BS and the UT is studied. LOS θ and φ are 90° and 0° respectively. Considering fixed values of LOS θ and φ indicates transmission to a specific direction through the channel. The user terminal is kept fixed, i.e., UT speed is zero and there is no question of moving direction. Base station height is kept at 10 m considering small cell deployment and UT height is 1.5 m which is comparable to standard height of human. The horizontal distance or two dimensional (2D) distance of the UT from the BS is varied starting from 50 m and up to 300 m to study the link performances for different cell sizes. There are one transmitting antenna and one receiving antenna and both are exhibiting isotropic pattern. Path loss (PL) is specified externally in the channel model block, and the oxygen absorption is included. The free space path loss (FSPL) is given by

$$\text{FSPL} = 20\log_{10}(d) + 20\log_{10}(f) + 20\log_{10}\left(\frac{4\pi}{c}\right) + G_t + G_r. \tag{2}$$

where d is the distance between BS and UT, f is the signal frequency, G_t and G_r are the transmitting and receiving antenna gain. In the small-scale scenario, i.e., UMi model, the actual path loss will be almost equal to the FSPL. The oxygen absorption is considered to be 15 dB/km [2] as already mentioned. Penetration loss and shadow fading are taken zero because it is clear LOS communication. The effect of noise is studied including a constant noise density. The noise density is considered to be −174 dBm/Hz. At the standard system temperature of 290 °K (16.85 °C), the noise density is 4.00388587e-21 Watts/Hz (−173.975 dBm/Hz).

5 Results and Discussions

The link performance is studied on the basis of measuring throughput and bit error rate (BER). BER is measured before and after decoding. LDPC codes are used for channel coding in 5G NR in down link configuration. The BER and throughput performance of the 5G NR downlink is simulated for four different distances between base stations and user terminals. The distances under consideration are 50, 100, 200 and 400 m. These BT-UT distances are chosen considering the small cell scenario

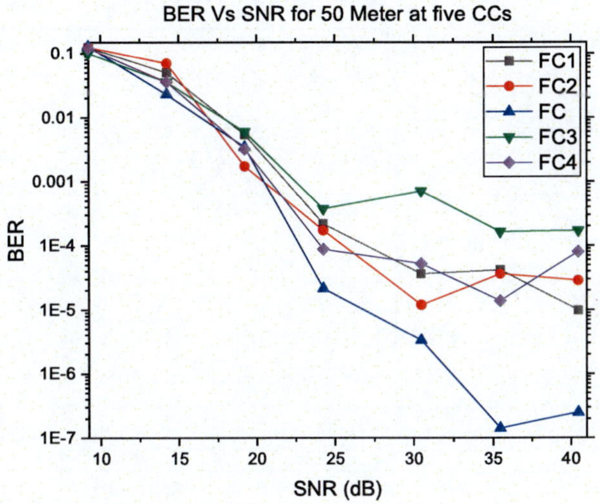

Fig. 4 Pre-decoder BER verses SNR for 50 m at all CCs

keeping in mind. The performance studies in these distances probably will be very helpful in designing practical small cell cases.

The pre-decoder BER vs SNR performances at all the component carriers are shown in Fig. 4 individually for the distance 50 m. The component carriers are FC1 = 59.21 GHz, FC2 = 59.61 GHz, FC = 60 GHz, FC3 = 60.39 GHz, FC4 = 60.79 GHz and the complete bandwidth is almost 2 GHz. At such very high bandwidth signal and very high central carrier frequency of 60 GHz, the performance variation throughout the band is very much important in view of the practical implementation of these designs. For this reason, the BER performances are studied at all component carrier separately and the overall BER performance is also given later in Figs. 5 and 6.

At a distance of 50 m, the BER is varying in the range of the order of 10^{-7}–10^{-4} for SNR above 35 dB for five different component carriers, where all these BERs are measured before the channel decoders, i.e., pre-decoder BERs. Post-decoder BERs are also measured, and it is shown later in Figs. 8 and 9. It is very important to observe that BERs are not same at all the carrier components for a certain value of SNR. This fact reveals that at such very high frequencies, the channel response is very much frequency dependent.

The overall pre-decoder BER and throughput variation with SNR is shown in Fig. 5. The result is very hopeful. BER is approaching 10^{-5} above 35 dB of SNR and throughput is achieving 7.17 Gbps. This BER is the average of individual BERs of all the five CCs, and the throughput is the sum of the throughputs achieved in all the individual CCs. This is necessary to mention here that maximum throughput achieved in each component carrier is 1.43 Gbps. The overall pre-decoder BER and throughput variation with received signal power is shown in Fig. 6. The BER is approaching 10^{-5} above the received signal power level −45 dBm and throughput is reaching 7.17 Gbps. This throughput is achieved in overall BW of 1950.703 MHz.

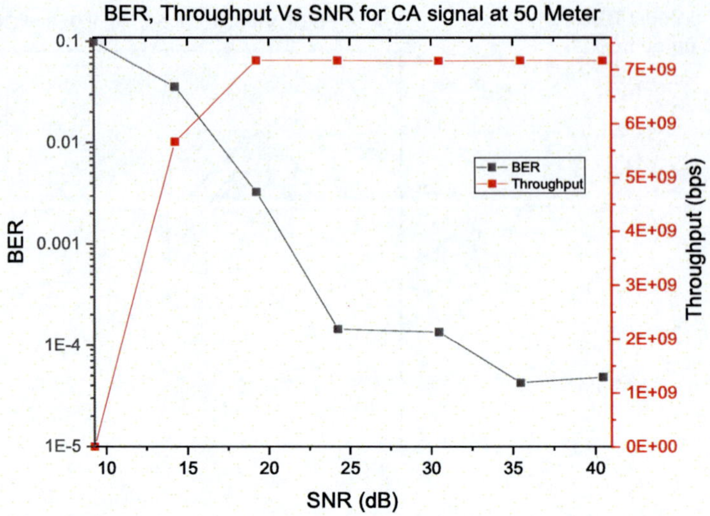

Fig. 5 Pre-decoder BER, throughput verses SNR for CA signal at 50 m

Fig. 6 Pre-decoder BER, throughput verses signal power for CA signal at 50 m

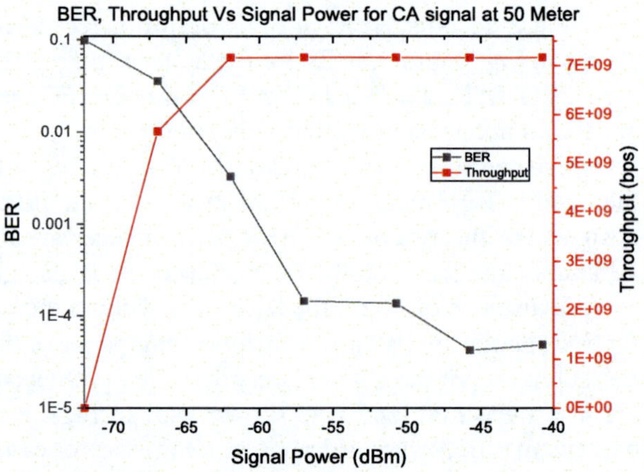

Figure 7 shows very interesting behavior of pre-decoder BER and throughput performance at five component carriers at fixed value of SNR, 40.48 dB. It is observed that throughputs achieved in all the CCs are maximum 1.43 Gbps and BERs are in the range of 10^{-7}–10^{-4}. This shows that the transmission channel is very frequency selective. It is giving different BERs at different component carriers.

The post-decoder BER and throughput variation with SNR and received signal power are shown in Figs. 8 and 9, respectively. It is very clear from these figures that BER measured after the channel decoder reaches almost zero level, i.e., almost error free above the SNR level about 20 dB and received signal power level of −60 dBm. The throughput also achieves the maximum level of 7.17 Gbps at the same level

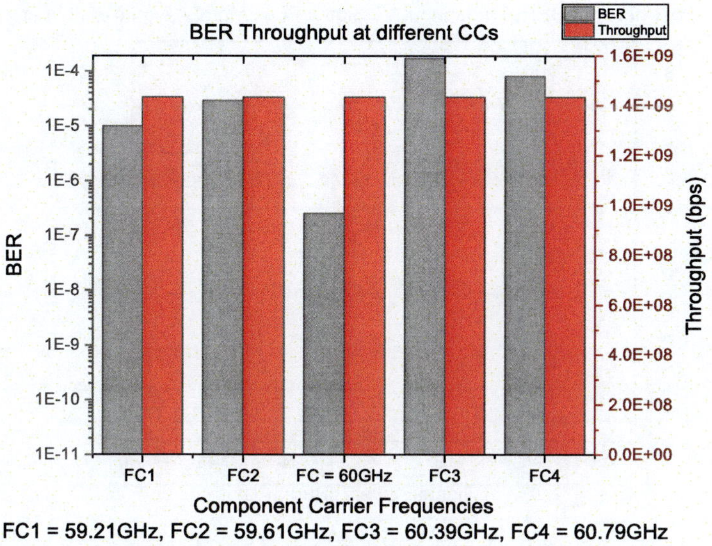

Fig. 7 Pre-decoder BER throughput at different CCs for 50 m

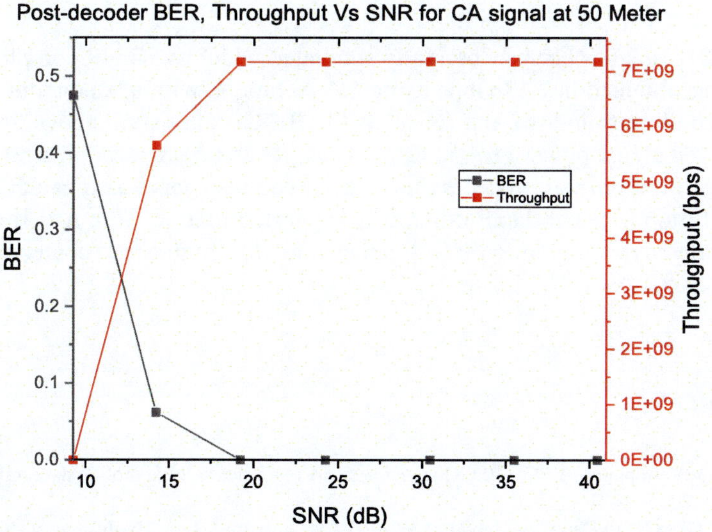

Fig. 8 Post-decoder BER, throughput verses SNR for CA signal at 50 m

of corresponding SNR and signal power. These results give the solid foundation of deploying this design of carrier aggregation in practical fields. Also, it ensures that there should be no hesitation to deal with the BERs up to the order of 10^{-4} in small cell applications as shown in Fig. 4.

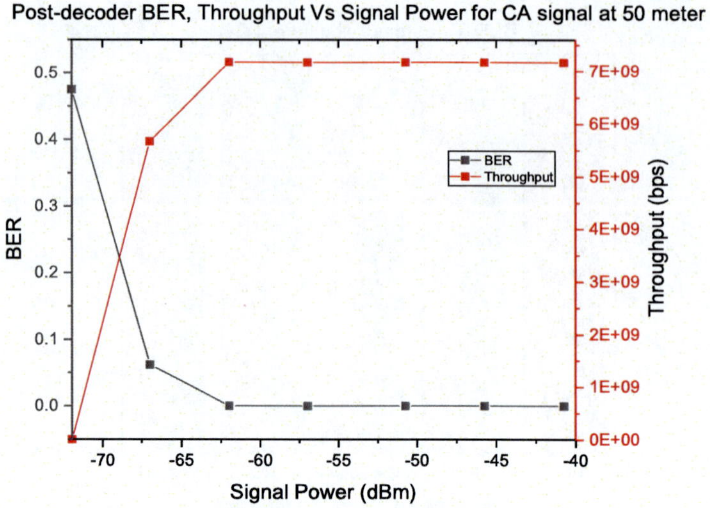

Fig. 9 Post-decoder BER, throughput verses signal power for CA signal at 50 m

6 Conclusions

The 60 GHz link has been successfully simulated which is 5G NR compliance. The throughput obtained is 7.17 Gbps using CA technique with a bandwidth of almost 2 GHz. The link found is quite robust as the BER is almost nil above SNR 20 dB and received signal power of −60 dBm. Thus, it can be proposed that 60 GHz link can be used not only as an access link, but it can also serve as a backhaul link in small cell mobile networks. Further, it is suggested that 60 GHz APs if combined with mobile edge computing (MEC) can enable ultra-high speed content download with low latency.

References

1. CISCO Whitepaper (2017) CISCO visual networking index: global mobile data traffic forecast update, 2016–2021
2. IWPC Ultra High Capacity Networks White Paper Version 1.1 on ‘Evolutionary and disruptive visions towards ultra high capacity networks’ April, 2014
3. 3GPP TR 38.901 v15.0.0, 2018–06, Study on channel model for frequencies from 0.5–100 GHz
4. Lu X et al (2019) Integrated use of licensed-and unlicensed-band mm wave radio technology in 5G and beyond. IEEE Access 7:24376–24391
5. Chang B, Chang (2019) Cost-reward-based carrier aggregation with differentiating network slicing for optimizing radio RB allocation in 5G new radio network. In: 2019 IEEE 10th annual information technology, electronics and mobile communication conference (IEMCON), Vancouver, BC, Canada, pp 0813–0819

6. Chavarria-Reyes E, Akyildiz IF, Fadel E (2016) Energy-efficient multi-stream carrier aggregation for heterogeneous networks in 5G wireless systems. IEEE Trans Wireless Commun 15(11):7432–7443
7. Wu T, Chen T (2018) Dual carrier modulation in channel aggregation for 60 GHz WLAN systems. IEEE Wireless Commun Lett 7(5):792–795
8. Jia J, Deng Y, Chen J, Aghvami A, Nallanathan A (2017) Availability analysis and optimization in CoMP and CA-enabled HetNets. IEEE Trans Commun 65(6):2438–2450
9. Jin B, Kim S, Yun D, Lee H, Kim W, Yi Y (2017) Aggregating LTE and Wi-Fi: toward intra-cell fairness and high TCP performance. IEEE Trans Wireless Commun 16(10):6295–6308
10. Liao H, Chen P, Chen W (2014) An efficient downlink radio resource allocation with carrier aggregation in LTE-advanced networks. IEEE Trans Mob Comput 13(10):2229–2239
11. Lin X, Andrews JG, Ghosh A (2013) Modeling, analysis and design for carrier aggregation in heterogeneous cellular networks. IEEE Trans Commun 61(9):4002–4015
12. Sengupta A, Davydov A, Wang G, Pawar S, Morozov G (2019) Low PAPR DM-RS design for 5G systems operating in high frequency bands. In: 2019 IEEE 90th vehicular technology conference (VTC2019-Fall). Honolulu, HI, USA
13. Stefanatos S, Foukalas F (2017) A filter-bank transceiver architecture for massive non-contiguous carrier aggregation. IEEE J Sel Areas Commun 35(1):215–227
14. Sakaguchi K, Haustein T, Barbarossa S, Strinati E, Clemente A, Destino G, Pärssinen A, Kim I, Chung H, Kim J, Keusgen W, Weiler R, Takinami K, Ceci E, Sadri et all (2017) Where, When, and How mmWave is Used in 5G and Beyond. In: IEICE Transactions on Electronics. E100.C. https://doi.org/10.1587/transele.E100.C.790

A Comparative Study of Parametric Spectrum Estimation Techniques for Cognitive Radio Using Testbed Prototyping

Debashis Chakraborty and Salil Kumar Sanyal

Abstract Cognitive radio (CR) has become an emerging field to rescue wireless communication applications from the spectrum scarcity problem. Spectrum estimation (SE) has been a key ingredient for faster and efficient network implementations using the concept of CR. In this work, we have performed a comparative study of SE technique for the CR systems by employing the null hypothesis approach. Autoregressive (AR), autoregressive moving average (ARMA) and autoregressive integrated moving average (ARIMA) modelling based on optimal data length and goodness of fit (GoF) has been utilized for optimal spectrum modelling. The optimization of the modelling has been achieved through the Akaike information criteria (AIC) and Bayesian information criteria (BIC). Validation and optimization of the time-series data samples have been accomplished using Fit (%) along with χ^2 test GoF. The entire process of SE along with the validation of data samples has been verified on the RICE University's FPGA-based WARP radio testbed in association with MATLAB. A thorough statistical analysis of variance and Standard Error (SER) of the received samples has been carried out for the optimization of sample time-series data length for optimal performance of the receiver or users. It is noteworthy that, we could achieve a much accurate and frugal SE with a data length of 250 only using ARIMA (3,1,2) model of the data samples with a significant improvement in power spectral density (PSD) compared to other conventional approaches. Extensive experimental work has been incorporated to establish the work.

Keywords Spectrum Estimation · Goodness of Fit · AIC and BIC · Statistical analysis · Autoregressive · WARP board

D. Chakraborty (✉)
Department of ECE, Future Institute of Engineering and Management,
Sonarpur, Kolkata, WB 700150, India

S. K. Sanyal
Narula Institute of Technology, Agarpara, Kolkata, WB 700109, India
e-mail: s_sanyal@ieee.org

M. Mitra et al. (eds.), *Computational Advancement in Communication, Circuits and Systems*, Lecture Notes in Electrical Engineering 786,
https://doi.org/10.1007/978-981-16-4035-3_30

1 Introduction

The sparsity of the wireless spectrum has been a growing concern to ensure the optimum usage of an allotted bandwidth. Emerging wireless services and smart devices are currently competing for available bandwidth for seamless connectivity. To avoid this "race-around" condition about the availability of spectrum, cognitive radio (CR) [1] has become a promising paradigm. CR is currently at the centre stage of wireless communication and networking research, where spectrum sensing is a core functionality of it. The goal of spectrum sensing is to detect spectral occupancy along with the estimate of power levels under various sensing constraints. Therefore, the job of the CR control system is to check and verify the availability of the designated spectrum for transmission and reception by the primary user (PU) [2]. If the spectrum remains unused, then it is the job of the CR system to allocate the band to the secondary user (SU) for use. To achieve this, the typical approaches have been to explicitly scan the entire spectrum and execute the subsequent sensing algorithms. Different spectrum sensing methods have been evolved in recent years, considering from narrowband energy detection to wideband spectrum analysis [3].

The premise of CR is that most of the band remains idle or eventually measured spectra are typically sparse. Most works on spectrum sensing are practised by identifying the activity of the PU and SU in the spectrum by full scanning of the entire available bandwidth. Spectrum sensing technique as it explores reduction in the sampling rate by exploiting the low-order correlation model, without having the problem of spectrum sparsity. Therefore, the primal line of these works is that linear power measurement using the autocorrelation function. The assumption of the nonparametric model for the spectrum and estimating the power was formulated as a linear programming (LP) problem [4]. Apriori knowledge of the signal admits a representation in terms of a parametric model of a certain order which falls into the category of classical parametric SE methods, being more accurate as compared to the nonparametric methods [5]. A parametric model provides a more parsimonious representation of the spectrum since it requires the estimation of fewer parameters as compared to a nonparametric model [6] (Fig. 1).

A different approach exploiting spectrum sparsity has been proposed in, where wideband filters are used to detect the occupancy in channels. The situation is very different in that type of network, where spectrum sensing arrangement is deployed with the use of scattered low-end sensors with limited communication and computational capabilities. It was possible to find a satisfactory estimate of the ambient power spectrum using fewer bits with these low-power and low-cost sensors. However, questions arise about its system complexity and computational overhead. Besides, adaptive sensor polling, hidden terminal problem, significant message—passing between sensors and inconsistent sensor readings are the key issues of these approaches.

In this work, it is primarily based on the concept of *GoF*. The interesting aspect of this approach has been the requirement of fewer samples for accurate spectrum sensing and detection over *blind sensing* methods like *energy detection* (ED). Moreover, the method could be adopted for any type of noise distribution environments

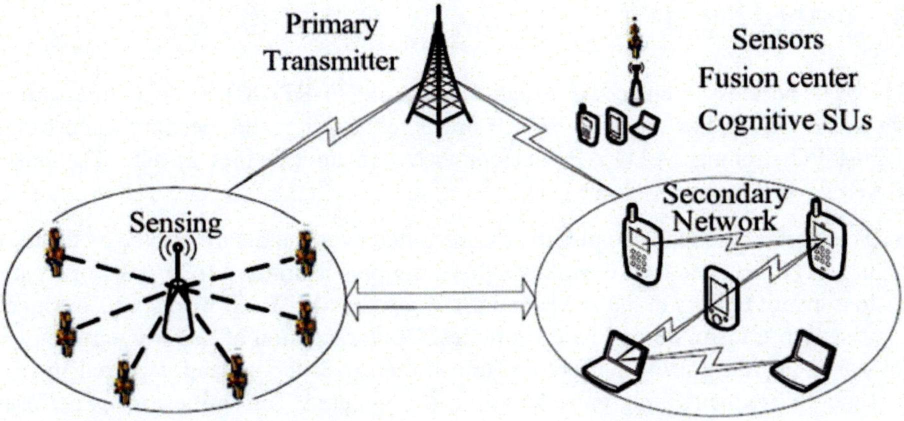

Fig. 1 Cognitive radio (CR) environment

like AWGN and Rayleigh. In the literature, different *GoF*-based tests are available like Anderson–Darling (AD), Kolmogorov–Smirnov test, Cramér–von Mises criterion, Shapiro–Wilk test, and Chi-squared test. [7]. In our setup, the AD test has been chosen because it employs very simple hypothesis testing or significance test using few samples to decide whether the samples are actually from random noise or they have certain alternative *cumulative distribution function* (CDF) F_o. As per the concept of *Null Hypothesis* testing [8], we find the following conditions for the detection of available spectrum. If the designated Hypothesis parameter $H = 0(H_0)$, then it indicates the presence of noisy signal only indicating the absence of PU or availability of spectrum hole. Alternately, if $H = 1(H_1)$, then it indicates the presence of meaningful signal with Gaussian or normal distribution along with noise representing the presence of PU. However, in the absence of a meaningful signal, the AD test produces a hypothesis value $H = 0(H_0)$ with a non-Gaussian distribution indicating a pure noise. Thus, the concept of *GoF* supported with AD test ultimately represents a CR system to detect spectrum holes.

After the confirmation of a meaningful presence of a signal, we proceed for modelling of the received signal for SE. Autoregressive (AR), Autoregressive Moving Average (ARMA) and Autoregressive Integrated Moving Average (ARIMA) [9] models have been chosen for their easy parameter estimation and statistical analysis benefits. In the beginning, it is started with default values of *data length* and *lag order* of the models. In the first stage, *data length* has been optimized using statistical analysis approach of estimating *Fit (%)*. In the second stage, we have optimized lag orders of the AR/ARMA/ARIMA models employing very conventional Akaike Information Criteria (AIC) and Bayesian Information Criteria (BIC) [10]. We have computed the AIC and BIC values of the models for different combinations of AR and Moving Average (MA) lag orders. The model which yields minimum values of AIC and BIC is chosen to be optimum with minimum lag order.

2 Testbed Setup

The basic wireless open access research platform (WARPLAB) node architecture is shown in Fig. 2. For multiple WARP nodes, connections are accomplished using a host PC running MATLAB and connected to an Ethernet switch. The basic WARPLAB flow is as follows [11].

- The digital baseband samples to be transmitted over-the-air are created by the user in MATLAB. The user downloads these samples from the MATLAB workspace to transmit buffers in the WARP nodes. From the MATLAB workspace, a trigger is sent to transmitter and receiver nodes. Upon reception of this trigger, samples stored in transmitting buffers are transmitted over-the-air and captured in real time. Subsequent paragraphs, however, are indented. Transmitter radios perform real-time digital-to-analogue conversion of baseband samples and real-time up-conversion. Receiver radios perform real-time down-conversion from analogue RF-to-analogue baseband and real-time conversion from analogue baseband-to-digital baseband.
- The captured (received) digital baseband samples are stored in receive buffers in the WARP nodes. The user reads captured baseband samples from the receive buffers in the WARP nodes to the MATLAB workspace, and then baseband samples in the MATLAB workspace are processed offline.
- As shown in Fig. 2, the testbed is configured as serial-in serial-out (SISO) configuration, with RB#1 transmitting 1 MHz–2V_{pp}, sinusoidal signal and RB#3 acts

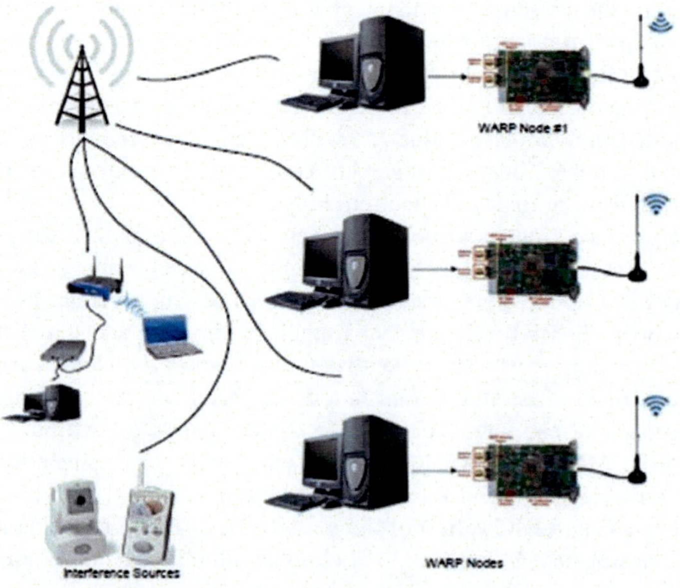

Fig. 2 WARP board network configuration

as a receiver. The Xilinx FPGA IC is configured by the desktop PC using the integrated simulation environment (ISE 14.3) [12] before running the algorithms. The samples are received over real-time during the transmission phase and sent for further analysis using WARPLAB. The node uses a carrier frequency of 2.4 GHz and a bandwidth of 40 MHz. The baseband signals are band-limited from 9 kHz to 10 MHz. In most of the applications, the WARP node uses quadrature or orthogonal signal transmission.

3 Mathematical Background

3.1 *Goodness of Fit (GoF) Test*

A GoF test is a blind nonparametric hypothesis test which can be used to detect the presence of signals in a noisy environment by determining whether the received samples are drawn from a distribution with a cumulative distribution function (CDF) F_0. The hypothesis to be tested can be formulated as follows [8]:

$$\begin{Bmatrix} H_0 : F_n(x) \neq F_0(x) \\ H_1 : F_n(x) = F_0(x) \end{Bmatrix} \tag{1}$$

where $F_n(x)$ is the empirical CDF of the received samples. The CDF could be computed as:

$$F_n(x) = |\{i : x_i \leq x, 1 < i < n\}/n| \tag{2}$$

Considering the classical binary hypothesis test [4]:

$$\begin{Bmatrix} H_0 : x_i = w_i \\ H_1 : x_i = r_i + w_i \end{Bmatrix} \tag{3}$$

where r_i are the received complex samples of the transmitted signal, filtered by the channel, and w_i is the complex noise. We now consider the non-negative random variable, $Y_i = |x_i|^2$, which corresponds to the received energy and construct the empirical CDF F_n as given in Eq. (2). We assume that, F_0, representing the hypothetical noise energy distribution, is known or can be estimated.

3.2 *Variance Analysis*

For a random variable with N number of scalar observations, the variance is described as [13]

Table 1 Variation of CI with mean and SER

Range	CI (%)
$\mu \pm 1\text{SE}$	68.27
$\mu \pm 1.96\text{SE}$	95.00
$\mu \pm 2\text{SE}$	95.45
$\mu \pm 2.5758\text{SE}$	99.00
$\mu \pm 3\text{SE}$	99.73

$$V = \frac{1}{N-1}\sum_{i=1}^{N}|A_i - \mu|^2 \tag{4}$$

where μ is mean of the observation vector A, expressed as:

$$\mu = \frac{1}{N}\sum_{i=1}^{N} A_i \tag{5}$$

SER is the standard deviation of the sampling distribution of statistical phenomena. The SER helps in testing whether the difference between observed and expected frequencies could arise due to chance and also gives an idea about the reliability and precision of a sample. The smaller the SER, the greater is the uniformity of sampling distribution, and hence, greater is the reliability. This also enables us to specify the limits within which the parameters of the data samples are expected to be distributed with a specified degree of confidence. Such an interval is usually defined as a confidence interval (CI). The following gives the relationship between the percentage of CI against the mean value (μ) and SER (Table 1).

3.3 *Power Spectrum Density*

The signal model description is given by considering a sequence of discrete-time WSS samples, given by $x(n)$, whose autocorrelation sequence is expressed as [14]

$$r(l) := E\left[x(n)x^*(n-l)\right] \tag{6}$$

where

$$r(l) = r^*(-l), \forall l, \text{ and } r(0) \geq 0 \tag{7}$$

where the vector

$$r_x = [r(1-k), \ldots, r(-1), r(0), r(1), \ldots, r(k-1)]^T \tag{8}$$

indicates a finite K-lag autocorrelation, then after applying windowing to the power spectrum estimate we obtain:

$$\hat{S}_X(\omega) = \sum_{l=-k+1}^{k-1} r(l)e^{-j\omega l} \tag{9}$$

However, this truncation cannot always guarantee non-negativeness at all frequencies. After discretizing the frequency axis, an approximate estimate of the power spectrum would be:

$$\widehat{s_x} = Fr_x \tag{10}$$

$$\widehat{s_x}(f) = \hat{S}_x\left(\frac{2\pi f}{N_F}\right)\text{for} f = 0, \ldots, N_F - 1 \tag{11}$$

4 Results and Analysis

As discussed in Eqs. (1, 3), the hypothesis testing for the received samples using (χ^2) test has been listed in Table 2. Hypothesis test results are returned as a logical value. The inferences drawn are guided by the following rule.

If$h = 1$, this indicates the rejection of the null hypothesis at the significance level ($\alpha = 0.05$).

If $h = 0$, this indicates a failure to reject the null hypothesis at same ($\alpha = 0.05$).

In our experiment, the returned value $H = 1$ (Column 2 in Table 2) indicates that both AD and (χ^2) tests reject the null hypothesis at the default significance level ($\alpha = 0.05$). Therefore it indicates, that the samples are not drawn from any noisy source having a normal or AWGN distributions, etc. Also by comparison of *critical value* (CV) with χ^2, we find from Table 2 that Chi-square value (11.1857) is much greater than the CV. This signifies that the test rejects null hypothesis having a normal distribution. This is crucial for a CR system for the detection of available spectrum.

Table 2 Hypothesis test results

Chi-Square (χ^2) test	Hypothesis value (H)	Probability value (p-value)	Test statistics value	Critical value (CV)
Presence of meaningful data samples	1 or H_1	3.8563e–36	11.1857	0.7507
Absence of meaningful data samples	0 or H_0	0.1854	0.5194	0.7470

Table 3 Model parameters (ARMA)

Parameter	Value	Standard error (SE)	t-Statistic
Constant	−3.22458e−05	0.000114257	−0.28222
AR(1)	1.60902	0.00893585	180.063
AR(2)	−0.989014	0.00864853	−114.356
MA(1)	−0.940719	0.0259692	−36.2244
MA(2)	−0.743997	0.0390489	−19.053
MA(3)	0.736106	0.0246089	29.9122
Variance	0.00194118	0.000122919	15.7924

Therefore, detection of non-Gaussian distribution is the first step of spectrum sensing using this *GoF* tests.

After employing the parametric modelling techniques, we could find the best optimum model parameters as available in Table 3.

The *GoF* is indicated as percentage of fit (%) in Fig. 3. It could be visualized that the AR(3) method offers very poor fitness, whereas the ARIMA(3, 1, 2) technique has the highest value of Fit (%). Interestingly, Fig. 3 suggests that if we select lesser data length, then it would have better Fit (%) at around data length of 200. AR approach has poor Fit (%), whereas the ARIMA method wins over it. It scores over 83% fit (%) in the WARP testbed analysis. In the final spectrum plot at the end, the accuracy of ARIMA over other methods has been highlighted. A comparison regarding the power spectrum variation in time-domain and frequency-domain has been shown in Fig. 4 for SNR analysis.

Variance analysis has always been an important parameter to evaluate the amount of error present in the fitted model with the estimated one *Variance* analysis of the selected AR(3), ARMA(3, 2) and ARIMA(3,1,2) with variations in *Data Length* has been elaborated in Fig. 5. It is interesting to note that, for the number of data samples between 20 and 200 the *variance* is very high. However, beyond the *data length* of

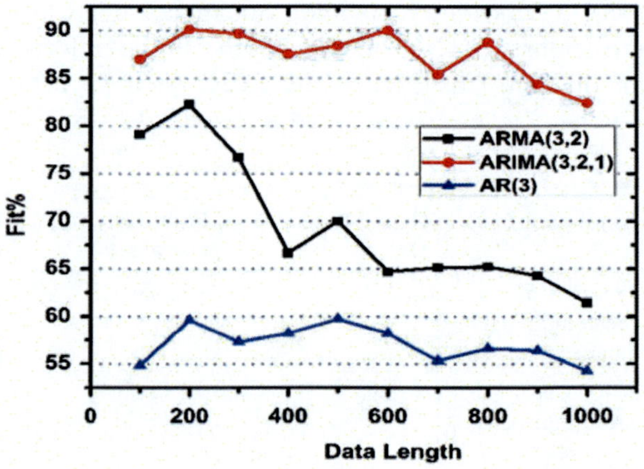

Fig. 3 Fitness (%) of various modelling approaches

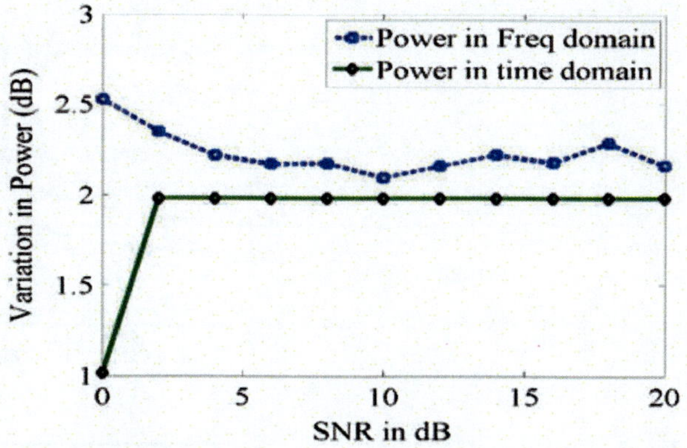

Fig. 4 SNR performances

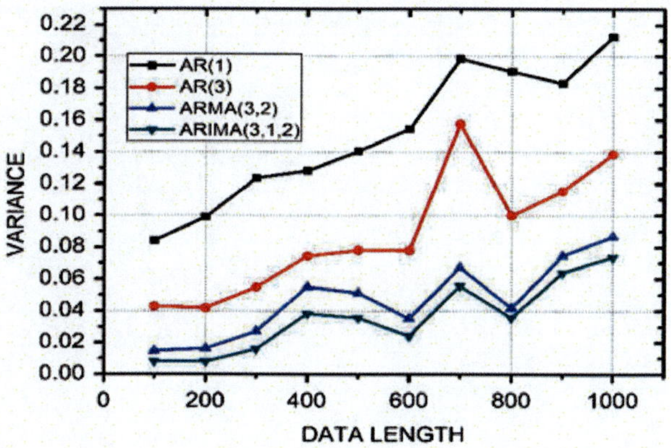

Fig. 5 Variance analysis of different modelling approaches

250, there is no significant improvement of *variance,* suggesting *data length* of 250 is optimum for this model. Even though the selection of higher *data length* above 500 gives us better fitment, this increases transmission duration of the samples resulting in slower detection of the spectrum, which is a poor performance characteristic for a CR system.

Finally, after optimizing all the parametric models under consideration are employed for final SE. The transmitted signal is having a frequency at 1 MHz and has been detected by the WARP board radio boards. The detection of the signal and the corresponding PSD has been plotted in Fig. 6. This indicates the spectrum obtained by the ARIMA method gives a better result. The above result of PSD as a variation of data length (N) has been shown in Fig. 7 for optimum data length analysis. As observed in Fig. 7 the optimum data length could be established at 200 approximately. Therefore, a frugal SE has been achieved by our work.

Fig. 6 Estimated spectrum

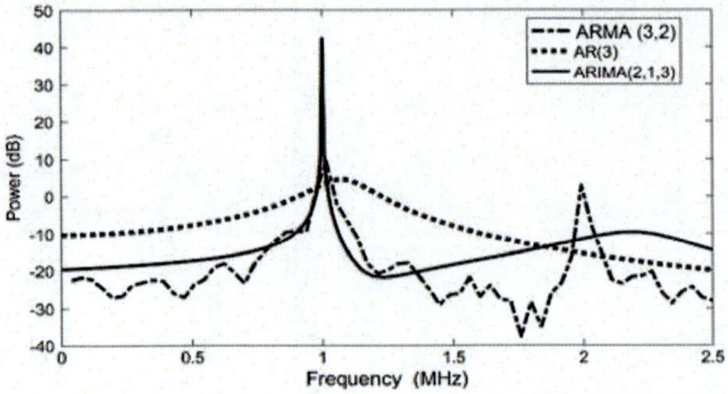

Fig. 7 Variation of PSD with data length

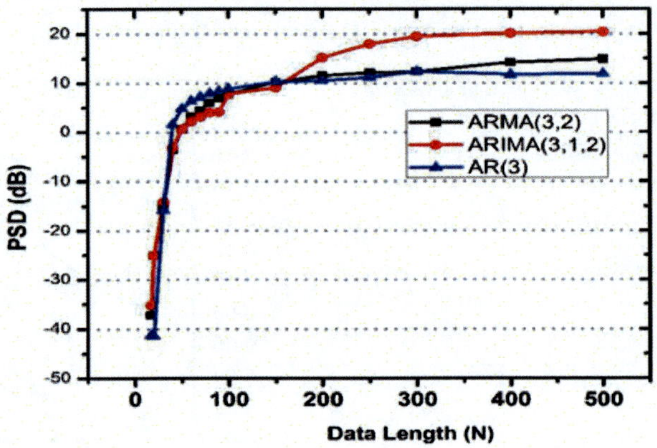

5 Conclusion

In this work, we have provided a unique approach of spectrum exploration and exploitation method using a real testbed like WARP. Sequential and fast SE technique aims at minimizing the time spent in sensing with the highest frugality. Analysing the samples using χ^2 the test gives us the notion of GoF. The SA using variance and Fit (%) have been our main ingredients for the entire exercise of SE with the most optimum value of data length. We have achieved frugality in data length of around 250 with a transmission duration or detection window time of 6.25 μS only. State-of-the-art SE techniques including parametric and nonparametric approaches have also been reviewed and verified empirically. A serious achievement of gain in PSD over the worst performer (AR) by ARIMA (3, 1, 2) model has been the key contribution of the work, as adjudicated by various analysis. It is noteworthy, we have been successful in establishing the bridge between the classical theory and empirical verification for the determination of the best regression model as well. The present work also motivates us for potential development and refinement of the demonstrations. Hence, the integration of the FPGA–WARP embedded module within the presented architecture could be explored further for MIMO applications. In conclusion, throughout this work thorough understanding of the theory and conception up to the implementation, and efficient reconfigurable architecture has been acquired, demonstrated, and discussed.

References

1. Mitola JI (2000) Cognitive radio: an integrated agent architecture for a software-defined radio, pp 0474–0474
2. Ali A, Hamouda W (2016) Advances on spectrum sensing for cognitive radio networks: theory and applications. IEEE Commun Surv Tutor 19(2):1277–1304
3. Arjoune Y, Kaabouch N (2019) A comprehensive survey on spectrum sensing in cognitive radio networks: recent advances, new challenges, and future research directions. Sensors 19(1):126
4. Chakraborty D, Sanyal SK (2020) Time-series data-optimized AR/ARMA model for frugal spectrum estimation in cognitive radio. Phys Commun 44:101252. https://doi.org/10.1016/j.phycom.2020.101252
5. Chakraborty D, Sanyal SK (2016 Sep) Performance analysis of different autoregressive methods for spectrum estimation along with their real-time implementations. In: Second IEEE International Conference on Research in Computational Intelligence and Communication Networks (ICRCICN). https://doi.org/10.1109/ICRCICN.2016.7813646
6. Chakraborty D, Sanyal SK (2019 Jul) WARP test bed implementation of lag order and data length-optimized AR Spectrum estimation algorithm. https://doi.org/10.1007/978-981-13-8687-9_18
7. Chakraborty D, Sanyal SK (2021) Test bed implementation of a scalable ARIMA model for spectrum estimation in cognitive radio-a null hypothesis approach. IETE J Res. https://doi.org/10.1080/03772063.2021.1944336
8. Javanmard A, Montanari A (2014) Confidence intervals and hypothesis testing for high-dimensional regression. J Mach Learn Res 15(1):2869–2909
9. Rawat DB, Yan G (2011) Spectrum sensing methods and dynamic spectrum sharing in cognitive radio networks: a survey. Int J Res Rev Wireless Sens Netw 1(1):1–13
10. Kadane JB, Lazar NA (2004) Methods and criteria for model selection. J Am Stat Assoc 99(465):279–290
11. Amiri K et al (2007) WARP, a unified wireless network testbed for education and research. In: 2007 IEEE international conference on microelectronic systems education (MSE'07). IEEE
12. Chu PP (2011) FPGA prototyping by VHDL Examples: xilinx spartan-3 version. Wiley
13. He J et al (2018) Quantifying expected uncertainty reduction and value of information using ensemble-variance analysis. SPE J 23(2):428–448
14. Hayes MH (2009) Statistical digital signal processing and modeling. Wiley

Study of Micro-Strip Patch Antenna for Applications in Contact-less Door Bell Looking at the COVID-19 Pandemic Situation

Arpita Santra, Arnima Das, Maitreyi Ray Kanjilal, and Moumita Mukherjee

Abstract As the technology advances, the modern trend of lifestyle also advances. The doorbell has an important responsibility in home safety; it is one of the competent and steady systems needs to be developed for better safety which could be access at a low cost. In this era, there are many doorbells systems doing different operation. This paper focuses on touchless type automatic doorbell systems which will ring the bell automatically when a visitor approaches near the door. This system is intended to people, and due to the spread of COVID-19 pandemic situation, it would be one of the safety steps that can be taken against corona. People are now more careful about their everyday work and their family. In the year 2020, the whole world is trapped in unprecedented COVID-19 pandemic. The situation takes away all our normal lifestyle, and all the researches are going on in controlling the situation and finding a new way of life. In this work, the author is trying to establish a contactless door alarm for the household application. Motivation behind the work is that due to the Corona virus spread around the world, we have to take utmost care in every step of our life. If we use the normal door alarm, then there will be the issue of contact for every people who will arrive in. But if there will be a replacement of the conventional door alarm with the help of antenna technology, then it can solve the issue with a contactless alarm. In this paper, the author have used the HFSS software for the proposed antenna.

Keywords Micro-strip · Gain · Radiation pattern · HFSS

A. Santra (✉) · A. Das · M. R. Kanjilal
Narula Institute of Technology, Agarpara, Kolkata 700109, India
e-mail: arpita.santra@nit.ac.in

A. Das
e-mail: arnima.das@nit.ac.in

M. R. Kanjilal
e-mail: maitreyi.kanjilal@nit.ac.in

M. Mukherjee
Adamas University, Kolkata, West Bengal 700126, India

M. Mitra et al. (eds.), *Computational Advancement in Communication, Circuits and Systems*, Lecture Notes in Electrical Engineering 786,
https://doi.org/10.1007/978-981-16-4035-3_31

1 Introduction

Many doorbell systems exist; doorbells ringing without touching it is a challenging task. Conventional doorbell systems work on the traditional way when a visitor presses the switch bell that rings inside the house. Due to the sudden spread of the novel corona virus, the current trend of thinking has been changed with immediate effect. All have to be very careful while moving from one place to another and not only that all the places of human contact as well as the use of touch-less methods are being taken into account by almost everyone. In this paper, the authors have proposed one door-alarm system for domestic use by avoiding the human contact. The drawbacks of the existing manual doorbell system include: (a) searching for switch, (b) some people may not reach the bell switch and (c) lack of safety due to pandemic. The proposed system will overcome all of the mentioned drawbacks.

Here is the working principle of the system

1.1 Block Diagram

As per the proposed block diagram, the working of the touchless alarm due to the signal sensed as the reflected signal from human body will overcome the drawbacks of conventional doorbell where any unknown person need to locate the door alarm first and then have to touch the alarm button by hand. Due to the pandemic situation, the conventional method of door alarm could be avoided by using the proposed system. Here, the author has performed analysis of the required micro-strip patch antenna [1, 2] on Ansoft HFSS software platform. The working principle of the system is a complete process of transmission and reception of the reflected signal by the antenna, and the signal received by the antenna which will trigger the ringing of the doorbell. The transmitted signal will be received by the people in front of the door, and the reflected signal from the people's body will be received by the receiving antenna at the outside of the door. The received signal value should reach a threshold value to trigger the ringing of the doorbell. The simulated antenna [3, 4] is a simple rectangular micro-strip patch [5] antenna with micro-strip line feeding [6] (Figs. 1, 2 and 3).

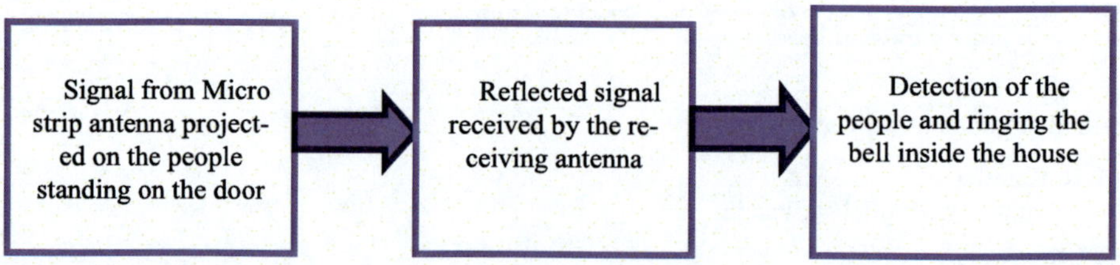

Fig. 1 Block diagram of the contactless door alarm system

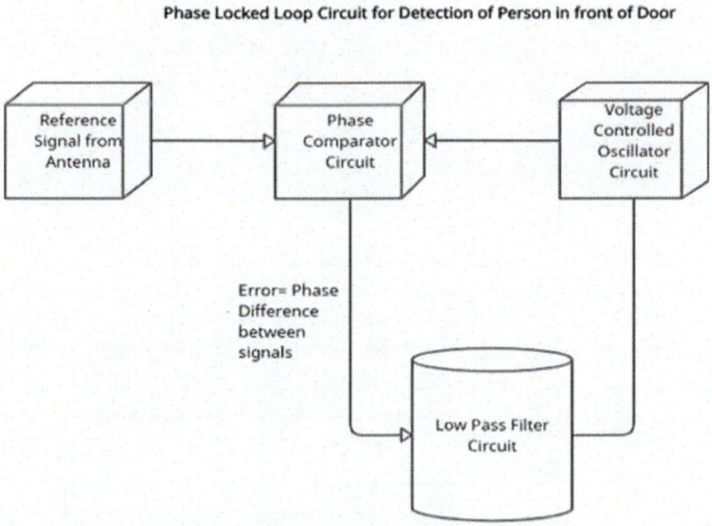

Fig. 2 Block diagram of the contactless door alarm system initiation with the help of phase locked loop circuit

Fig. 3 Micro-strip patch antenna with micro-strip feeding

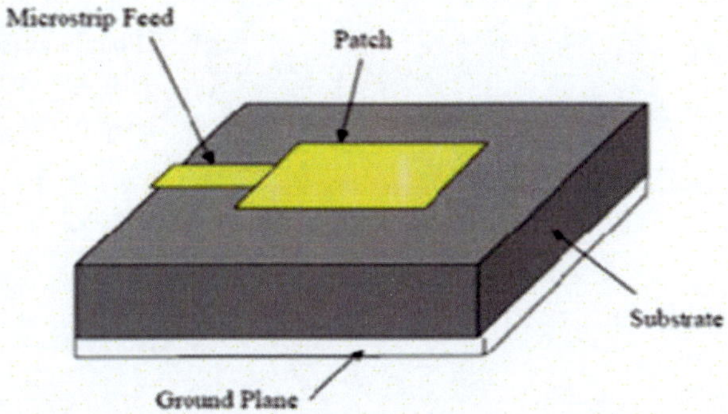

The antenna selection is based on certain features of the simulated antenna which include the return loss [7] is found to be less, resonant frequency is more, VSWR < 1.5, ease of fabrication is simple, impedance matching [8] is easy, reliability is better, bandwidth is 2–5%.

Flowchart: Here is the flowchart of the proposed antenna-based touchless door alarm system (Fig. 4).

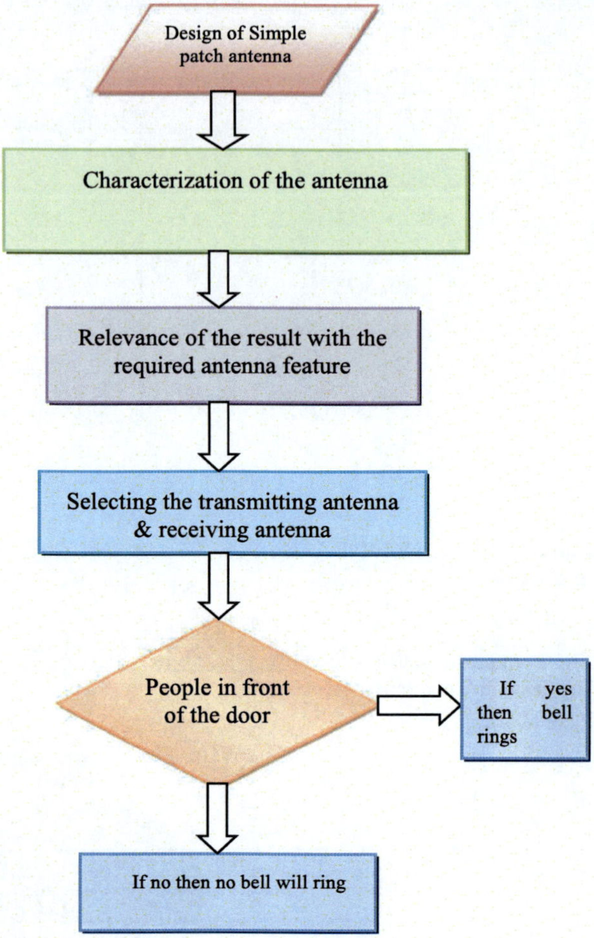

Fig. 4 Flowchart of the proposed system

2 Results and Discussion

The author found the results using the HFSS software. The simulated antenna is shown in the following Figs 5, 6 and 7, and the antenna radiation pattern gain is shown.

3 Conclusion

This paper is focused on an automatic doorbell which works when a visitor stands in front of the door, and then, the sensed signal from the human body enables the circuit and rings the bell automatically. Doorbells provide information that someone

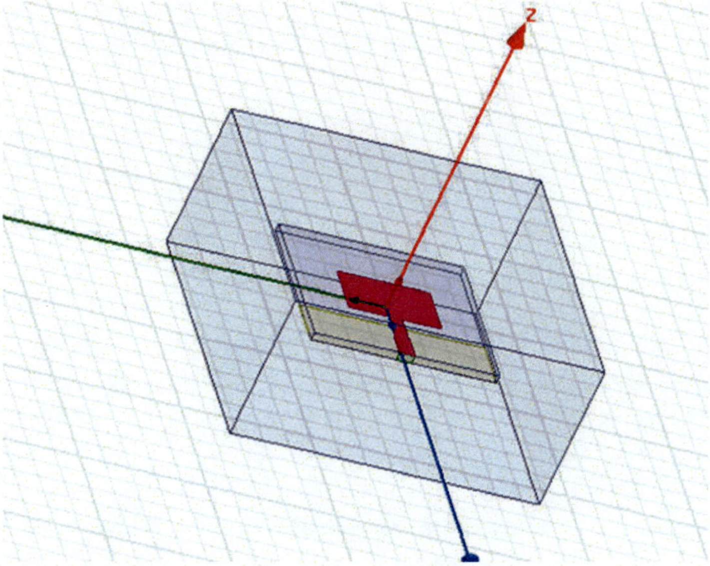

Fig. 5 The micro-strip patch antenna simulated on the HFSS platform

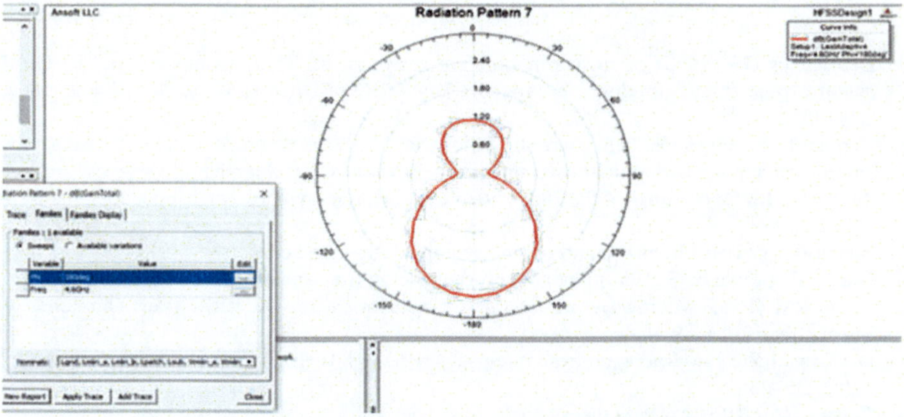

Fig. 6 The radiation pattern at 180° of the micro-strip patch antenna simulated on the HFSS platform

is standing in front of your door so that the person present in the house can open the door. The simulated results show relevance to the process as proposed in the touchless doorbell system. The authors also looking forward to complete the system design with the help of hardware and software mapping and a study of the comparison is to be done in future.

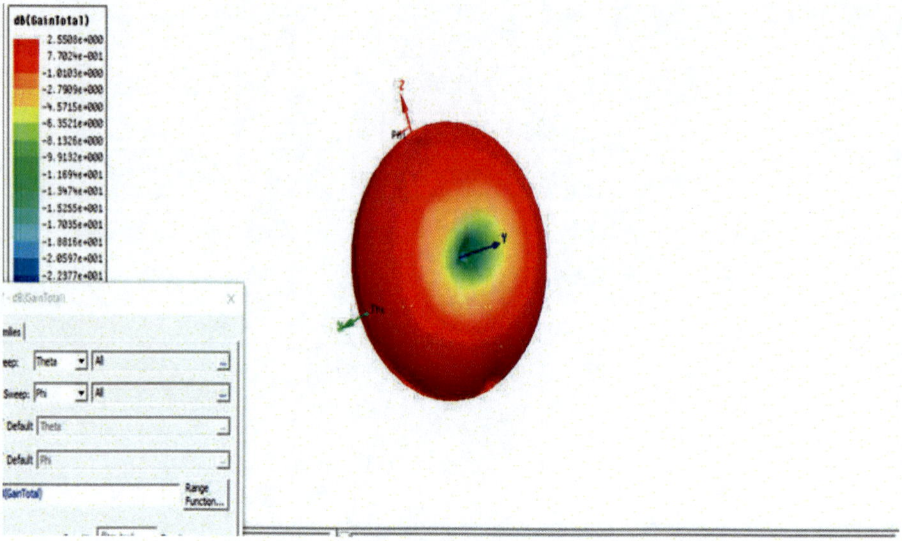

Fig. 7 The gain of the micro-strip patch antenna simulated on the HFSS platform

References

1. Deschamps GA (1953) Microstrip microwave antennas. In: Third symposium on the USAF antenna research and development program. University of IIlinois, Moticello, IIlinois, 18–22 Oct 1953
2. Bernhard JT, Mayes PE, Schaubert D, Mailoux RJ (2003) A commemoration of Deschamps' and Sichak's 'Microstrip microwave antennas': 50 years of development, divergence, and new directions. In: Proceedings of the 2003 antenna applications symposium. Moticello, IIlinois, pp 189–230
3. Bancroft R (2019) Microstrip and printed antenna design, 2nd edn
4. Gutton H, Baissinot G (1955) Flat aerial for ultra-high frequencies. French Patent no.703113
5. Barret RM (1955) Microwave printed circuits-a historical survey. IEEE Trans Onmicrowave Theory Tech 3(2):1–9
6. Denlinger EJ (1969) Radiation from microstrip radiators. IEEE Trans Microwave Theory Tech 17(4):235–236
7. Balanis CA (2015) Antenna theory analysis and design, 3rd edn
8. Lo YT, Solomon D, Richards WF (1979) Theory and experiment on microstrip antenna. IEEE Trans Antenna propags AP-27:137–149

Impacts of COVID-19: A Comprehensive Study Using Linear Regression Analysis in a Predictive Approach

Shreyashree Mondal, Soumya Bhattacharyya, Puspak Pain, Sujata Kundu, Shyamapriya Chowdhury, Neha Dey, and Ankush Baran Basu

Abstract Human history is observing a very strange time fighting an invisible enemy; the novel COVID-19 is the greatest challenge to humankind since the Second World War. The current outbreak of COVID-19 coronavirus infection among humans in Wuhan (China) and its spreading around the globe is heavily impacting global health and mental health. Novel coronavirus (n-CoV) is a generic name given to severe acute respiratory syndrome coronavirus 2(SARs-CoV-2). It has rapidly spread around the world posing enormous mental, social, economic, and environmental challenges to the entire human population. This paper evolved from an overview of the coronavirus and its effect on public health and economics. The main focus of this paper is to survey the various species and types of COVs. The overall statistics of the count around the world and an inclusive survey of its impact on society is being discussed in this paper. In this paper, the linear regression analysis of different vaccines commissioned around the world in COVID-19 and manifold updated information across India has been analyzed in a statistical approach.

Keywords COVID-19 · Socio-economic impact · SARS-CoV-2 · Coronavirus · Vaccine · Regression analysis

1 Introduction

A coronavirus is an enormous group of RNA viruses, mostly causing diseases in birds and mammals. Coronaviruses belong to the family of Coronaviridae [1]. They are generally spherical, single-stranded RNA genomes enclosed within a proteinaceous envelope having bulbous spike-like projections nearly 20 nm long and are especially

S. Mondal · S. Bhattacharyya · S. Kundu · S. Chowdhury (✉) · N. Dey
Department of Information Technology, Narula Institute of Technology, Kolkata, W.B. 700109, India
e-mail: shyamapriya.chowdhury@nit.ac.in

P. Pain · A. B. Basu
Department of Electronics and Communication, Narula Institute of Technology, Kolkata, W.B. 700109, India

M. Mitra et al. (eds.), *Computational Advancement in Communication, Circuits and Systems*, Lecture Notes in Electrical Engineering 786,
https://doi.org/10.1007/978-981-16-4035-3_32

zoonotic. These viruses can cause trivial to fatal infections in humans. Some of the deadly varieties caused severe acute respiratory syndrome coronavirus (SARS), Middle East respiratory syndrome coronavirus (MERS), and now COVID-19. Novel denotes "new pathogen of a known family" of coronavirus or (n-CoV) appeared first in Wuhan, China, seafood market in December 2019. As of May 2020, 212 countries and territories around the world have reported for n-CoV cases. Novel coronavirus has put the whole world health institution to workday night. China informed the outbreak of n-CoV to the World Health Organization (WHO) Country office in China on December 31, 2019, and WHO defined the ongoing pandemic as a Public Health Emergency of International Concern on January 30, 2020. WHO introduced the disease as COVID-19 [2–10].

1.1 Classification

Based on these virus phylogenic, it can be divided into four genera entitled, alpha-COV, Beta-COV, delta-COV, and Gamma-COV [3]. The alpha and beta coronaviruses generally infect only mammals. The beta and delta types infect birds but also mammals [5]. Out of hundreds of coronaviruses, there are only seven of them which affect humans. Some common human coronaviruses are 229E (alpha), NL63 (alpha), OC43 (beta), HKU1 (beta).

The more severe types of coronaviruses are that are originated from animals and which are more dangerous to human being are

- MERS-CoV, a beta virus that causes, the Middle East respiratory syndrome (MERS).
- SARS-CoV, a beta virus that causes severe acute respiratory syndrome (SARS).
- SARS-CoV-2, which causes COVID-19.

COVs are positive-sense single-stranded RNA genome ranging from 26.4–31.7 kilobases.SARS-CoV-2 which is a beta type CoV is the cause of the pandemic. Phylogenetic relationships among 2019-n-CoV and another beta-Cov's showed that there are 89% nucleotide similarities of 2019-n-Cov with the bat SARS. [3–12] The survival stability of SARS in the human body is quite strong, but it has been observed that the SARS-CoV was quite inactive when heated for 30 min at 60 °C. Viruses are always changing or mutating after studying various samples researchers found out that there are two types of strains called "L" and "S" in which "L" type is more severe and has spread the most than the "S" type. These two types of strains have a genetic difference of only 4%.

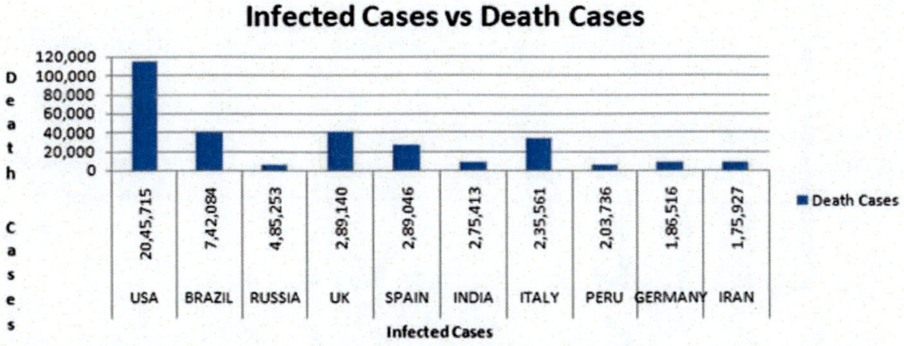

Fig. 1 Graphical representation of COVID-19 infected verses death cases in top ten countries as per Table 1 [10]

1.2 *Epidemiology*

Coronavirus, the current pandemic, is transmitted from human to human and has swept through all over the world very rapidly affecting at least 212 countries and territories with significant mortality and morbidity rates. The number of cases of COVID-19 is rapidly increasing all over the world. More than 3.58 M people are infected by this virus worldwide. The epidemiology curve can be divided into four phases: The local outbreak from the seafood market in Wuhan, China, in December 2019 led to the pandemic. The second phase started on January 13 which led to the rapid expansion of the virus through hospitals, family, or close contact. In this phase, the epidemic spread from Wuhan to other areas too and also to various other countries. The third phase started on January 26 where the number of cases increased 240-fold more than before. The number of cases and fatality rate increased rapidly all over the world. The fourth phase is when the pandemic has affected a large number of people worldwide and is increasing day by day with a cluster of cases as it happened in China. It becomes uncontrollable and the disease becomes pandemic (Figs. 1 and 2; Table 1).

1.3 *Transmission and Symptoms*

The National Institute of Health (NIH) recommended that few classes of people have the highest risk of fatality due to n-Cov. They are:

i. An infant or young children
ii. Elderly people
iii. People with a low immune system
iv. Pregnant women

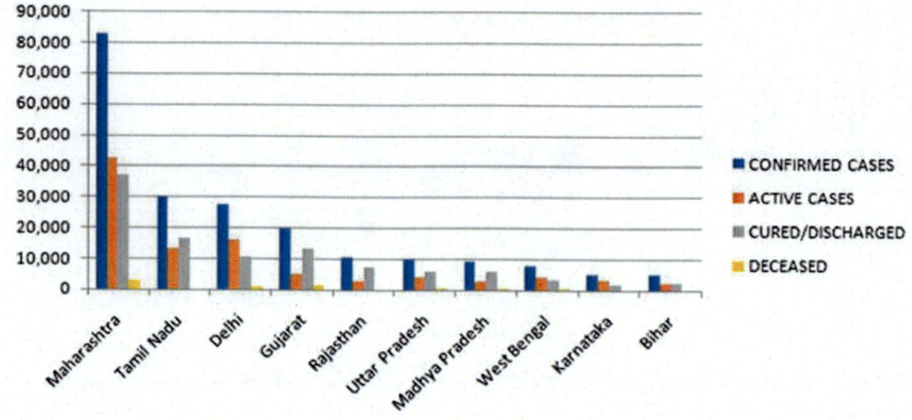

Fig. 2 Graphical representation of COVID-19 (a) Confirmed cases (b) Active cases (c) Cured/discharged (d) Deceased at top ten states in India [10]

The incubation period on an average is 5.2 days, though it varies from patient to patient. From the confirmed Chinese study, the first 425 patients were the median age people nearly 59 years, and 56% were male. Major cases (55%) before January 1, 2020, had a direct link with the Huanan Seafood Wholesale Market. There are asymptomatic patients as well as septic shock patients. The common signs seen in patients include fever having the highest rate (98.6%), infected person becoming lethargic forming (69.6%), shortness of breath, and dry cough [8]. The disease levels may be categorized into mild ones that include mild fever, headache, sore throat, dry cough, etc. These affect the majority (81%) of SARs-CoV-2 cases where there is a probability that the patients may quickly decline to a severe level [3–6].

1.4 Diagnosis, Prevention, and Treatment

Diagnosis of COVID-19 in the areas where the outbreak is so vast is necessary. It is reported based on the symptoms that COVID-19 can be diagnosed using real-time reverse transcription-polymerase chain reaction (RRT-PCR) that detects the presence of RNA fragments of the virus from the infected respiratory secretions using a nasopharyngeal swab. Other diagnosis processes include computed tomography (CT) image of the chest, which is faster as well as more sensitive, but accuracy is not close concerning PCR. It is recommended for the health workers to have proper personal protective equipment (PPE) kit while having close contact with COVID-19 patients. It is advised not to have close contact with individuals, animals and also for immune compromised people to avoid public gatherings. Frequently washing hands with soaps and use of alcohol-based hand sanitizers becomes a good practice. Clinics, hospitals, and medical departments must maintain hygiene to control the infections. The use of N95 masks, FFP3 masks are strictly recommended to healthcare workers

Table 1 (a) Total no. of cases and death rate of COVID-19. (b) Death cases in top ten countries throughout the world according to Worldometers. "COVID-19 Coronavirus Pandemic." (2020) [10] as on June 6, 2020

(a)

Country	Infected cases (in person)	Deaths (in person)
USA	2,045,715	114,151
BRAZIL	742,084	38,497
Russia	485,253	6142
UK	289,140	40,883
Spain	289,046	27,136
India	275,413	7719
Italy	235,561	34,043
Peru	203,736	5738
Germany	186,516	8831
Iran	175,927	8425

(b)

Name of state/UT	Confirmed cases	Active cases	Cured/discharged	Deceased
Maharashtra	82,968	42,609	37,390	2969
Tamil Nadu	30,152	13,503	16,395	254
Delhi	27,654	16,229	10,664	761
Gujarat	19,617	5074	13,324	1219
Rajasthan	10,337	2605	7501	231
Uttar Pradesh	10,103	3927	5908	268
Madhya Pradesh	9228	2721	6108	399
West Bengal	7738	4236	3119	383
Karnataka	5213	3179	1973	59
Bihar	4831	2504	2298	29

along with the use of face and eye-protective equipment, gloves, gown, head cover, and boots. Recently, convalescent blood plasma transfusions are being trialed to treat. Results of this mechanism led to normalizing the body temperature within a short span of days [7–15].

2 Socio-Economic Impacts

2.1 Impact on Economy

The outbreak of the pandemic threatened the worldwide economy to a large extent. Due to the considerably rising of COVID-19 cases across the globe, the stock market fell on February 24, 2020. [9] According to the US stocks, they felt the sharpest fall of 1191 points within a single day making it the largest since 2007–2008. [10–12] According to the estimations made by the "United Nations Economic Commission for Latin America," this pandemic might lead to extreme poverty for 14–22 million Latin American people. The Organization of the Petroleum Exporting Countries (OPEC) felt a sharp decline in the oil prices in early February 2020 due to low consumption and demand from worldwide countries [11]. The International Monetary Fund (IMF) announced that this economic deterioration is "far worse" than that of the Great Recession in 2009 [12]. Based on an estimated report made by the United Nations (UN), on March 12, India will be facing a trade impact of US$348 million, making it one of the 15 worst economies around the globe [13]. The drop of 8.18% or 2919 points of BSE SENSEX is the lowest in the past decades while NIFTY dropped 9% or 950 points [14].

2.2 Education

The ongoing pandemic has not only affected the world's economy but also has led to a total shutdown of the educational system like schools, colleges, and universities [15]. From the data gathered on April 27, 2020, nearly 1.725 billion learners are presently staying back at their homes due to the complete closure of the educational institutions. Nearly, 186 countries across the globe are maintaining closures according to the report given by UNICEF. Shutting down the school has not only affected students and the families but also the teachers [16–19]. COVID-19 made a severe impact on underprivileged children, letting them facing disturbed learning, negotiated nutrition, and ultimately resulting in the rise of economic cost to the daily worker families.

A countrywide lockdown of schools and colleges was announced by the union government on March 16 [20]. On March 19, the Central Board of Secondary Education (CBSE) and Joint Entrance and Examination (JEE) were postponed till March 31, [21, 22], and were to reschedule after the lockdown ends. The Union Public Civil Service Commission (UPSC) suspended the interview for Civil Services Examination 2019 and only to be held after the lockdown. Same for the SSC exams in Tamil Nadu [23, 24].The Indian Space Research Organization (ISRO) also postponed the GISAT-1 mission launch in April 2020.

2.3 *Entertainment and Sports*

Shutting down of the cinema halls was made compulsory in most of the Indian state governments. Film organizations determined to stop producing films, daily soaps. The International Indian Film Academy Awards, [25, 26] which was planned to take place on March 27, was canceled to control the outspread. The Wagah-Attari border ceremony announced to conduct the ceremony without any audiences since March 7. The Padma Awards ceremony that was supposed to be held on April 3 was also postponed [27, 28]. The Board of Control for Cricket in India (BCCI) declared for the first time postponement of the opening ceremony of the Indian Premier League on March 13 and later BCCI suspended the tournament till the pandemic comes to end [29, 30].

2.4 *Transportation*

The Government of India on March 19 stated that no international flight will have the permission to land in India from March 22 [31]. On March 23, the union government announced the postponement of all the domestic flights from March 25, 2020 [32].

During this situation, on March 17, the Western railway system of India increased the platform tickets from Rs. 10 to Rs. 50 for 250 stations. Central Railways terminated 23 trains. Taking the considerations on the day of the Janata Curfew on 22 March, [33] where complete lockdown was maintained across the country, Indian railways cancelled 3700 trains across the country [34–36].

2.5 *Migration of Workers*

Reports from a survey implied that about 139 million laborers work in different cities and towns in India. Due to the shutdown of their workplace like factories, they are left with no source of income. During the starting days of the lockdown, there was a clear image of long processions of workers walking miles to get back to their native state [37, 38].

3 Predictive Analysis: Discussion

Virologists and scientists across the globe are attempting to develop vaccines to prevent the spread of severe COVID-19 disease. Till date, a few vaccines have been devised and trialed on humans. Several dosage limits of the vaccines have been directed. In this paper, with the gathered data of the infected cases in India [39, 40],

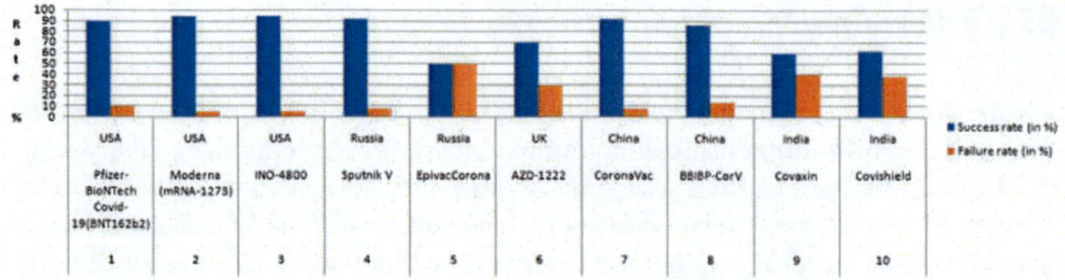

Fig. 3 Vaccine trial success rate in most populated countries in the world

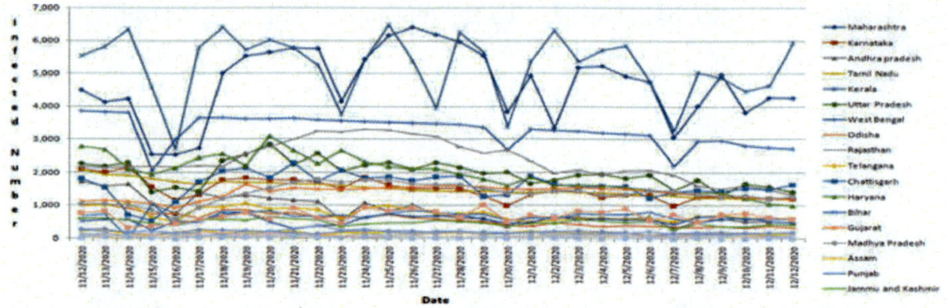

Fig. 4 Displaying trend over time for past 30 days (12.11.2020—11.12.2020) and manifold updated information of COVID-19 pandemic throughout the 29 states in India

a linear regression which is one of the fundamental supervised-machine learning algorithms has been demonstrated and analyzed. Here, no threshold value is required and the value is predicted by an integer number. It can also be concluded from the yielded graphical presentation that the spreading is weakening as a whole (Figs. 3 and 4; Table 2).

In this survey work, a sample dataset (e.g., Bihar) for prediction with linear regression model during the period of 12.11.2020 to 11.12.2020 has been examined. The predicted number of infected persons in the 30th day (11.12.2020) also justifies the actual data collected from authorized data source [39, 40]. The accuracy percentage of the predictions for the number of future infected people using the linear regression model has been achieved within the range of 85.57 to 97.18% in Google-CoLab python library platform (Figs. 5 and 6).

4 Conclusion

In this article, we studied the manifold impact of COVID-19 throughout the world. Isolation, hygiene, and social distancing are the only mode to minimize the transmission of the virus for further spreading. In this work, an intensive survey study on the open-source sample dataset in a predictive approach during a stipulated period and

Table 2 Vaccine trial [41–43] success rate in most populated countries in the world

Sl. No.	Name of the vaccine	Name of country	Type	Doses	Success rate (in %)	Failure rate (in %)
1	Pfizer-BioNTech COVID-19(BNT162b2)	USA	mRNA-genetic code of virus	Two injections-28 days apart	90	10
2	Moderna (mRNA-1273)	USA	mRNA-genetic code of virus	Two injections-1 month apart	94.5	5.5
3	INO-4800	USA	Plasmid vaccine	Two doses-4 weeks apart	94(phase 1)	6
4	Sputnik V	Russia	Non-replicating virus	Two doses-3 weeks apart	91.4	8.6
5	EpivacCorona	Russia	Peptide vaccine	Twice 14–21 days	50.4	49.6
6	AZD-1222	UK	Replication-deficient adenovirus	Two injections-1 month apart followed by full dose	90 (half dose)	30
					62 (full dose)	
					70(avg dose)	
7	CoronaVac	China	Formalin-Alum adjuvant	Two dozes-14 days gap	92	8
8	BBIBP-CorV	China	Inactivated vaccine	Two doses-21 days apart	86	14
9	Covaxin	India	Inactivated vaccine	Two doses-28 days apart	60	40
10	Covishield	India	Replication-deficient adenovirus	Replication-deficient adenovirus	62	38

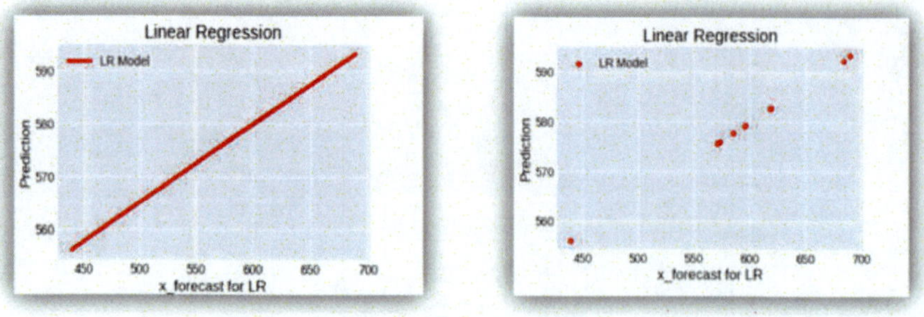

(a) Bihar (b)

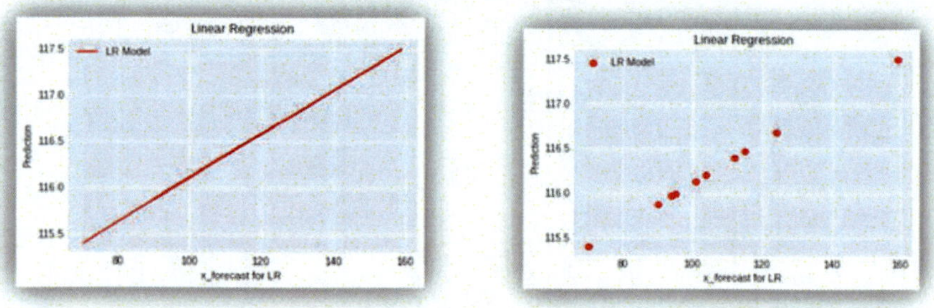

(c) Goa (d)

Fig. 5 Linear regression analysis of the sample datasets **a–b** for Bihar, and **c–d** for Goa in line graph and scatter plot approach during the period of 12.11.2020 to 11.12.2020

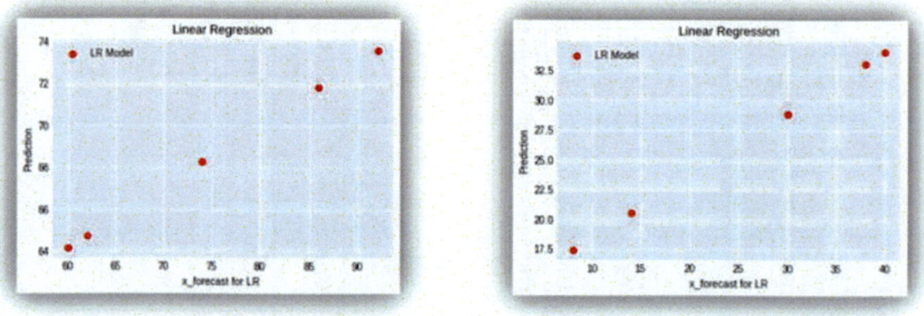

(a) (b)

Fig. 6 **a** Success rate. **b** Failure rate as scatter plot approach using linear regression analysis of different vaccines commissioned round the world in COVID-19 and manifold updated information as per the data of Table 2 the impact of the COVID-19 in several dimensions has been examined. The accuracy percentage of those predictions for the number of affected people through the linear regression model has been achieved to proximate reality using the Google-CoLab python library that justifies the actual data collected from that said source.

References

1. Fehr AR, Perlman S (2015) Coronaviruses: an overview of their replication and pathogenesis. Meth Mol Biol 1282:1–23. https://doi.org/10.1007/978-1-4939-2438-7_1
2. Lu R, Zhao X, Li J et al (2020) Genomic characterisation and epidemiology of 2019 novel coronavirus: implications for virus origins and receptor binding. Lancet 395:565–574. https://doi.org/10.1016/S0140-6736(20)30251-8
3. Cascella M, Rajnik M, Cuomo A, Dulebohn SC, Napoli RD (2020) Features, evaluation and treatment coronavirus (COVID-19). StatPearls Publishing, Treasure Island, FL
4. Drexler JF, Gloza-Rausch F, Glende J, Corman VM, Muth D, Goettsche M, Seebens A, Niedrig M, Pfefferle S, Yor-danov S, Zhelyazkov L, Hermanns U, Vallo P, Lukashev A, Muller MA, Deng H, Herrler G, Drosten C (2010) Genomic characterization of severe acute respiratory syndrome-related coronavirus in European bats and classification of coronaviruses based on partial RNA-dependent RNA polymerase gene sequences. J Virol 84:11336–11349
5. Chan JF, To KK, Tse H, Jin DY, Yuen KY (2013) Interspecies transmission and emergence of novel viruses: lessons from bats and birds. Trends Microbiol 21:544–555. https://doi.org/10.1016/j.tim.2013.05.005
6. Ghinai I, McPherson TD, Hunter JC et al (2020) First known person-to-person transmission of severe acute respiratory syndrome coronavirus 2 (SARS-CoV-2) in the USA [Epub ahead of print]. Lancet. Accessed 21 March 2020. https://doi.org/10.1016/S0140-6736(20)30607-3
7. Li Q, Guan X, Wu P et al (2020) Early transmission dynamics in Wuhan, China, of novel coronavirus-infected pneumonia [Epub ahead of print]. N Engl J Med. https://doi.org/10.1056/NEJMoa2001316
8. Wang Y, Wang Y, Chen Y, Qin Q (2020) Unique epidemiological and clinical features of the emerging 2019 novel coronavirus pneumonia (COVID-19) implicate special control measures [Epub ahead of print]. J Med Virol. https://doi.org/10.1002/jmv.25748
9. FTSE 100 plunges 3.7 percent as Italy confirms sixth coronavirus death. City A.M. London. 24 Feb 2020
10. Worldometers (2020) COVID-19 coronavirus pandemic
11. Reed S (2020) OPEC scrambles to react to falling oil demand from China. The New York Times. ISSN 0362-4331. Retrieved 14 Feb 2020
12. Larry Elliott Economics editor (2020) 'Great Lockdown' to rival great depression with 3% hit to global economy, says IMF | Business. The Guardian. Retrieved 27 April 2020
13. Trade impact of Coronavirus epidemic for India estimated at 348 million dollars: UN report". The Economic Times. Retrieved 5 March 2020
14. Sensex Crashes 2,919 Points, Nifty Ends at 9590 In Worst Day For Markets Ever. NDTV.com. Retrieved 13 March 2020
15. COVID-19 Educational Disruption and Response. UNESCO. 4 March 2020. Retrieved 12 April 2020
16. Adverse consequences of school closures. UNESCO. 10 March 2020. Retrieved 15 March 2020
17. Lindzon J (2020) School closures are starting, and they'll have far-reaching economic impacts. Fast Company. Retrieved 22 March 2020
18. Distance learning solutions. UNESCO. 5 March 2020. Retrieved 23 March 2020
19. 290 million students out of school due to COVID-19: UNESCO releases first global numbers and mobilizes response. UNESCO. 4 March 2020. Retrieved 6 March 2020

20. Schools Closed, Travel To Be Avoided, Says Centre On Coronavirus: 10 Points. NDTV.com. Retrieved 18 March 2020
21. CBSE 10th and 12th Board exams postponed due to coronavirus epidemic. The Times of India
22. CBSE Class 10, 12 Board exams postponed due to coronavirus; to be rescheduled after March 31. indiatv. 18 March 2020
23. Coronavirus outbreak: UPSC defers civil services interviews. Economic
24. TN SSC 10th board exams postponed in Tamil Nadu, Puducherry due to coronavirus. Hindustan Times. 21 March 2020. Retrieved 23 March 2020
25. IIFA 2020 cancelled due to coronavirus scare, fresh date to be announced soon. Hindustan Times. 6 March 2020. Retrieved 6 March 2020
26. Jha L (2020) Coronavirus scare in India: All film, TV, web shoots cancelled. Livemint. Retrieved 15 March 2020
27. Coronavirus: Attari border ceremony to be conducted without public presence. Livemint. Retrieved 6 March 2020
28. Singh BJ (2020) Upcoming Padma Awards ceremony postponed in wake of coronavirus outbreak. India Today. Retrieved 14 March 2020
29. With India in lockdown, IPL 2020 suspended indefinitely. ESPNcricinfo. 15 April 2020. Retrieved 15 April 2020
30. It's official: IPL 2020 postponed to April 15 due to coronavirus. The Times of India. Retrieved 13 March 2020
31. No international commercial flight will be allowed to land India from March 22. indiatvnews. 19 March 2020
32. Coronavirus: India suspends domestic flights from 25 March. Livemint
33. Delhi Metro to be closed on Sunday As Part of PM's "Janata Curfew" Move. ndtv
34. Janata curfew: 3700 trains cancelled on Sunday. The Times of India
35. COVID-19 Janta curfew: Railways cancels 3700 trains on Sunday. Livemint. 20 March 2020
36. Train services suspended till May 3, no ticket booking till further orders. Livemint. 14 April 2020. Retrieved 16 April 2020
37. Ara I (2020). No work, no money': thousands stranded on Anand Vihar Bus Stand. The Wire
38. Jha S (2020). Fighting Covid-19: After the long walk, jobless migrants head home by bus. Business Standard India
39. https://www.covid19india.org/. Accessed on 11 Dec 2020
40. https://www.mygov.in/covid-19/. Accessed on 11 Dec 2020
41. https://www.raps.org/news-and-articles/news-articles/2020/3/covid-19-vaccine-tracker
42. https://www.indiatoday.in/coronavirus-outbreak/story/oxford-coronavirus-vaccine-70-per-cent-effective-against-covid-19-pfizer-moderna-sputnik-v-vaccines-success-rate-efficacy-1743318-2020-11-23
43. https://www.precisionvaccinations.com/vaccines/bnt162-sars-cov-2-vaccin

Word Estimation in Continuous Colloquial Bengali Speech

Suman Das

Abstract Word segmentation is a crucial part in any speech to text conversion. Many works have been done on popular languages, especially on English, but a very few work has been carried out on Bengali language, especially on colloquial speech. In our work, we present a simple pitch profile-based technique to find the words in a Bengali speech. We extract the feature of the existence of words based on the pitch profile of a speech. To find the pitch profile, we have used the state phase technique. A simple deviation of a 20 ms window is studied to find the pitch. In order to reshape the pitch, power profile of the speech is used. Then apply morphology to make the profile more robust. Finally, we cluster the data and use silhouette index to select the number clusters present in the data which in turn estimate the word boundaries. The algorithm is tested over continuous colloquial Bengali speech.

Keywords Signal processing · Word boundary detection · Pitch profile

1 Introduction

Speech is the most commonly used mode of communication among human beings; it is natural to expect people to carry out spoken dialogue with computers. Speech synthesis has already attained a respectable level in some of the Indian language. On the other hand, speech recognition, the process of translating a speech signal into a sequence of words, is yet to make its mark on Indian languages. Although recognition of phonemes and small vocabulary words in some of the Indian languages has been attempted, yet the recognition of continuous speech in Indian languages is still awaiting serious attention. Lack of information regarding word boundary implies that the automatic speech recognizers have to assume that at each and every moment, a new word may begin. This assumption results in many superfluous word hypotheses, particularly in a huge search space in which many hypotheses have to be considered

S. Das (✉)
Narula Institute of Technology, Kolkata, India

Springer Heidelberg, Tiergartenstr. 17, 69121 Heidelberg, Germany

M. Mitra et al. (eds.), *Computational Advancement in Communication, Circuits and Systems*, Lecture Notes in Electrical Engineering 786,
https://doi.org/10.1007/978-981-16-4035-3_33

simultaneously. The computation of the probabilities for every path through this large search space slows down the processing of the speech. If one could find reliable information about word boundaries in the speech stream, it might be possible to speed up the recognition process. Speech recognition can be considerably improved if lexical knowledge is incorporated in the recognition system. It is necessary to detect the word boundary to effectively use lexical knowledge. It is estimated that lexical knowledge can enhance recognition rate for spoken word up to 95% with a phoneme recognition rate of around 80%. Word boundary detection can also be used for Spoken Document Retrieval (SDR) and Key Word Spotting.

Spoken language interface to computers is a topic that has lured and fascinated engineers and speech scientists for more than five decades. For many, the ability to converse freely with a machine represents the ultimate challenge to artificial intelligence, as the production and perception of human speech constitute the highest form of human cognition. In addition to being a challenging topic, spoken language interfaces are fast becoming a necessity. In the near future, interactive networks will provide easy access to a wealth of information and services that will fundamentally influence how people work, play and conduct their daily affairs. Today, access to such networks is limited to people who can read, a relatively small section of the population in the developing countries. Advances in human language technology are needed for the average citizen to communicate with networks harnessing natural communication skills using everyday devices, such as telephones. Without fundamental advances in speech-based user-centred interfaces, a large portion of society will be debarred from participating in the age of information, resulting in further stratification of society leading to tragic loss of human potential.

2 Methodology

In (Fig. 1), we have shown a typical sentence in Bengali. Is it possible to estimate how many words are there? To answer this question, we need the analysis of audio signal.

In (Fig. 2), it is noted that the audio signal can be classified in three categories, [1]

1. Silence (S).
2. Voice (V).
3. Unvoiced (U).

The segment where no voice part is present referred as silence (S). Now, the difference between voice and unvoiced is, for the voice part, our vocal cord vibrates if there is vowels (Fig. 3) present and it has quasi-periodic (Fig. 4) nature.

But for unvoiced section, our vocal cord does not vibrate when consonant (Fig. 5) present and it has quasi-random (Fig. 6) nature. The period of quasi-periodic signal is commonly known as pitch. So our idea is to extract the pitch of a signal to estimate the presence of word.

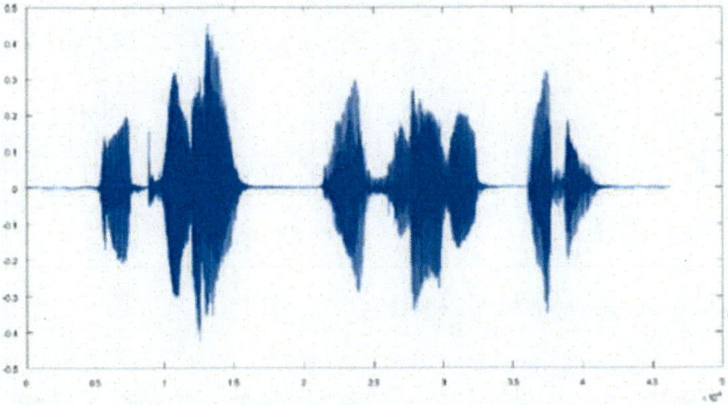

Fig. 1 Bengali speech

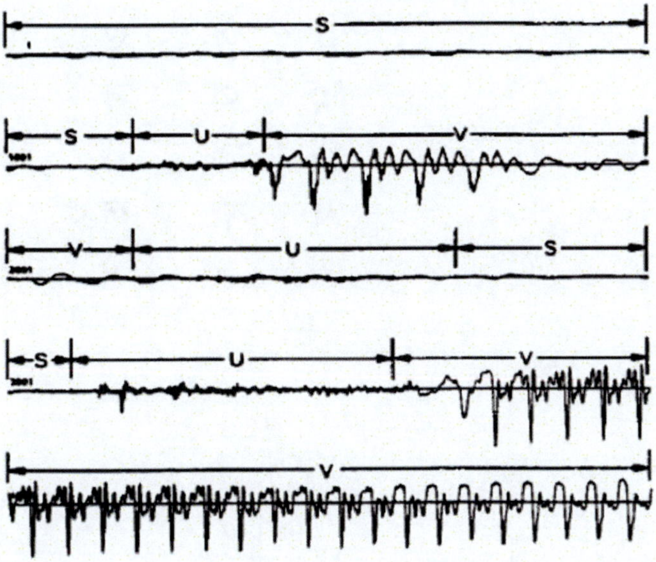

Fig. 2 Position of S, U and V on a signal [1]

Fig. 3 Vowels of Bengali speech

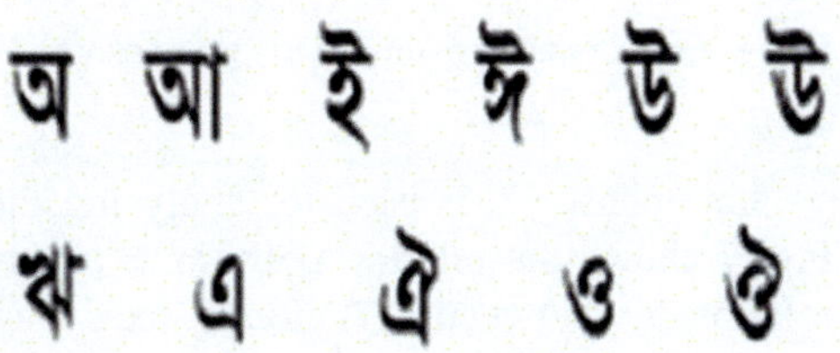

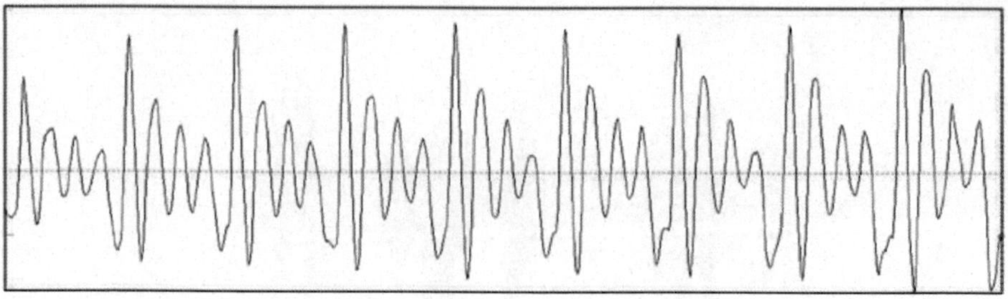

Fig. 4 Quasi-periodic nature of Bangla vowel 'আ'

Fig. 5 Consonant of Bengali speech

ক	খ	গ	ঘ	ঙ
চ	ছ	জ	ঝ	ঞ
ট	ঠ	ড	ঢ	ণ
ত	থ	দ	ধ	ন
প	ফ	ব	ভ	ম
য	র	ল		
শ	ষ	স	হ	

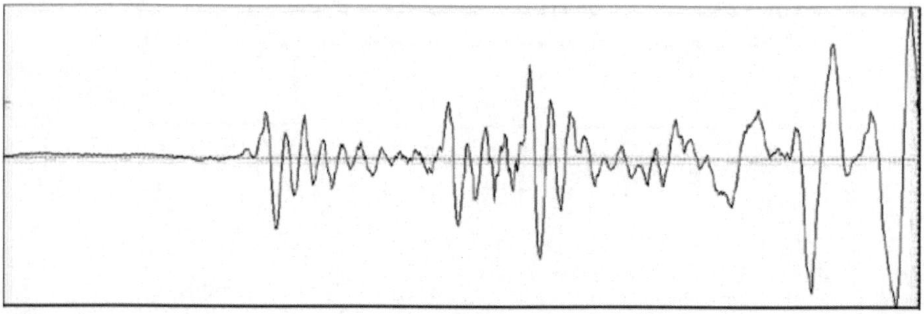

Fig. 6 Quasi-random nature of Bangla consonant 'ক'

State phase [2] is used for analysing the signal. Let $x\ (n)$ is a time series, then Fig. 7 shows the relation between $x\ (i)$ and $x\ (i + L)$, where L is a delay. If the relationship between $x\ (i)$ along x-axes and $x\ (i + L)$ along y-axes is periodic, then we get a straight line with a 45° along the x-axes [3].

Here, in our work, we cannot get a perfect symmetrical nature. The line which we get is not symmetrical but close to symmetrical that means the nature of the signal

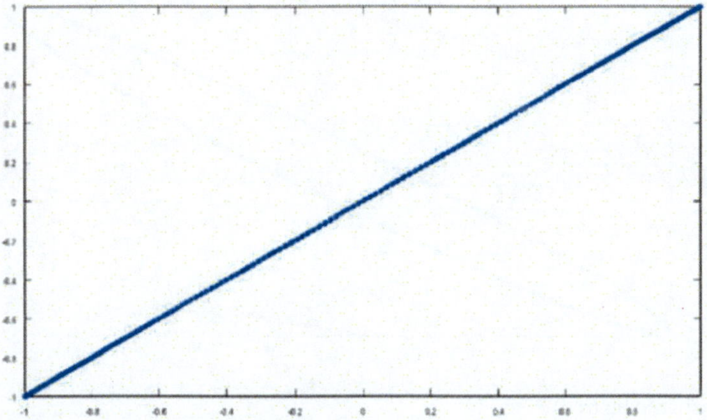

Fig. 7 Relation is periodic, get a symmetrical line

is quasi-periodic. It has been seen that the voiced speech is not completely periodic, it is quasi-periodic Fig. 8,

In quasi-periodic signal [3], pattern of the signal is regular but interval of that signal is not fixed. And for a periodic complex signal, the displacement at two points with phase difference of 2pi will have the same value. This implies that in the state-phase diagram, the points representing such pairs will be lying on a straight line with a slope of pi/4 with respect to the axes.

In Fig. 9, quasi-random [3], as the delay increases, the points are scattered over a broad region. The scattered region attains maximum width for a phase difference of pi/2. It again collapses into a straight line for the phase difference of 2pi. A similar thing also happens for a quasi-periodic signal. It may be observed that as the delay increases, the points are scattered over a broad region. The scattered region attains

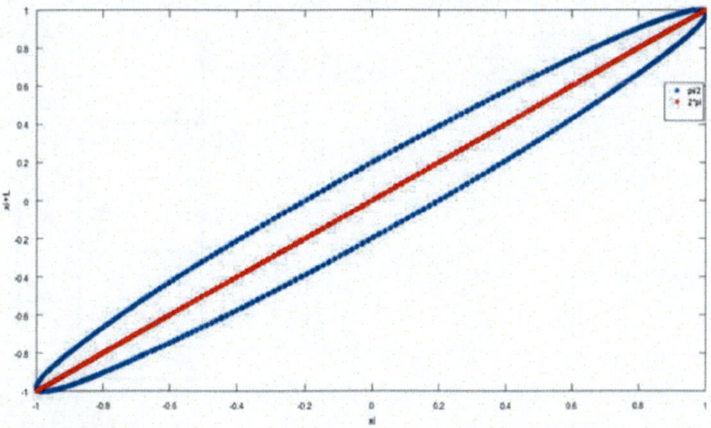

Fig. 8 Quasi-periodic (Blue)

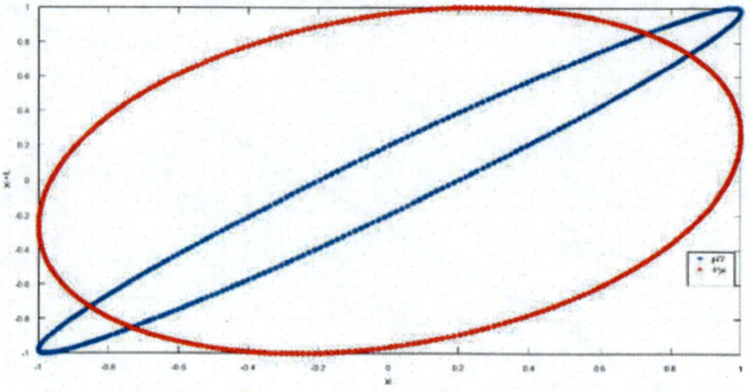

Fig. 9 Quasi-random (red)

maximum width for a phase difference of pi/2. It again collapses into a straight line for the phase difference of 2pi. The corresponding delay gives the period of the signal.

So intuitively to differentiate between quasi-random and quasi-periodic we have to estimate the deviation from the straight line with a slope of pi/4. More in the deviation, the signal is more quasi-periodic in nature.

In Fig. 10 the dotted line indicates the line having a slope of pi/4. XP is the perpendicular from the point *X* on the dotted line and represents the deviation of the point *X*. XA is the perpendicular from the point *X* to the axis *x* (*i*) and it intersects the dotted line at point *Q*. Or in another way, it can explain as, the *X* point is deviated from the straight line with amount of *p* point, and now with the help of the Eq. (4.1), it measures that the amount of point "*X*" is deviated from the straight line.

2(XP)2 = (XQ)2 = [*x*(*i* + *L*) − *x* (*i*)]2 The square of the deviation XP is proportional to square of *x* (*i* + *L*) − *x*(*i*).

Fig. 10 State phase diagram

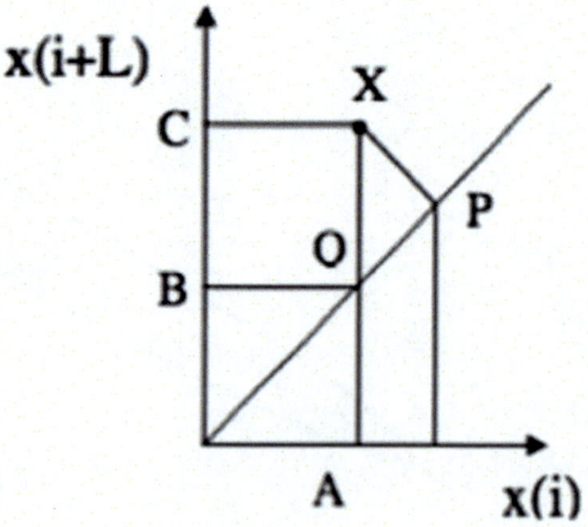

To estimate the average magnitude difference function (AMDF), we have used the $s = (1/N)\sum_{i=1}^{N}(x(i + L) - x(i))$formula as suggested in [4].

Where N = Number of point present in each section, $x(i + L)$ = delay point, L = delay.

The speech signal consists of basically three different kinds of waveforms, namely the quasi-periodic, the quasi-random and the silence. The pitch detection algorithm (PDA) is required only in the quasi-periodic portion of the signal. Therefore, it is necessary that the PDA includes a voice detection algorithm (VDA). Again, the quasi-periodic region consists of signal representing vowels, laterals, nasals and voice bar. The graphs (Figs. 11, 12 and 13) show the nature of the deviation vs delay for quasi-periodic, quasi-random and silence.

If we compose Figs. 11, 12 and 13, it is noted that for quasi-periodic signal, the minima of AMDF is much close to zero. So the first minima (Lm) from the origin gives the period of the quasi-periodic signal. The corresponding pitch can be calculated using Pitch = F_s/L_m.

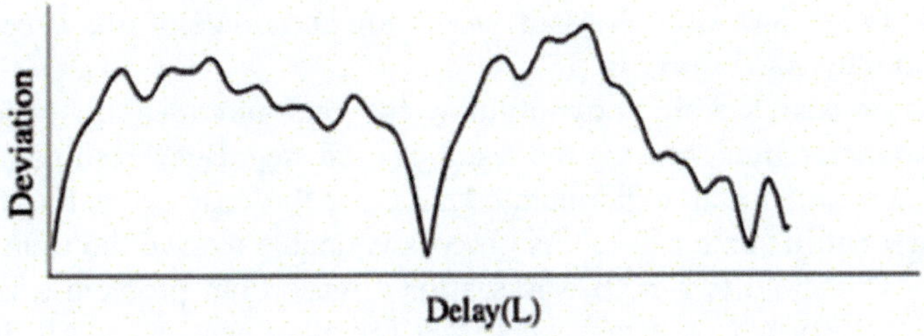

Fig. 11 Deviations against delay for quasi-periodic signal

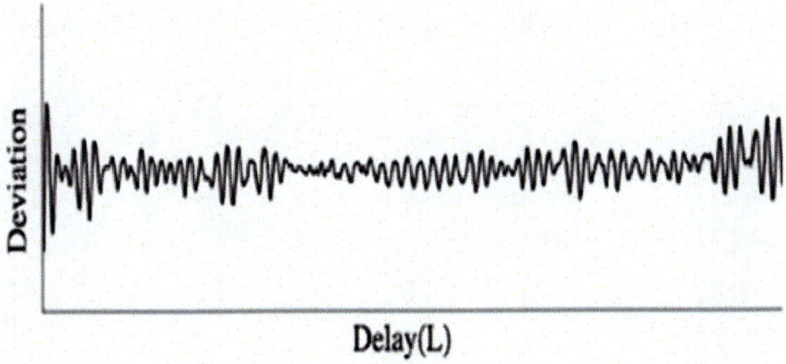

Fig. 12 Deviations against delay for quasi-random signal

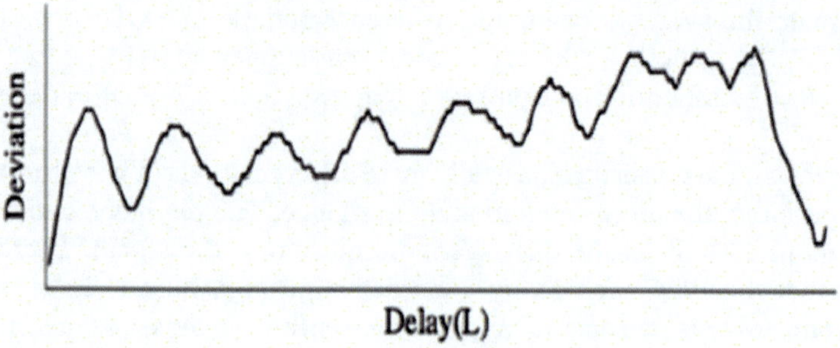

Fig. 13 Deviations against delay for silence in the speech signal

3 Result Analysis

From Fig. 14 (red), the pitch range is between 120 and 220 Hz. The mean of this range is $m1 = 177$ and standard deviation is $s1 = 12$.For double, $m2 = 354$, and range is set between 342–366. For half, $m3 = 88$, and range of this is between 88 and 100. Final pitch shows on Fig. 15.

But in this process, we are not continuing for pitch modification, because here we are set parameter after showing the result of each step. But threshold (120–220) parameter is not same for all experimented data. For this case, we get perfect result (Fig. 16, (red)) but in some cases, this process is unable to give the right pitch for the sentence. Threshold (75–500) was set for a reason but problem arise for the threshold (120–220), where it may be varied for other sentences. All the time it cannot be possible to tuning the threshold (120–220).

The silhouettes constructed below are useful when the proximities are on a ratio scale (as in the case of Euclidean distances) and when one is seeking compact and clearly separated clusters [5]. Indeed, the definition makes use of average proximities

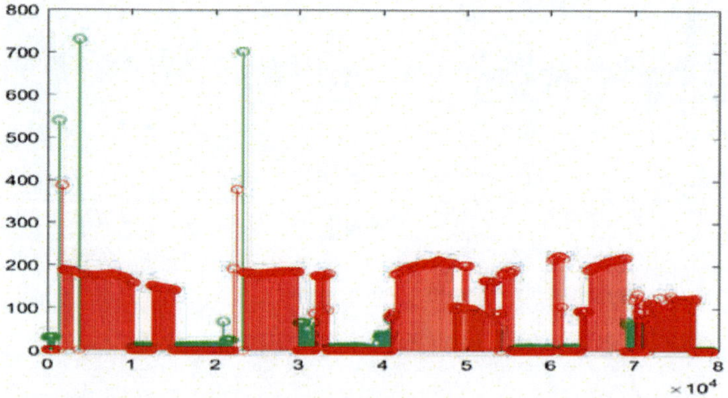

Fig. 14 Pitch profile [red (75–500)] and raw pitch [(green)]

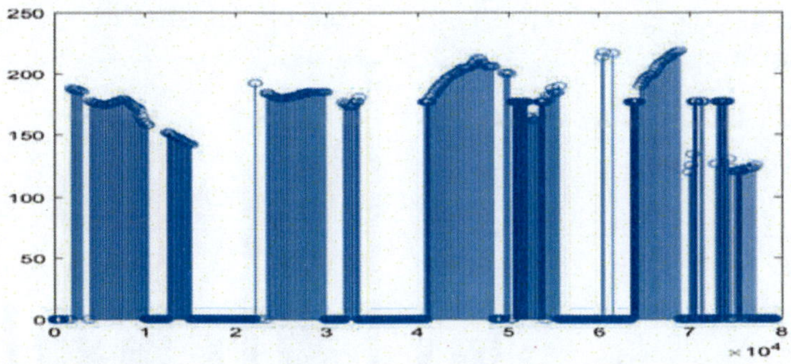

Fig. 15 Final pitch (range between 120 and 220)

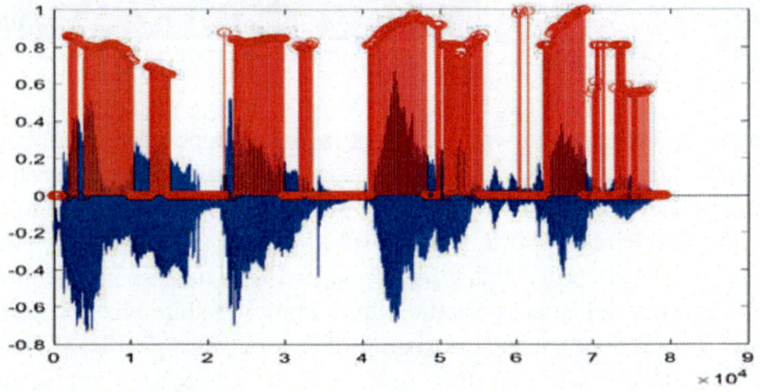

Fig. 16 Final pitch profile (red)

as in the case of group average linkage, which is known to work best in a situation with roughly spherical clusters.

In order to construct silhouettes, we only need two things: the partition we have obtained (by the application of some clustering technique) and the collection of all proximities between objects.

In Fig. 17, along *x*-axis, it represents the total number of sentences, and along *y*-axes, it represents the cluster number belong to each sentence. We have taken $k = 10, 15, 20\%$ and structural element = 70, 90, 110, 130, 150; in total, we get 15 combinations, out of this combination which sentence get how many right cluster has been shown in Fig. 17.

4 Conclusion

Very few work has been done on Bengali colloquial speech as a speech recognition system. Our aim is to enhance the research on this direction. Doing so, the most

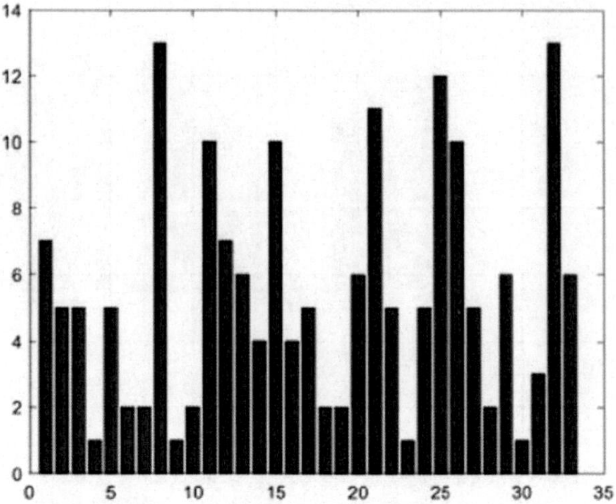

Fig. 17 Overall result for all experimented sentences

crucial part is to identify the words present in the sentence. The challenge in this work is to detect the word boundary. As the speech is colloquial, the word boundary is not prominent. In this thesis, we take the pitch profile as the feature to detect the presence of words. In every word, there must be a vowel, and to utter the vowel, there is a vibration of vocal cords. This vibrating signal is in quasi-periodic in nature. In our thesis, we track this quasi-periodic nature applying state phase technique. The pitch profile we get is noisy and consisting of several false positives. To overcome that, we use the power profile of the speech. Then we binarize the pitch and apply morphological operation to detect the word boundary. Finally, a cluster validity index is used to find the word boundaries. The process is simple, yet robust. In our study, we set 15 different combination of parameters to study 60 different speech signal. It is observed that 30 times we get correct result. In rest of the cases our algorithm missed by one word. We admit that there is a need of more research work to tune the parameters. In future, we have a plan to continue the work with more data sets and develop other features, such as consonant–vowel (CV) transitions to detect the words, and a manner-based lexicon database will be created to recognize the Bengali colloquial speech.

References

1. Rabiner L, Juang BH, Yegnanarayana B (2009) Fundamental of speech recognition. Pearson
2. Chowdhury S (2006) Concatenative text-to-speech synthesis: a study on standard colloquial Bengali, Ph.D. thesis. Statistical Institute, Kolkata
3. Acoustics of Bangla Speech Sounds, Asoke Kumar Datta. Springer Publications
4. Das Mandal SK (2007) Role of shape parameters in speech recognition: a study on standard colloquial Bengali (SCB), Ph.D. thesis. Jadavpur University
5. Rousseeuw PJ (1987) Silhouettes: a graphical aid to the interpretation and validation of cluster analysis. J Comput Appl Math 20:53–65

Author Index

M. Mitra et al. (eds.), *Computational Advancement in Communication, Circuits and Systems*, Lecture Notes in Electrical Engineering 786,
https://doi.org/10.1007/978-981-16-4035-3